ISBN 978-0-282-25008-9
PIBN 10845551

1 MONTH OF
FREE
READING

at

www.ForgottenBooks.com

By purchasing this book you are eligible for one month membership to ForgottenBooks.com, giving you unlimited access to our entire collection of over 1,000,000 titles via our web site and mobile apps.

To claim your free month visit:

www.forgottenbooks.com/free845551

Copyright, 1905, by INTERNATIONAL TEXTBOOK COMPANY.

Entered at Stationers' Hall, London.

Electric Power Stations: Copyright, 1905, by INTERNATIONAL TEXTBOOK COM-
PANY. Entered at Stationers' Hall, London.

Telegraph Systems: Copyright, 1905, by INTERNATIONAL TEXTBOOK COMPANY.
Entered at Stationers' Hall, London.

Telephone Systems: Copyright, 1905, by INTERNATIONAL TEXTBOOK COMPANY.
Entered at Stationers' Hall, London.

Applied Electricity: Copyright, 1905, by INTERNATIONAL TEXTBOOK COMPANY.
Entered at Stationers' Hall, London.

PRINTED IN THE UNITED STATES.

47

23142

CONTENTS

---

iv   .CONTENTS

CONTENTS

PUBLISHERS' STATEMENT

The sections on electric-power stations included in this volume were written by J. H. Vail, Consulting Engineer, Philadelphia, Pa. The parts relating to telegraphy and telephony were prepared by H. S. Webb, M. S., and that on applied electricty by R. B. Williamson, M. E. In the preparation of the sections on power stations the technical press has in some cases been consulted, particularly with reference to descriptions of various power stations, and the publishers wish to express their indebtedness to *The Electrical World and Engineer*, *The American Electrician*, and *Power*, for such material as may have been made use of.

ELECTRIC POWER STATIONS

(PART 1)

INTRODUCTION

1. In addition to the study of the various systems of electrical distribution and the types of dynamos and auxiliary electrical apparatus required in the power station for carrying out the distribution, it is necessary to consider the various appliances, used in the power station, which are not electrical in character and also to take up the power station as a whole. The first involves a study of those types of steam boilers, steam engines, waterwheels, etc. that are best adapted for the purpose. It also includes a consideration of the features that govern the location of the plant, type of building, fuel, water supply, etc.

2. The electric power station of the present day being built for generating electricity as a commercial product, follows in the legitimate line of its predecessors and contemporaries, the waterworks and the gasworks. In each is concentrated the apparatus necessary to manufacture or force through circulating or distributing mediums its particular product for the supply of individual consumers in quantities as required.

The electric power station involves the assembling within a properly designed building, situated in a suitable location, such a combination of selected apparatus as will facilitate the extracting of the heat units contained in the fuel, and their transmission over a given territory; or, if a water-power station, the conversion of the foot-pounds of energy developed by the falling water, into electric energy for similar distribution.

Of the original energy contained in the fuel, if coal is used, it is surprising how much is lost in conversion and transmission; this is demonstrated under the heading Efficiency of Power Stations. These enormous losses serve to emphasize the importance of the most judicious selection of every appliance requisite for an electric power station, to the end that the loss may be reduced to a minimum.

LOCATION

3. The standard of excellence for the location of the station will be described with a view to attaining it as far as practicable, when property is about to be selected. The most desirable plot will meet the following conditions:

(*a*) The ground should have a suitable elevation above high-water mark to avoid danger from floods, it should be nearly level, and should have ample area for construction of the contemplated station and its future growth to more than four times its immediate estimated capacity. There should also be an abundance of space for a storage warehouse that will contain all supplies, and yard room for poles, cross-arms, underground materials, coal, teams, etc.; in fact, a plot of ground amply large for the entire future business of the company. Any subdivision of working force, either in actual station operation, supply department, or offices, means an added future expense of supervision over what would be the minimum cost in the ideal location.

(*b*) The selection of a location for building a power station involves a careful analysis of its environment, because there are several points which, if not given due attention, may be the cause of an avoidable expense of greater or less amount, either in cost of construction, maintenance, operation, or future legal entanglements.

(*c*) The ideal site will not involve any annoyance to the surrounding neighborhood, due to the noise of machinery in operation, the delivery of coal, the handling of ashes, smoke, dust, smell of oil, and the other numerous unavoidable incidents that continually attend on central-station operations and are objectionable when near districts containing schools, churches, residences, or similar buildings.

(d) The fire-risk must be considered by careful investigation of all the surroundings with a view to danger from external fire. A power station is liable to as great danger from fire in adjacent buildings as within its own walls.

(e) The least cost of construction will be obtained on a site where excavations are readily made at minimum cost, where extra piling and blasting are not necessary, where firm hard pan, clay, gravel, or rock are found within a few feet of the surface, and where no more grading is required than the usual filling in and grading off around foundations. The security of the structure depends on solid underlying soil or rock on which may be built foundations of unquestionable solidity. Piling on soft ground is expensive, and causes more or less anxiety about the security of the structures.

(f) There should be an abundant, never-failing supply of pure water for boiler-feed purposes, and, if possible, for condensing purposes also, free of cost except for piping; therefore, a waterside location is preferable, and the elevation for pumping should not exceed 18 feet. A variation from this will add to what will be normal cost for these items under ideal conditions. If a condensing station is desired, the free supply of water for this purpose is essential even if water for boiler feed must be purchased.

(g) The fuel supply should be absolutely reliable and delivered at the premises by railroad on elevated trestle, or boat alongside, at the lowest rates. The storage bunker should be of such capacity as to permit securing a full stock during the season of lowest prices, and sufficient for 4-months' use to carry the winter load, or over any long strike at coal mines or on transportation lines, and the arrangement should be such that inclement weather may not, at any season, delay delivery and thereby imperil station operation. If oil is used for fuel, the storage tanks should be of ample capacity, located in vaults below ground, and properly ventilated.

(h) The electrical center of the entire district to which

power is supplied is the most desirable location for the station, other things being equal, so that the normal cost of copper for feeders may not be exceeded, and the whole cost of the system of distribution kept within the lowest limits. The relative advantage of different sites as regards cost of copper can readily be ascertained by estimating, for the maximum load, the pounds of copper under a given loss from several locations, and comparing costs.

Ideal locations possessing every requisite are not always secured, but each variation therefrom should receive its estimate of cost. The value of one location should be compared with another on the cost of the several items of construction and operation, and thus an intelligent conclusion can be reached. Extra cost of operation at one location over another should be capitalized on a 6-per-cent. basis to show the relative values. That is, if one location represented an additional yearly cost of operation of $300 over some other location, the first location should be charged with an additional investment of $5,000 capital in order to make a fair comparison of the first cost in the two cases.

4. Reliability.—An electric power station is constructed to supply its product on instantaneous demand and in any amount irrespective of other notice from the consumers to the producer. The consumers make a contract for electrical energy, and an abundant supply sufficient for their requirements must be available on demand, the same as water or gas. Therefore, special stress must be laid on the point of insuring absolute reliability of service throughout every detail of the whole equipment and system of distribution. No station can be considered a reliable source of supply unless every feature of its equipment liable to derangement or accident is at least duplicated, and so interconnected throughout that in the event of trouble the load may be transferred from one set of apparatus to another, without causing the slightest inconvenience to consumers, or interfering with the continuity of the service. If the service from the station is irregular or subject to interruption, this quickly creates in the minds of the consumers a lack of confidence in the ability of the company to fulfil its obligations in supplying its product, and such lack of confidence on the part of the public will be a serious impediment to the growth of the business. For these reasons the duplication of apparatus, and the complete interchangeability of connections should always be insisted on as of first importance.

EFFICIENCY OF POWER STATIONS

5. In an electric power station it is important to know as accurately as possible, what degree of economy is obtained from the entire equipment operated as a unit; or, taking the total energy supplied in the form of heat units held in the fuel, what percentage is lost and what percentage is recovered in the form of useful work in light or power. The commercial success or failure of an electric power station is largely determined by the efficiency and reliability of the equipment, and the actual efficiency is therefore a subject of special interest to the central-station manager and the engineer.

6. The **efficiency** of the entire plant in the aggregate must necessarily depend on the net efficiency of each unit of apparatus composing the equipment, since their accumulative losses account for a large part of the total energy wasted in the system, thus readily affecting the final efficiency. Emphasis is given to these facts to the end that, knowing of certain fixed and irreparable losses, one may fully realize the importance of using such equipment as will keep these losses down to the minimum. In the electric power station operated by steam the heat units in the fuel represent the original amount of energy applied, and which are generally accepted as 14,600 British thermal units per pound of coal, it being understood that much coal averages less than this, and an excess is exceptional. The efficiency of fuel may be stated as the total amount of heat it is capable of generating. The proportion of the generated heat that may be utilized, depends on the efficiency of the boiler. The efficiency of a coal depends not only on its chemical composition and theoretical value in heat units contained per pound, but also on the percentage of moisture, ash, and non-combustible material contained, as well as its size and condition for use. In the burning of coal there are certain unavoidable losses (such as heat lost in moisture, heat lost in excess of air supplied, heat lost in products of combustion, heat lost in unburnt coal), the aggregate of which will rarely fall short of 15 per cent., and more frequently exceed that amount.

7. Boiler Efficiency.—The efficiency of a boiler may be defined as the ratio of the heat utilized in evaporating the water to the total heat supplied by the combustion of the fuel. It will vary according to the relative ratios of heating surface to grate area, the cleanliness of the boiler, method of setting, thickness of plates, etc. Exhaustive tests show that **boiler efficiency** is sometimes as low as 21 per cent., and seldom reaches 88 per cent.; with the best types of boilers, with grate area and heating surface carefully proportioned, an efficiency of 75 to 80 per cent. is attainable with a clean boiler very carefully managed. There is such a great difference between a competent and an incompetent fireman that the efficiency of the best-designed boiler under the most favorable conditions may be greatly reduced by bad management.

8. Engine Efficiency.—The best engine, apart from its boiler, has about five-sixths the efficiency of a perfect engine, the other sixth being lost through waste of heat by radiation, conduction, cylinder condensation, and friction. Its efficiency as a heat engine is from 10 to 15 per cent.; that is, of the energy represented by the heat stored in the steam, only from 10 to 15 per cent. is converted into useful work. Engine friction is a factor of waste in all engines, and the size, type, and condition of the engine will affect the resistance and the effort necessary to overcome it. This will vary from 1 pound per square inch of piston, in large, well-designed engines, to 3 or 4 pounds in inefficient machines. The net useful work is represented by the indicated work minus the engine friction, and in good practice this will be 85 to 90 per cent., which represents the mechanical efficiency as a machine. That is, of the actual amount of indicated work done in moving the piston in the cylinder, 85 to 90 per cent. is available at the engine shaft, the balance being lost in friction between the various rubbing surfaces.

Of the boiler and engine combined, Dr. R. H. Thurston states "that of all the heat derived from the fuel, about seven-tenths is lost through the existence of natural conditions over which man can probably never expect to obtain control, two-tenths are lost through imperfections in apparatus, and only one-tenth is utilized in good engines." In this combination of waste probably two-tenths at least of the heat derived from the fuel is lost in the boiler and steam pipes.

9. Generator Efficiency.—In electric generators ranging in output from 50 to 1,000 kilowatts the efficiency at full load will range from 90 to 94 per cent.; with half load, somewhat less. In this statement no distinction is made between direct-current and alternating-current generators. Outside of the generator, other sources of loss are found; in electrical conductors and connections, and at any point where an abnormal rise of temperature can be detected, it may be certain that a loss is occurring.

10. Switchboard losses may reach from .125 per cent. upwards, and may be prevented by exercising great care in the selection of switches, instruments, etc., and particularly in details of bus-bar work and connections. All bus-bars and electrical conductors and connections must be liberal in capacity; every joint should have true surfaced contacts of at least twice the area of the conductor, and all bolting, soldering, or brazing must be very carefully done.

11. Outside Losses.—Losses outside of the station will largely depend on the characteristics of current and system. The actual loss in distributing conductors can be accurately predetermined for specified loads. With the direct-current system the maximum loss will take place during the hours of maximum load, and the loss is reduced as the load goes off; with the alternating-current system, the line loss is comparatively small, but there is considerable loss in the transformers even when the useful load is very light. The magnetizing current of the transformers causes a certain amount of copper loss and the core losses are practically constant at all loads.

12. Selecting Station Equipment.—Economy consists in avoiding all unnecessary expenditures and losses, and in making a profitable disposition of what would otherwise be wasted. In aiming for economy of fuel the whole combination of boilers, engines, condensers, piping, pumps, heaters, etc. must be considered individually and collectively. A poor boiler will not demonstrate the virtues of a good engine, and likewise an engine extravagant in steam may render useless all the economy obtained with a good boiler. The contract requirements for a boiler, engine, generator, or any piece of apparatus should be based on the highest economies demonstrated in its class; these should be clearly set forth in the contract and specifications, and should be rigidly adhered to. The impossible should not be demanded. The man who designs and erects a station should set aside all personal preferences and prejudices, and should be broad enough to select his equipment on established merit only. If an article secured from a manufacturer of doubtful reputation is found unsatisfactory after it is delivered, erected, and started, the purchasing company will have no satisfactory redress. The removal and replacement by some other apparatus may make good the deficiency, but can never compensate for the annoyance, loss of time, and money involved.

LOAD CURVES

13. Determination of Probable Load.—In advance of building a station it is possible to construct a load diagram that will closely represent the load curve or daily output of the station, from the following data: First, a careful canvass of the lighting and power to be obtained within the area of the district to be supplied. Second, the subdivision of each kind of service under different classifications, such as motive power, street lighting, store lighting, hotel lighting, residence lighting, theater lighting, church lighting, etc., and an estimate of the percentage of connected load that will be operated during certain hours, and for a given period.

The combination of these several classified loads will overlap at certain periods, and unite to form the load line of the station. The load line will vary during the different seasons of the year, and under normal conditions the maximum load will be met from December 1 to January 1 in cities and towns. In summer resorts, the maximum load is at the height of business during the summer season. The station must always be prepared to take care of this maximum load, and to be so equipped frequently requires investment in expensive apparatus that has an earning capacity during three summer months only and must lie idle for the remainder of the year. For this reason the apparatus so employed does not require to be of so costly or economical a character as that used daily during the entire year.

14. Example of Load Diagram.—The load diagram shown in Fig. 1 combines arc lighting, incandescent lighting, and motive power supplied from a single station. The arc-lighting load represented by the full line is all-night lighting, starting at 6 P. M. and carrying the total load until 5 A. M. The load for arc lighting is represented at 75 kilowatts for 100 burning lights. The output for incandescent lighting is indicated by the dash line and represents a combination of lighting in private residences, stores, saloons, hotels, theaters, etc. It will be noted that the load is least between 11 P. M. and 5 A. M. There is a slight peak at 6 A. M. which again diminishes, and as business lighting starts up in the various establishments, the load increases. Assuming that the day is a stormy one, there is a gradual increase of load until the peak is reached at 6 P. M., after which many stores and places of business close up. There is an increase again after 6 P. M. on account of additional lighting used in theaters and places of amusement. The lighting on the incandescent system gradually diminishes until midnight. The motive-power load is represented by the dotted line, which shows the aggregate of current required by motors used for manufacturing purposes, elevators, etc. This load is almost constant between 7 A. M. and 12 o'clock noon, at which time a number of motors in manufacturing establishments are shut off; these are started again at 1 P. M. and gradually closed off between 5 and 6 P. M. The total load on the station, which is a combination of all of the above, is represented by the dot-and-dash line and shows the demands on

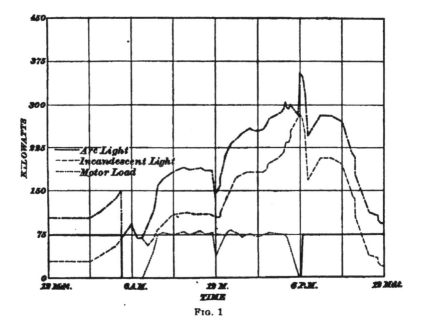

FIG. 1

the station by the combination of the various kinds of service supplied. This line is found by adding the ordinates of the three lines representing the various loads. If it is assumed that the arc lighting is supplied from the alternating-current system, the motive power from the direct-current system, the incandescent lighting in the business district and part of the residence lighting from the direct-current, and the remainder of the incandescent lighting in outlying districts from the alternating current, it becomes evident that a station supplying current for this variety of business can conveniently do so by the use of double-current generators, the arc lighting and alternating-current incandescent lighting being served from the alternating side of the generator, and the motive power and business district lighting from the direct-current side.

15. If this station were equipped with two double-current generators each of 100 kilowatts and one double-current generator of 300 kilowatts, it would have such a combination of apparatus as could be economically manipulated for the daily load during week days. On Sunday, the incandescent load and motive load is expected to be at the lowest point between 6 A. M. and 6 P. M.; therefore, during these hours on Sunday, which would represent 624 hours per year, the load could be carried on a single 100-kilowatt generator running at about one-half load. This illustrates a typical case, and in such a station provision should be made for an early increase in the number of generating units.

16. Provision for Future Extensions.—It is a matter of history, that during the past 15 years, 90 per cent. of the stations have been designed on too small a scale to meet successfully and economically the rapid growth and expansion of the business; constant extension, moving, and rebuilding is evidence of the enormous cost entailed by this lack of foresight. For these reasons the plant should be such as to admit readily of extensive additions, and the future should be anticipated to such a large degree that changes and improvements necessary to accommodate an enlarged equipment can be readily and economically harmonized with the original design without destroying first construction or causing expensive alterations.

STORAGE OF COAL

17. It is very important that every electric power station using coal should be provided with a large storage capacity to insure a continuous supply of fuel during extended periods of bad weather, delays in transportation, strikes, etc., and also to permit taking advantage of low market rates. Coal-storage bunkers should be so arranged as to admit of the most economical methods for handling the coal, and a thoroughly reliable system of conveyers should be installed when the plant is sufficiently large to warrant the expense.

In designing a coal bunker, no fixed rule can be given as to its dimensions, except to make it as large as possible according to the space available. The approximate weights of coal per cubic foot are as follows:

Anthracite {
Buckwheat . . . . 55 pounds per cubic foot
Pea . . . . 50 pounds per cubic foot
Broken . . . . . . 60 pounds per cubic foot
}
Bituminous . . . . 45 to 50 pounds per cubic foot

18. Pressure Exerted by Coal.—If bituminous coal is piled up, it will assume a slope of about 35°, or in other words, its angle of repose is 35°. With anthracite coal, the angle of repose is about 27°. For example, in Fig. 2, if ab represents a retaining wall for holding back bituminous coal, the coal will, if simply piled up, assume the slope am. If the space is filled level with the top of the wall along the line bm, there will be a certain pressure exerted per each foot length of wall due to the mass of coal abm. If the coal is heaped up in the bunker until it assumes the slope bc, which is the maximum slope it can assume without running off, it is plain that the pressure exerted per lineal foot of the retaining wall will be considerably greater than tbat when the surface of the coal was level. Table I, prepared by the Link Belt Engineering Company, shows the pressure exerted for each lineal foot of wall for walls of different heights, when the coal is filled level along the line bm, Fig. 2, and when it is piled up along the line bc. For example, if the depth of the wall ba were 10 feet and the coal filled flat with the top along the line bm, the total pressure exerted on a strip of the wall of height ba = 10 feet and breadth 1 foot will be 637 pounds. The table also gives the pressure on the lowest foot of wall, or in other words, the maximum pressure to which any square foot of wall is subjected; in this case

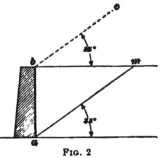

FIG. 2

the pressure on the lowest foot is 121 pounds. If the coal were piled along the line bc, the total pressure would have been 1,000 pounds per lineal foot and 190 pounds on the lowest foot.

Table II is similar to Table I, but shows the pressures exerted by anthracite when piled up to the line bm, Fig. 3, or heaped up to the slope bc. The pressures exerted by anthracite are fully 50 per cent. greater than those exerted by bituminous, partly because of the greater weight per cubic foot of the anthracite and partly because of its smaller angle of repose, which is shown in Fig. 3.

19. Labor-Saving Appliances.—In designing an electric power station and its equipment, especial attention

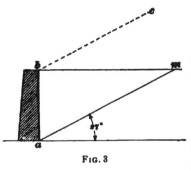

FIG. 3

should be given to the largest practicable reduction of labor throughout the station; convenience of arrangement, size of units, and many other features will, if rightly selected, accomplish this purpose. Appliances for handling coal and ashes have been brought to a high degree of perfection and are great labor savers. At a reasonable cost it is possible to arrange for automatically handling coal from the time it

TABLE I

HORIZONTAL PRESSURE EXERTED BY BITUMINOUS COAL AGAINST VERTICAL RETAINING WALLS PER FOOT OF LENGTH

Link Belt Engineering Company

Depth ba in Feet	Horizontal Surface bm		Sloping Surface bc		Depth ba in Feet	Horizontal Surface bm		Sloping Surface bc	
	Total Pressure Pounds	Pressure on Lowest Foot Pounds	Total Pressure Pounds	Pressure on Lowest Foot Pounds		Total Pressure Pounds	Pressure on Lowest Foot Pounds	Total Pressure Pounds	Pressure on Lowest Foot Pounds
1	6.4	6.4	10	10	26	4.305	325	6,760	510
2	25.	19.	40	30	27	4,641	338	7,290	530
3	57.	32.	90	50	28	4,993	350	7,840	550
4	102.	45.	160	70	29	5,358	363	8,410	570
5	159.	57.	250	90	30	5,732	376	9,000	590
6	229.	70.	360	110	31	6,122	389	9,610	610
7	312.	83.	490	130	32	6,523	401	10,240	630
8	407.	96.	640	150	33	6,935	414	10,890	650
9	516.	108.	810	170	34	7,362	427	11,560	670
10	637.	121.	1,000	190	35	7,778	440	12,250	690
11	770.	134.	1,210	210	36	8,253	452	12,960	710
12	917.	146.	1,440	230	37	8,754	465	13,690	730
13	1,076.	159.	1,690	250	38	9,193	478	14,440	750
14	1,248.	172.	1,960	270	39	9,682	490	15,210	770
15	1,433.	185.	2,250	290	40	10,192	503	16,000	790
16	1,630.	197.	2,560	310	41	10,669	516	16,810	810
17	1,840.	210.	2,890	330	42	11,236	529	17,640	830
18	2,063.	223.	3,240	350	43	11,797	541	18,490	850
19	2,298.	236.	3,610	370	44	12,331	554	19,360	870
20	2,548.	248.	4,000	390	45	12,968	567	20,250	890
21	2,809.	261.	4,410	410	46	13,478	580	21,160	910
22	3,083.	274.	4,840	430	47	14,100	592	22,090	930
23	3,369.	287.	5,290	450	48	14,679	605	23,040	950
24	3,669.	299.	5,760	470	49	15,275	618	24,010	970
25	3,981.	312.	6,250	490	50	15,925	631	25,000	990

TABLE II

HORIZONTAL PRESSURE EXERTED BY ANTHRACITE COAL AGAINST VERTICAL RETAINING WALLS PER FOOT OF LENGTH

Link Belt Engineering Company

Depth b a in Feet	Horizontal Surface b m		Sloping Surface b c		Depth b a in Feet	Horizontal Surface b m		Sloping Surface b c	
	Total Pressure Pounds	Pressure on Lowest Foot Pounds	Total Pressure Pounds	Pressure on Lowest Foot Pounds		Total Pressure Pounds	Pressure on Lowest Foot Pounds	Total Pressure Pounds	Pressure on Lowest Foot Pounds
1	9.78	9.78	14.2	14.22	26	6,611.1	498.78	9,612.8	725.21
2	39.12	29.34	56.9	42.66	27	7,129.5	518.35	10,366.	753.67
3	88.02	48.90	127.1	71.10	28	7,667.6	537.90	11,149.	782.10
4	156.48	68.46	227.5	99.54	29	8,225.0	557.46	11,988.	810.54
5	244.50	88.02	355.5	127.98	30	8,802.0	577.01	12,797.	839.0
6	352.08	107.58	511.9	156.42	31	9,398.5	596.59	13,665.	867.41
7	479.22	127.14	696.8	184.86	32	10,015.	616.14	14,561.	895.86
8	625.92	146.70	910.1	213.30	33	10,650.	635.70	15,486.	924.30
9	792.18	166.26	1,151.8	241.74	34	11,306.	655.26	16,439.	952.7
10	978.00	185.82	1,422.0	270.18	35	11,980.	674.81	17,420.	981.19
11	1,183.38	205.38	1,720.6	298.62	36	12,675.	694.39	18,429.	1,009.6
12	1,408.32	224.94	2,047.7	327.06	37	13,389.	713.94	19,467.	1,038.1
13	1,652.82	244.50	2,403.2	355.50	38	14,123.	733.50	20,533.	1,066.5
14	1,916.88	264.06	2,787.1	383.94	39	14,875.	753.07	21,629.	1,095.0
15	2,200.50	283.62	3,199.5	412.38	40	15,648.	772.63	22,752.	1,123.4
16	2,503.68	303.18	3,640.3	440.82	41	16,440.	792.20	23,904.	1,151.8
17	2,826.42	322.74	4,109.6	469.26	42	17,252.	811.74	25,084.	1,180.3
18	3,168.72	342.30	4,607.3	497.70	43	18,083.	830.73	26,293.	1,208.7
19	3,530.58	361.86	5,133.4	526.14	44	18,934.	850.86	27,530.	1,237.2
20	3,912.00	381.42	5,688.0	554.58	45	19,804.	870.41	28,793.	1,265.6
21	4,313.00	400.98	6,271.0	583.26	46	20,695.	889.99	30,090.	1,294.0
22	4,733.5	420.54	6,882.5	611.46	47	21,605.	909.54	31,412.	1,322.3
23	5,173.7	440.10	7,522.5	639.90	48	22,533.	929.10	32,763.	1,350.9
24	5,633.3	459.67	8,190.7	668.35	49	23,482.	948.66	34,143.	1,379.4
25	6,112.6	479.22	8,887.5	696.79	50	24,450.	968.21	35,550.	1,407.9

reaches the station by car or boat, until the ashes are removed from the station. For example, the coal may be delivered by car or boat and rapidly transferred, by traveling conveyers, to an elevated coal bunker at a cost of from 3 to 6 cents per ton for power and attendance; it may be delivered direct from the coal bunker, through chutes, to each mechanical stoker, each chute being fitted with a recording apparatus for weighing the coal delivered to each hopper and stoker.

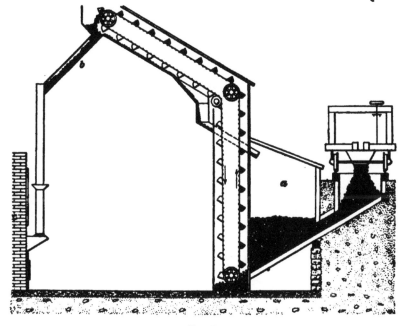

Fig. 4

20. Fig. 4 shows an efficient type of coal conveyer made by the Link Belt Engineering Company; it is of the continuous-running type and takes the coal from the storage bin *a* and delivers it to the reservoir chute *b* from which it goes to the hopper of the mechanical stoker, which feeds it automatically under the boiler. The conveyer consists of a number of buckets on an endless chain, the buckets filled with coal being shown black and the empty buckets white. It is usually driven by an electric motor or small steam engine.

21. Fig. 5 shows a coal-storage and conveying arrangement installed by the C. W. Hunt Company. The coal is here brought to the station in the barge *a* and is removed by the bucket *b*, which is drawn up the incline and discharged into the storage bin directly above the boilers *c*. The coal passes through the weighing device *g* to the chute *d* and is fed, by the automatic stoker, on to the inclined grate bars *e*. The ashes drop into a carrier at *f* and are conveyed to a

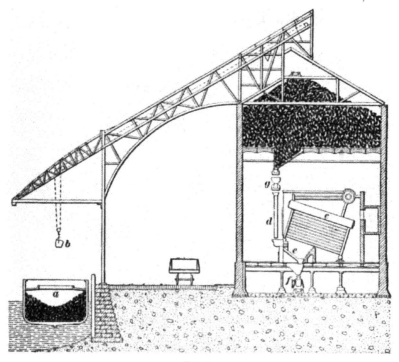

FIG 5

hopper in the second story of the building, from which they are readily loaded into wagons for removal. No shoveling is required for handling either coal or ashes. Before passing into the spout *d*, the coal is run into the suspended hopper *g* where it is weighed, thus allowing an accurate record to be kept of the amount of coal burned under each boiler.

22. Fig. 6 shows an arrangement for a coal-storage plant, where the coal is delivered either by water or rail;

47—3

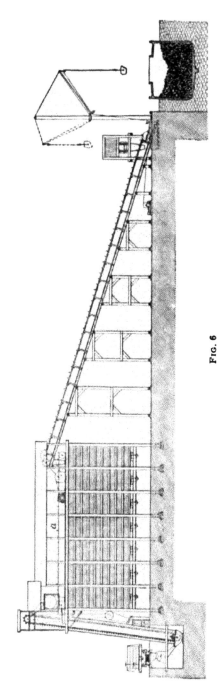

Fig. 6

the storage bin is capable of holding 1,000 tons. The coal, after being unloaded, is passed through crushers to reduce the large lumps to suitable size and is then carried by the conveyers to the storage bin. A horizontal conveyer that can be run in either direction passes across the top of the bin, as indicated at *a*; it takes the coal from either of the other conveyers and delivers it in any desired part of the storage bin, thus allowing the bin to be evenly filled. The coal is delivered from chutes arranged at the bottom of the bin. The operation of the other parts will be plain from Fig. 6, so that further comment is unnecessary. Fig. 7 gives the general appearance of the bin, showing the end where coal is unloaded from the cars. Fig. 8 shows the unloading of coal from barges for the same installation.

There are many kinds of coal conveyers in use, but it is impossible to mention more than a few typical examples here. One type that is largely used for station work is the belt

FIG. 7

FIG. 8

conveyer, of which the Robins, Fig. 9, is a prominent
example. This consists simply of a continuous rubber belt
running up an incline, and resting on inclined rollers that
make it assume a trough shape. The coal lies in the trough
formed by the belt and is carried up the incline by the motion
of the belt. Of course, the pitch of the incline has to be less
than that which would cause the coal to slide on the belt.

Fig. 9

23. Miscellaneous Devices.—In addition to the labor-
saving devices for handling coal and ashes, there are a
number of appliances that, while not so important, effect a
considerable saving and make the station much more convenient to operate. A traveling crane is a great convenience
for lifting heavy parts quickly and will save in time and
labor a large percentage of its cost, which will be from $600
up, according to the type and load capacity. Oil tanks, oil
filters, oil piping and circulating pumps not only reduce

labor but increase the cleanliness, safety from fire, and effect an economy of from 25 to 40 per cent. in the amount of oil used. Recording meters should be used to record the current generated and supplied to each feeder, including that used for station supply; these meter readings assist in accuracy of records and aid in calculating station efficiency. Every department of the station service should be so systematized and its cost accounted for, that the total cost per kilowatt-hour can be accurately determined and the cost of each item reduced to a kilowatt-hour basis.

FOUNDATIONS

24. That portion of a substructure that serves as a base on which to erect a superstructure, is designated as the **foundation.** Its object is to obtain a solid base on which to erect the machinery or building; the stability of the entire structure is threatened by a foundation of insufficient size, improperly built, or of unsuitable materials. Where supporting heavy machinery, the foundation should, without the slightest vibration, withstand the incessant daily shocks to which it will be subjected. The least yielding may cause fractures in walls or spring the frame of the machinery, causing destructive strains, heating of bearings, and loss of power by increased friction. Suitable foundations are therefore of first importance to the life and efficiency of whatever they support, and deserve the most careful consideration.

All foundations for moving machinery should be constructed absolutely independent of the walls of the building, but where two or more foundations for individual units are within a short distance of each other, it is better that they should be bonded together, and thus form a monolithic structure, as the combined mass and increased area will render them the more capable of withstanding greater strains than when individually constructed.

25. Prior to purchasing the ground and in advance of the preparation of the plans and specifications for an electric power station, the nature of the soil should be determined, as explained later, because the knowledge of the strata thus gained will aid in determining what class of foundation must be provided. The power of a soil to sustain weight varies according to its nature and condition. The building laws of the larger cities limit the load per square foot on soils, and Table III shows the allowance in a few cities.

TABLE III

ALLOWABLE LOAD PER SQUARE FOOT ON VARIOUS SOILS

Kind of Soil	Load in Tons (2,000 Pounds) per Square Foot			
	Philadelphia	Chicago	Buffalo	Milwaukee
Solid natural earth of dry clay . . .	$2\frac{1}{2}$	$1\frac{3}{4}$	$3\frac{1}{2}$	4
Clay, moderately dry.				2
Clay, soft				1
Gravel and coarse sand, well cemented				8
Gravel and sand.	4		4	
Clay and sand.		$1\frac{1}{2}$		
Dry sand confined.		2		

Rock, according to its kind, may sustain from 18 to 180 tons per square foot, but the load placed per square foot should not exceed one-eighth of the crushing limit. Where the soil is of a yielding nature, piles or grillage must be used to support the foundations. This kind of work adds considerably to the cost of building, and should be taken into consideration by a careful estimate of cost when selecting property for a station.

26. Excavation.—This is a feature of construction work so uncertain in its requirements and cost, and so important in its effects on the future of the structure to be reared, that it is wise before building to make careful tests over the ground. Test pits should be dug to ascertain the character of the subsoil and consequent depth of excavation required; it is not wise to rely on the experience of others who have built in the vicinity. The earth often varies considerably in short distances, and such contingencies as water, quicksand, rocks that must be blasted away, or soft subsoil requiring piling should be known in advance.

The specifications must provide, and a price be fixed, for each extra cubic foot of excavation found necessary over that originally specified. The keeping of the sides safely shored, and pumping the pit clear of water, should be definitely required. This leaves the responsibility with the contractor, and gives him no chance to demand disproportionately large amounts for work not anticipated. When the ground is not uniform in character, it is necessary to go to a sufficient depth to reach firm dry ground, or the walls of a building erected on it are almost certain to crack because of uneven settlement. Equally serious results follow in the case of a foundation for heavy machinery. It is advisable to pay for this class of work by the cubic yard of *excavation*, not by the cubic yard *excavated*.

27. Cost of Excavation.—The cost of excavation varies so much with the locality, on account of the character of the soil, cost of labor, etc., that close figures suitable for general use cannot be given, and the various items should in any particular case be open to competitive bids. The following figures are merely averages of actual bids published in the "Engineering Record" for work in various parts of the United States, and can serve only as an approximation in any particular case: Loam, 32 cents per cubic yard; clay or very stiff soil, 60 cents; rock, $4.20 per cubic yard. The maximum and minimum of the bids on which the foregoing figures were based are as follows: Minimum, loam, 29½ cents per cubic yard; stiff clay, 50 cents per cubic yard; rock, 50 cents per cubic yard. Maximum, loam, 43 cents per cubic yard; stiff clay, 65 cents per cubic yard; rock, $8 per cubic yard.

The reason for the great disparity in the figures is due to the nature of the soil, the amount of excavation, the distance to the dump, and the cost of labor.   There is also a difference in the meaning of the terms used.   For example, loam may be very light and sandy or heavy and thick, while rock may mean loose stones, or a practically solid stratum that would have to be removed by blasting, possibly under adverse conditions.   The above figures are for depths of about 6 feet. The prices at greater depths increase at the rate of about 50 per cent. for every 6 feet additional depth, over the price for the first 6 feet.   For example, if the first 6 feet cost 1, 12 feet would cost 1.5; 18 feet, 2¼; and so on.

28.   Great care must be taken when excavating near buildings not to undermine or in any manner weaken or injure them; underpinning should be provided.   This is especially the case when any quantity of water is encountered, or where great depths of foundations, or blasting is necessary.   In the case of rock excavation it must always be borne in mind that the force exerted on surrounding objects by explosives depends on the degree to which the resulting gas is confined.   An experienced foreman will proportion the charges to suit the circumstances.   The contract should clearly specify that the contractor is to assume all responsibility for danger to adjacent property, and he should furnish a satisfactory bond for this purpose.   A sketch showing the shape and all dimensions of the excavation, and its relative position with regard to fixed objects from which the contractor can take his measurements, should always be made a part of the specifications and should be so marked and referred to as to be easily identified.   Special examination should be made as to whether water will be encountered, and how the subsoil is affected by it; the specifications should not fail to require all pumping and drainage to be done by the contractor.   It is always desirable to extend the excavation to firm soil if consistent with cost.   The refilling around finished foundations should be completed to desired grade line, the earth selected to be free from large stones or broken bricks, must be rammed as filled in, and puddled and well settled in place.

29. Piles.—In case a firm subsoil cannot be obtained by excavation, foundations are supported on **piles**—long posts, driven into the ground by means of pile drivers until they are capable of supporting the load imposed on them. A heavy weight or hammer is raised to a considerable height and allowed to drop on the head of the pile and the process repeated until the pile is driven to the desired depth. The tops of the piles are then evened off to a uniform height and the foundation erected on top.

30. Various formulas have been devised for calculating the safe load that a pile will support, but these rules differ considerably because no rule can apply to all conditions. The following formula is very largely used:

$$L = \frac{2\,WH}{p+1} \qquad (1)$$

where L = safe load, in net tons, that the pile is capable of
　　　　　supporting;
　　W = weight of hammer, in tons;
　　H = drop of hammer, in feet;
　　p = penetration, in inches, due to last fall.

This is known as *The Engineering News* formula and gives results that are on the conservative side. It is used in the building laws of a number of cities, for specifying the load on piles. The formula gives the safe load and it is not necessary to allow a factor of safety.

EXAMPLE.—In driving piles for a power-station chimney foundation, the piles were driven until the drop of the hammer produced a movement of .5 inch for each successive drop. The weight of the hammer was 1,500 pounds and the drop 20 feet. What safe load would each pile support assuming that they were driven in firm soil?

SOLUTION.—In formula 1, we have $H = 20$, $W = \frac{1500}{2000} = .75$ ton, $p = .5$; hence, the safe load would be

$$L = \frac{2 \times .75 \times 20}{.5 + 1} = 20 \text{ tons. Ans.}$$

The driving requirements for piles in municipal work will

be given in the local building laws, as are also the dimensions, loading, and spacing. Table IV shows the load and spacing for piles in a few cities.

TABLE IV

ALLOWABLE LOAD FOR PILES

Cities	Piles	Allowable Load Net Tons per Pile
Philadelphia .	Small end, 5 inches; head, 12 inches. Spaced not over 30 inches center to center.	20
New York . .	Small end, 5 inches. Spaced not over 30 inches center to center.	20
Buffalo	Small end, 6 inches. Spaced not over 36 inches center to center.	

31. Grillage Work Under Foundations.—In designing foundations to rest on subsoils of a yielding nature, and where piling is unnecessary or impracticable, provision should be made for uniform settlement. This is particularly important where large generating units are to be erected, as unequal settlement is liable to destroy the alinement. The load may be distributed over a greater area by the use of beams embedded or resting on concrete, making an arrangement usually called **grillage**, as shown in Fig. 10. The columns *a, a* carrying the load rest on the beams *b*, which are, in turn, supported by the beams *c* resting on the concrete *d*. By this means the load is spread over a large area and uniform settlement secured.

32. Setting Foundation Bolts.—When foundations are built for generators, engines, or other heavy apparatus, it is necessary to locate accurately the holes of the holding-down bolts. It is seldom necessary to make the expensively framed templets often used for this purpose. A far more economical and accurate method is the following: After completing the excavation and laying out of the footing course for the foundation, determine the exact floor level for the finished work, and the exact center lines of the machines to

be set.   Lay 2″ × 6″ or 2″ × 8″ timbers on edge, to span the area at, say, 24 inches or 30 inches between centers and having their top edge $\frac{7}{8}$ inch below the floor line.   These timbers are laid across the shortest span similar to beams for a floor, and cross-
bridged if necessary to retain them in
a fixed position.   Having located the
center lines of the engine or generator unit to be erected, overlay the
beams on the lines for the bolt holes
with $\frac{7}{8}″$ × 6″ or $\frac{7}{8}″$ × 8″ strips nailed
to the beams, being careful to see
that the beams shall not interfere
with the bolts.   If some bolts are
higher than others, block up accord-
ingly.   On these $\frac{7}{8}$-inch strips, the
exact center lines of all bolts can be
accurately laid off, and at the center
of each bolt a hole bored sufficiently
large to pass the bolt neatly.   Having
hung each bolt in position, slip on it
a wooden washer of the thickness of
the bedplate, and screw on the nuts.
Draw each bolt through the nut as
far as it will come on the finished
work and allow for the shims to be
used in leveling up the bedplate.

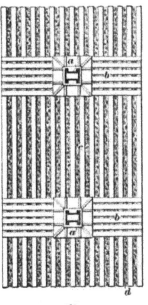

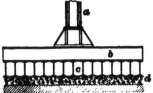

Fig. 10

After the foundation is finished the timbers are removed.
This kind of templet is cheap, strong, and, when well laid, more
accurate than the framed templet, which often gets twisted.

IMPORTANCE OF FIREPROOF CONSTRUCTION

33.   It is very essential that the station building should
be of fireproof construction, and also should be, as far
as practicable, securely protected against fire from adjacent
properties.   The cost of fireproof construction is not greatly
in excess of that which is not proof against fire, and the

saving of insurance premiums, the greater durability, and the sense of security obtained well repays the additional money expended.

34. The salient features of this construction are walls of brick, concrete, or stone, the interior face being smooth finished or pointed and the finished walls painted or calcimined with cold-water paint. Wood ceiling or sheathing should be entirely excluded. The roof trusses should be of structural steel, but the roof can be sheathed with 2-inch plank and overlaid with slate, or, for a large building, may have trusses of sufficient strength to carry a fireproof roof. The floors can be laid around the foundations with concrete made of stone, gravel, or cinders, and finished with a granolithic coating. A fire-wall should separate the boiler room from the engine room. The floor of the boiler room will be most durable if laid with vitrified brick on a concrete bed. All doors should be of Underwriters' pattern, made as follows: Two courses of $\frac{7}{8}$-inch matched pine, one course vertical the other diagonal, a layer of $\frac{1}{8}$-inch sheet asbestos between; the wood secured with wrought or wire nails clinched, all exterior surfaces covered with $\frac{1}{8}$-inch sheet asbestos, then entirely overlaid with tin, the joints lock-seamed, nailed, and hammered down over the nails. The doors should be self-closing by hanging on a sloping track and heavy wrought-iron hardware secured by through bolts should be used. Such doors will withstand a more intense heat than sheet-iron doors. Windows on sides exposed to fire from without should be protected with shutters constructed in a similar manner. All switchboard work should be thoroughly fireproof and all precautions taken to avoid short circuits and burn-outs. Wherever conductors pass through walls or floors they should be thoroughly protected by porcelain sleeves.

35. Lubricating oils should be stored in a separate fireproof enclosure. Everything about the station should be kept scrupulously clean. In a large station, similar lines of construction will be followed on a more elaborate scale, and the structural-steel frame will be an important feature. In designing a power station, it should be remembered that the convenient and secure placing of the equipment on the respective foundations is the first and essential feature; the building should be regarded as a suitable enclosure designed to protect the equipment and employes from the elements. Utility should be the first consideration; architectural design is secondary. Every feature that tends toward the economy of operation should be carefully studied out. Very frequently the inconvenient arrangement of the apparatus in a station, or location of floor levels, etc., may require the employment of extra labor, over what would be necessary with a better thought-out plan. The cost of such labor will seldom be less than 5 or 6 per cent. per year on an investment of $10,000.

36. In the matter of roofs, preference as to durability, first cost, and least cost of maintenance come in about the following order: Tile laid in cement on concrete, slate laid in cement on concrete, slate laid on plank, slate laid on lath, tin laid on plank, and corrugated iron laid on lath. Metal roofings should be avoided, as they entail a cost for painting and maintenance that is not incurred with a roof of slate, tile, or concrete.

WATER SUPPLY

37. Without a supply of water no electric power station can be operated. A gas-engine station needs it for cooling the engine cylinders; a water-power station for driving the wheels. For a steam-power station it is essential for boiler feeding, and condensing the exhaust steam in case condensers are used. Any deficiency in its purity, reliability, and abundance will affect the economy of the station. The ideal quality of water is rarely found and the best that can be done is to make sure of a constant and unfailing supply sufficient for the successful operation of the station; its deficiencies in quality must be overcome by the method best adapted to make it fit for boiler-feed purposes.

38.   Having determined on the best source of supply the next step is to ascertain what ingredients are contained in the water, since in its natural form almost all water contains, either in solution or suspension, more or less vegetable and mineral matters.   River and lake waters contain from 5 to 20 grains per gallon in solution, and from 10 to 15 grains in suspension.   Well and spring waters contain from 10 to 600 grains in solution and little in suspension.

39.   Practically considered, **boiler scale** is usually formed from carbonates of lime and magnesia, sulphates of lime and magnesia, chlorides, and silicious material.   The heat employed to generate steam sets free the impurities held in suspension, and if these are not eliminated before the water is fed into the boiler they are precipitated on the inner surface of the shell and on the tubes to which they firmly adhere in the form of scale.

EFFECTS OF SCALE

40.   Much has been written concerning the waste of heat caused by scale on boiler plates and elaborate theoretical tables have been prepared showing the extra expenditure of fuel required according to the increased thickness of scale; much of this data is unreliable and misleading.   Clean boilers are desirable, and thick scale is detrimental, but a scale of $\frac{1}{16}$ to $\frac{1}{8}$ inch does not require the expenditure of any considerable excess of fuel over and above what the same boiler would require with clean plates; the thin scale may prevent pitting of plates and tubes.   There can be no doubt but that scale has a detrimental effect on the heat-absorbing power of the plates, but not to the extent frequently claimed.   The objection to a thicker scale is that it may cause the metal of the boiler to become heated to such a high degree as to cause burning or bulging of the plates, or some other form of rapid deterioration.

Ordinary boiler scale of a more or less porous character has but trifling effect on the efficiency or capacity of a boiler, until the deposits become so thick as to interfere seriously with the transmission of heat. A hard, dense scale is more detrimental, depending on quality and thickness, but under scarcely any circumstances will the losses equal those stated in the published tables. Soot and fine ash dust are greater non-conductors of heat than scale and will seriously interfere with capacity and efficiency.

IMPURITIES IN WATER

41. The analysis of water should be given in grains per United States gallon of 231 cubic inches, which, in round numbers, contains 58,400 grains. To change the figures of an analysis given in parts per 100,000, to grains per gallon, multiply by .584 or, to obtain an approximate figure, by .6. To determine the pounds of scale-forming, or incrusting, solids per 1,000 gallons, divide grains per gallon by 7. Water containing more than 10 grains of incrusting solids per gallon is classed as *hard water*, and water containing from 40 to 50 grains or upwards is liable to foam or prime. The dirt in the water causes it to foam and froth on the water surface in the boiler, which action is apt to lead to *priming;* i. e., the carrying over of water, to the engine, mixed with the steam.

The following is a description of the properties and action of the substances found in waters requiring purification for boiler-feed purposes:

42. Carbonate of Lime.—Carbonate of lime is the commonest form in which lime occurs in water. It is but slightly soluble in chemically pure water, but when carbonic acid is present it is found in the form of bicarbonate of lime, which is quite soluble. Bicarbonate of lime, when carried into a boiler, is decomposed by the heat; the carbonic acid is driven off with the steam and normal carbonate of lime is formed, which is practically all precipitated in the boiler when the temperature reaches 290° F. Carbonate of lime alone does not form very hard scale, but it is responsible for a good deal of the mud that is found in boilers. However, it may form part of a very hard scale when materials that cement it to the sides and flues of the boiler are present.

43. Sulphate of Lime.—Sulphate of lime (commonly called gypsum or plaster of Paris) is a common constituent of natural water and is responsible for the hardest kind of boiler scale, this scale sometimes being as hard as porcelain. It is almost entirely precipitated when the boiler pressure is at 50 pounds, precipitation being in the form of heavy crystals that at once fasten themselves to the sides of the boiler. Sulphate of lime attaches itself to the sides of a boiler much more firmly than carbonate of lime.

44. Chloride of Calcium.—Chloride of calcium is sometimes found in natural water, in which it is very soluble. It is classed among the corrosive minerals found in water. It does not, of itself, form scale, but when other sulphates are present a transfer of acids takes place and calcium sulphate is formed.

45. Calcium Nitrate.—Calcium nitrate rarely occurs and is even then of but little importance, as the quantity is usually very small. It, of itself, does not form scale, but in the presence of sulphate of soda an exchange of acids takes place in the boiler and the nitrate is converted into sulphate of lime. Its action is corrosive.

46. Magnesium Carbonate.—Magnesium carbonate in its commonest form is used as a toilet preparation; it is then known as magnesia. It behaves in exactly the same manner as carbonate of lime, its bicarbonate being soluble, and its normal carbonate being practically insoluble. Magnesium carbonate is much used as lagging for boilers and is an excellent non-conductor of heat, but when in the form of boiler scale is on the wrong side of the shell.

47. Magnesium Chloride.—This is a very objectionable mineral when present in boiler water, it being very corrosive in its action, quickly pitting and grooving boilers that use water containing it.

48. Magnesium Sulphate.—Sulphate of magnesium (commonly known as Epsom salts) is a common constituent of natural waters, in which it is extremely soluble. It does not, of itself, form boiler scale, but is broken up by the lime salts when present in the water, and forms scale.

49. Sodium Sulphate and Sodium Chloride.—These salts (commonly known as Glauber's salt and common salt, respectively) do not form boiler scale nor corrode iron; they are not objectionable unless present in very large amounts, when they may cause foaming or priming in the boilers.

50. Sodium Carbonate.—This constituent does not form boiler scale nor corrode, but is objectionable when present in large quantity, as it causes foaming.

51. Iron.—Iron is generally present in water in the form of bicarbonate, but iron bicarbonate being a very unstable compound, quickly gives out its excess of carbonic acid, and absorbing oxygen is converted into iron rust, this being the cause of many waters turning red when standing exposed to the air for a short time. Carbonate of iron causes boiler scale. Waters from the vicinity of coal beds sometimes contain sulphate of iron, which is extremely corrosive.

52. Carbonic Acid.—Carbonic acid is found in all natural waters; it is the same gas as used in soda and seltzer waters and is responsible for the presence of many of the above minerals, as it holds them in solution in the water.

53. Silica.—Common sand is nearly all pure silica. Though contained in almost every water, silica is found to the greatest extent in warm waters. It is frequently in combination with alumina, and except in some few cases, is present in such small quantity that it has little to do with the formation of boiler scale.

54. Acids and Alkalies.—Water may be alkaline, neutral, or acid; it is generally alkaline. Some waters are acid, although this is quite rare, except where waters are drawn from coal mines or from the vicinity of coal beds. These waters may become quite acid with sulphuric acid, which is produced by the oxidation of the pyrites or sulphide of iron that is always found with coal.

47—4

55. Suspended Matter.—Organic or inorganic matter may be held in suspension in water and is very variable in quantity, depending on the source of supply of the water and the condition of the rainfall and the season. Suspended matter forms boiler scale only by being cemented to the boiler by other materials.

56. There should be no question in the mind of any intelligent engineer regarding the necessity of adopting the most effective methods of purifying and heating the water before it is fed into the boilers. This will not only economize in coal, but also in the cost of cleaning the boilers, and will materially add to their life and durability. The aim should be to collect the scale-forming matter and sediment from the water before it enters the boilers.

TESTING FEEDWATER .

57. To determine what objectionable ingredients are contained in the water, a careful chemical analysis is necessary. If a competent chemist is not accessible the engineer may be able to make the following simple tests. For this purpose the following chemicals and apparatus will be required: One $\frac{1}{2}$-pint bottle of soap solution, one 2-ounce bottle of lime water, one 2-ounce bottle of chloride of barium, one 2-ounce bottle of ferrocyanide of potassium, one 2-ounce bottle of hydrochloric acid, one 2-ounce bottle of nitric acid, one 2-ounce bottle of tincture of cochineal, one 2-ounce bottle of metallic mercury, one 2-ounce bottle of carbonate of ammonia (crystals), one 2-ounce bottle of chloride of ammonia, one 1-ounce bottle of oxalic acid (crystals), one 1-ounce bottle of phosphate of soda (crystals), slips of blue and red litmus paper, one 4-ounce flat-bottom clear-glass bottle, one wooden test-tube holder, one small spirit lamp, $\frac{1}{2}$ pint of alcohol, one test-tube brush, $\frac{1}{2}$ dozen of test tubes.

Fill a clean bottle with the water you desire to test and proceed as follows:

1. *Test for Hard or Soft Water.*—Take a clean test tube and pour into it about ¾ inch in depth of the soap solution, then add three or four drops only of the water; if the solution becomes milky or curdly the water is hard. Or dissolve a small quantity of soap in alcohol and put a few drops of the solution in a vessel of water; if the water turns milky, it is hard, if not it is soft.

2. . *Test for Acid or Alkali.*—Dip into a test tube half filled with the water, a strip of red litmus paper; if it turns blue the water is alkaline. Dip a strip of the blue litmus paper into the water; if it turns red the water contains acid.

3. *Test for Carbonic Acid.*—Pour about ¾ inch of water into a test tube and then pour in the same quantity of lime water; if carbonic acid is present the water will become milky; on adding a little hydrochloric acid, the water will become clear again.

4. *Test for Sulphate of Lime or Gypsum.*—Pour water to the depth of 1½ inches in a test tube and add a little chloride of barium; if a white precipitate is formed and will not redissolve when a little nitric acid is added, sulphate of lime is present.

5. *Test for Magnesia.*—Fill a test tube one-fourth or one-third full of water, hold it with the tube holder, and bring it to the boiling point over the spirit lamp; then add the point of a knife full of carbonate of ammonia, and a very little phosphate of soda; if magnesia is present it will form a white precipitate, but as it may not do so at once it is better to set it to one side for a few minutes.

6. *Test for Lead.*—Fill a test tube one-fourth full of water and add one or two drops of tincture of cochineal; if there be but a trace of lead in the water the solution will be colored blue instead of pink.

7. *Test for Copper.*—Add to some water in a test tube a little filing dust of soft iron and a few drops of chloride of ammonia; a blue colorization denotes the presence of copper.

8. *Test for Iron.*—To some water in a test tube add one drop of ferrocyanide of potassium; the water will become blue if iron is present.

58. The following simple tests may also be used and under some conditions may be found more convenient than those previously given:

1. *Earthy Matter, or Alkali.*—Dip litmus paper into vinegar; if on immersion in water the paper returns to its former shade, the water contains earthy matter or alkali.

2. *Carbonic Acid.*—Take equal parts of water and clear lime water.. If combined or free carbonic acid is present, a precipitate will be produced, in which, if a few drops of hydrochloric acid is added an effervescence will take place.

3. *Magnesia.*—Boil the water to one-twentieth part of its weight, drop a few grains of neutral carbonate of ammonia and a few drops of phosphate of soda into it; a precipitate will be formed if magnesia is present.

4. *Iron.*—Dissolve a little prussiate of potash and mix it with the water; the water will become blue if iron is present.

5. *Lime.*—Put two drops of oxalic acid into a glass of water and blow on it; if the solution becomes milky, it indicates the presence of lime.

6. *Acid.*—Immerse a piece of blue litmus paper in the water; if it turns red, the water is acid. If it forms a precipitate on adding lime water, carbonic acid is present.

IMPROVEMENT OF WATER BY CHEMICAL TREATMENT

59. Kennicott Water Softener.—A successful method now employed for the softening of water before it is fed into the boilers, is that brought out by the Kennicott Water Softener Company. The result is accomplished by chemical purification, the ingredients (lime, soda, or other neutralizing agents) being introduced into the "raw" water in accurate proportions. Fig. 11 (*a*) shows a cross-section of the apparatus and (*b*) a plan looking down on top.

The machine consists of a steel settling tank *a* provided with a conical bottom *b*. In the center of this tank and open at the top and bottom is a cone-shaped conduit or downtake *c* within which is a tank *d* for the preparation of the lime water; this tank is, of course, closed at the bottom. The

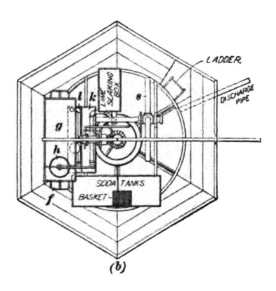

—TO SODA TANK

DISCHARGE PIPE

WOOD FIBER FILTER.

LIME HOIST.

LADDER.

DISCHARGE PIPE

LIME SLAKING BOX

8

_i_ _k_

_g_

_h_

SODA TANKS

BASKET

_f_

(b)

PERFORATED BAFFLE PLATES

_a_

UDGE DISCHARGE

87

(a)

FIG. 11

lime-saturator tank is enclosed in the conduit so that danger from freezing of water in small parts is obviated. The supply pipe *e* runs through the body of the apparatus for the same reason. On the top of the tank is placed a cast-iron overshot waterwheel *f*, which furnishes all the power necessary to operate the apparatus. The water from the source of supply is delivered to the hard-water box *g* above the top of the settling tank; from this it passes through a slot in the bottom, the size of the slot being adjusted according to the amount of water to be treated. Within this box is a float *h* with chains passing over pulleys and connected to hinged inlet pipes in the two boxes or small tanks *k*, *l* that contain soft water and the soda solution, respectively. The object of this arrangement is to vary the supply from those tanks in accordance with the supply of raw water. Thus, if the rapidity of the supply increases so as to maintain a depth of 6 inches in the box, or a head of 6 inches over the slot in the bottom of the box, the rising of the float will allow the ends of the hinged inlet pipes to drop 6 inches below the surface of the respective tanks, thus maintaining a constant uniform head. The ends of these hinged inlet pipes are closed with caps, each cap having a small slot of such size as to give the required proportion of solution to the water. This, of course, varies according to the character of the water to be treated.

60. In order to separate the lime and magnesia from the water a solution of lime water is added. The lime and magnesia are in the form of soluble bicarbonates, but when lime water is added these change to carbonates, which are insoluble and therefore settle out. This method of treatment was discovered by Doctor Clark, an English scientist. In order to remove the sulphates, chlorides, and nitrates of lime and magnesia, the water is treated with a solution of sal-soda or soda ash, which leaves in the water harmless sodium sulphate and sodium chloride. This method of treatment was discovered by Doctor Porter, another English scientist, and the whole process of treatment is often referred to as the *Porter-Clark* process of water softening. In the Kennicott softener, some of the water that has been softened is used to make up the lime solution, as it is found that less lime is required than if hard water is used.

Referring to Fig. 11 (*a*), the flow of water is indicated by the arrows. The hard water flows up through pipe *e*, discharges into box *g*, and flows over the waterwheel *f* into the top of the cone. At *n*, the soda solution is introduced in the proper proportions. In the upper part *p* of the lime-saturator tank the water is thoroughly mixed, by means of revolving paddles, with the lime water, which flows up through a perforated plate *m*. The water then enters the top of the settling cone, in which its velocity continually decreases, causing the particles held in solution to fall to the bottom of the cone, the larger particles carrying the smaller ones with them. On reaching the bottom of the tank, the current is reversed, and the water rises through a series of perforated conical baffle plates *o, o*; at the same time the velocity continues to decrease, owing to the increasing diameter of the water space. On reaching the top, the water passes upwards through a filter compartment filled with wood fiber, and enters a shallow soft-water tank *r* from which it flows to the storage tank for supplying the boilers. At the bottom of the settling tank is a conical hopper with valve for discharging the sediment. The soft water for making the lime solution is lifted from tank *r* to tank *k* by means of a lift wheel *s* driven by the waterwheel and provided with curved buckets that discharge the water through a hollow shaft into tank *k* from which the water flows to the bottom of the lime saturator through the pipe *t*. The elevation of tank *k* gives sufficient head to make the lime water in the saturator come up through plate *m* to be mixed with the hard water. The waterwheel is also arranged so that, by throwing in a clutch, it can be used for operating a hoist for raising the lime and soda to the top of the tank. The whole apparatus is automatic in its action and the only attendance necessary is that to keep up the supply of chemicals. It is claimed that the expense per thousand gallons of water purified is as follows: For the cost of chemicals, 3 cents; for interest on investment, ½ cent; for attendance, ½ cent; making a total cost of approximately 4 cents, which in especially adverse conditions might be increased to 5 cents.

61. Extra Cost Due to Impure Water.—While much has been written and many investigations made about the use of impure water and the resultant cost, the data connected therewith has probably never been so well assembled as by the American Railway Master Mechanics Association. A committee appointed on this particular subject has accumulated sufficient evidence to show that the cost of using impure waters in locomotive boilers is as follows:

Cost of extra cleaning per boiler per year . $ 50
Cost of extra repairs per boiler per year   . 360
Cost of extra fuel per boiler per year   . . 340

making a total of $750 extra cost per boiler per year due to the use of bad water.

PURIFICATION OF WATER BY LIVE STEAM

62. Many of the impurities are precipitated and form scale when the water is heated to the high temperature corresponding to the steam pressure maintained in the boiler. It follows therefore that if some device is provided in which the feedwater can be heated up to or near the temperature of the water in the boiler, the impurities can be precipitated before the water is fed into the boiler. These devices might be called *combined heaters and purifiers*, but since they usually use live steam they are often called **live-steam feedwater heaters.** They are made in a variety of forms; some are constructed with removable pans for collecting the scale-forming material, while others depend on the use of a blow-off and occasional opening up of the heater for the removal of the sediment.

63. Fig. 12 shows a **Hoppes purifier,** which is of the removable-pan type. It consists of a cylindrical shell *a* fitted on one end with a removable head *b*, shown in the figure as

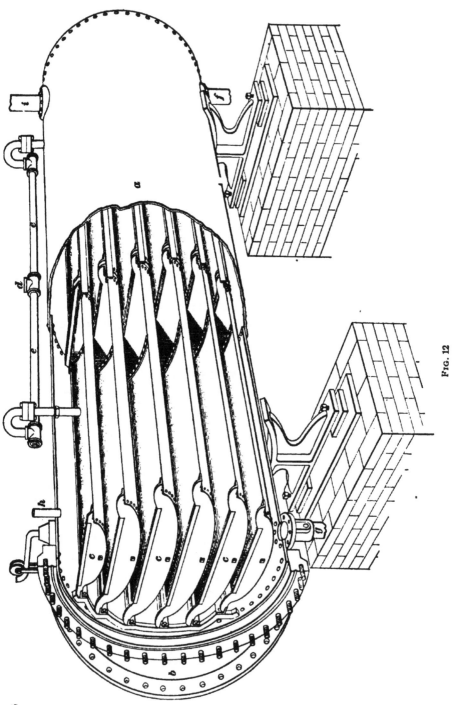

FIG. 12

taken off and swung out of the way; a series of shallow steel pans *c, c* are placed within the purifier. The feedwater enters at *d* and flows through the branch pipes *e, e* into the top pans, from which it flows in thin sheets over the edges of the pans and finally out of the purifier through the pipe *f*. Live steam from the boiler enters through the pipe *i* and heats the water in the purifier to a temperature nearly equal to its own, and in so doing precipitates those impurities that become insoluble by heat. Mud and earthy matter deposit on the inside of the pans while the scale-forming substances coat the outside; the pans are removed occasionally and cleaned. The feedwater flows through *f* into the boiler by gravity, the purifier being placed higher than the boiler and having a pressure equal to that in the boiler within it. A blow-off pipe is connected at *g* and a glass water-gauge connects to *g* and *h*.

Fig. 13 shows the arrangement of a live-steam purifier in connection with an exhaust-steam heater. The exhaust steam from the engines passes through the exhaust pipe *a* into the heater *b* where it meets the cold feedwater delivered from pipe *c*. The exhaust steam heats the water to a temperature of, say, 206° to 210°. From *b* the water is pumped by the boiler feed-pump *d* into the live-steam purifier *c*, which is supplied with live steam through pipe *f*; hence, the water must be pumped in against the boiler pressure. The purifier is arranged above the level of the boilers, so that the water can run in by gravity through pipes *g*.

Should the station be so situated that it must use, for boiler feed, water that is highly impregnated with incrusting materials, such as carbonates and sulphates of lime and magnesia, it may be decided to use a live-steam purifier, but it should only be used as a last resort since it is more economical to use, if possible, sources of waste heat for heating the feedwater.

When the water contains in solution such ingredients as can be neutralized or rendered harmless by the addition of substances, such as caustic soda, kerosene oil, or trisodium-phosphate, it is desirable to add the neutralizing agent before

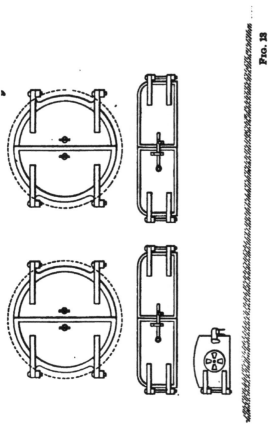

FIG. 13

the water enters the heater. The supply may be stored in a barrel or tank and fed regularly in the proper proportions by

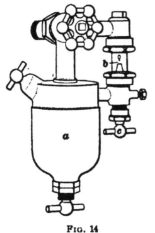

Fig. 14

a suitable appliance. Kerosene oil fed into a boiler has the effect of loosening scale and preventing its adherence to the tubes. The oil is fed in by a device very similar in appearance to the lubricators used for engine cylinders.

Fig. 14 shows a *boiler oil injector*, as it is called. The oil is contained in the reservoir *a* and passes through the glass tube in the form of drops on its way to the boiler. The attendant can therefore see the amount supplied and regulate it, as may be required, by means of the valve *c*.

HEATING FEEDWATER

64. The feedwater furnished to steam boilers must of necessity be heated from its normal temperature to that of steam before evaporation can commence, and if not otherwise accomplished, the heating will be done at the expense of fuel that should be utilized in making steam. This temperature at 75 pounds pressure is 320°, and if we take 60° as the average temperature of feedwater, we have 320 − 60 = 260 British thermal units of heat required to raise 1 pound of water from 60° to 320°. It requires 1,151 heat units to convert a pound of water at 60° into steam at 75 pounds pressure, so that the 260 units required for heating the water represents about 22.6 per cent. of the total. All heat, therefore, that can be imparted to the feedwater before it enters the boilers is just so much saved, not only in cost of fuel, but in capacity of boiler. It must be remembered that of the total heat generated in the furnace of the boiler, nearly 80 per cent. is lost and cannot be converted into mechanical power on account of the low efficiency of the steam engine as a heat engine.   Therefore, the sources of waste heat must be utilized, and the question of the proper selection of auxiliary appliances to obtain from this waste the largest possible benefit becomes of first importance.   The unused heat units have cost just as much in proportion to the price paid for the fuel as the useful units; hence, in the interest of economy, the largest possible percentage of the heat units that would otherwise be wasted must be returned to the boiler and thereby utilized.

65. Sources of Waste Heat.—The principal sources of waste heat are as follows: First, the exhaust steam from engines; second, the exhaust steam from pumps and auxiliary appliances; third, the heat carried off by the gases passing from the furnace through the flues and up the stack. In one case we have the heat rejected with the exhaust steam, which may be utilized in one or more of the numerous types of heaters, and in the other case we have the heat of a higher temperature rejected by the boiler furnaces.   Of the several methods of deriving benefit from these escaping heat units, the least in cost are those employing exhaust-steam heaters.   The economizer method, or the method of making use of the waste heat in the furnace gases, is the more expensive, but under certain conditions it is desirable.

SELECTION OF EXHAUST STEAM FEEDWATER HEATER

66.   The ingredients contained in the water will largely determine the type of exhaust-steam heater to be used in any given plant.   These heaters are divided into two general classes, known as *open heaters* and *closed heaters*.   An open heater may be defined as one in which the water is in contact with the atmosphere.   In a direct-contact open heater, the exhaust steam comes into contact with the water, which, by means of suitable devices, is broken into spray or thin sheets in order to readily absorb the heat of the steam.   In a coil heater, the exhaust steam passes through coils of pipe submerged in a suitable vessel containing the water to be heated and open on top.   **Closed, exhaust-steam,**

feedwater heaters may be defined as heaters in which the feedwater is not exposed to the atmosphere, but is subjected to the full boiler pressure. The steam does not come into contact with the water; the latter is heated through coming into contact with metallic surfaces, generally those of tubes, that are heated by the exhaust steam.

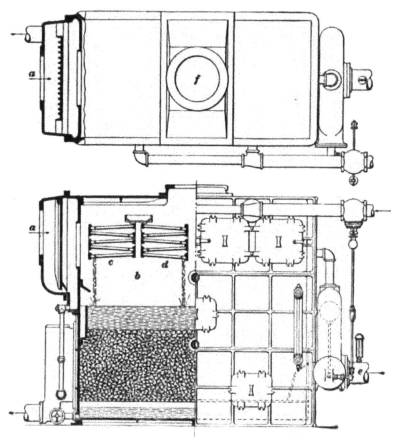

FIG. 15

67. Cochrane Heater.—Fig. 15 shows the **Cochrane heater,** which will serve as an example of the open type. This heater serves also as a purifier. The exhaust steam enters at *a* and passes into the chamber *b*, where it comes into contact with the cold feedwater, which flows over the trays *c d*;

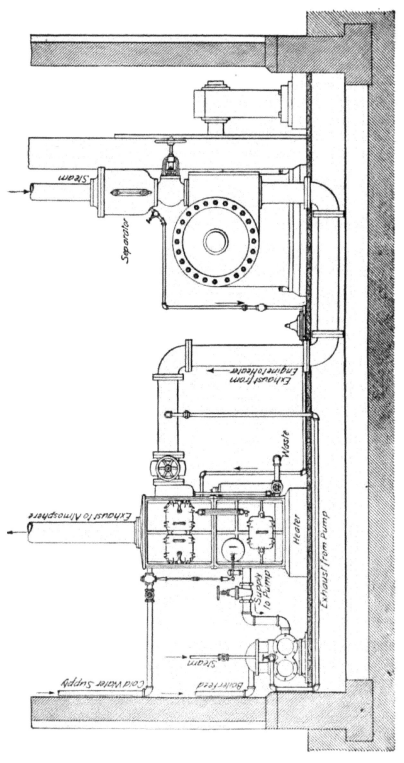

Steam

Separator

Exhaust from Engine to Heater

Waste

Exhaust to Atmosphere

Supply to Pump

Heater

Exhaust from Pump

Cold Water Supply

Steam

Boiler Feed

FIG. 16

this exposes a large surface of water to the steam. The heated water collects in the lower part, filters through a thick layer of coke, and is drawn off by the boiler feed-pumps attached to the outlet *e*. The exhaust steam passes out at the outlet *f* on top of the heater. Fig. 16 shows a Cochrane, open, exhaust-steam heater connected to an engine and boiler feed-pump.

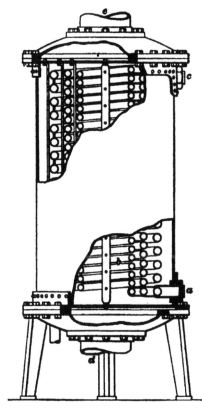

FIG. 17

68. Closed Heaters. Fig. 17 shows an American feed-water heater, which is of the closed type. The feed-water enters at *a*, passes through the triple pipe coil *b*, and out at *c*; the exhaust steam enters at *d* and passes out at *e* after passing between the coils of the pipe and heating the water contained therein. Fig. 18 shows a Berryman heater, which is of the closed type also. The heating surface is obtained by means of tubes *a, a* through which the exhaust steam passes. The steam enters the bottom *b* of the heater through the pipe *c*, and there being a steam-tight partition in the bottom, is caused to flow up through one leg of the tubes and down the other and thence through pipe *d* to the atmosphere. The condensed steam is discharged through the drip *e*. The feed-water enters the heater through pipe *f* and leaves at the top through the feedpipe *g*. A safety valve *h* is fitted to the heater, which prevents any overpressure due to a closing of the globe valve in the feedpipe before the pump is stopped. When the safety valve is open, the water discharges through

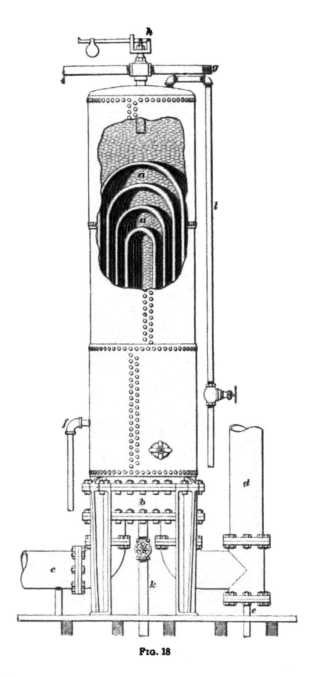

Fɪɢ. 18

i to the sewer. The heater is fitted with a bottom blow-off *k* and a surface blow-off *l* for removing settled and floating impurities in the water. It heats the water to a temperature between 200° and 212°, and in so doing precipitates the scale-forming substances that become insoluble at these temperatures.

Where the boiler feedwater is free from acids, salts, sulphates, and carbonates so that no scale is formed under a high temperature, the closed feedwater heater will be found satisfactory. This kind of water is not available in the majority of places. Heaters of the coil type may be used with pure water, but should not be used with water that will precipitate sediment or scale-forming matter of any kind, because they have no sediment chamber. The coil heater is very efficient as a heater, as the water circulating through the coils is a long time in contact with the surface surrounded and heated by the exhaust steam. Heaters of the closed type with straight tubes and sediment chamber can be more readily cleaned than those having curved tubes, but the curved tubes allow more freedom for expansion and contraction. Heaters of the tubular type should have ample sediment chambers and may be used with water that contains organic matter or earthy matter, but not with scale-forming ingredients. Carbonate of lime is liable to combine with earthy matter and form an exceedingly hard scale.

69. Heaters of the open exhaust-steam type have the advantage of bringing the exhaust steam in direct contact with the feedwater; some of the exhaust steam is condensed (thus effecting a saving in feedwater) and sediment and scale-forming ingredients (excepting the sulphates of lime and magnesia) are precipitated or will settle to the bottom of the heater. The oil in the exhaust steam must be intercepted by special oil extractors, filters, or skimmers, generally combined with the heater and, by automatic regulation, sufficient fresh feedwater is added to make up the total quantity required. The open heater is available for heating and purifying where the impurities do not require a temperature exceeding 175° for precipitation. When the system is properly arranged all live-steam drips and discharge from traps will be discharged into the heater.

70. Table V shows the percentage of the lime thrown down from feedwater at various temperatures. A temperature of 290° F. is necessary for the precipitation of all the lime.

TABLE V
PERCENTAGE OF LIME PRECIPITATED FROM WATER AT VARIOUS TEMPERATURES

Temperature Degrees Fahrenheit	Per Cent. of Lime Thrown Down	Temperature Degrees Fahrenheit	Per Cent. of Lime Thrown Down
217	50.0	245	77.4
219	52.3	250	81.7
221	56.8	255	86.0
227	60.5	261	90.3
232	64.5	266	94.0
236	69.0	271	97.7
240	73.3	290	100.0

HEATING SURFACE OF FEEDWATER HEATERS

71. When determining the size of closed heater required for any given plant, the following quantities should first be obtained: The maximum weight of water to be heated per hour, the natural temperature of the water entering the heater, the temperature of the steam, and the temperature to which the water must be heated. A standard boiler horsepower has been defined by the American Society of Mechanical Engineers as equal to the absorption of 33,330 British thermal units per hour. This is equivalent to the evaporation of 34.5 pounds of water per hour at 212° F. into steam at 212° F. or of 30 pounds of water at 100° F. into steam at 70 pounds pressure. The number of square feet of tube surface required for each boiler horsepower can be calculated from the formula

$$S = .248 \log \frac{T_s - T_i}{T_s - T_o} \qquad (2)$$

where S = square feet of tube surface per horsepower, or the surface required to heat 34.5 pounds of water per hour;

T_s = temperature of steam passing through heater;

T_i = temperature of water entering the heater;

T_a = temperature of water leaving the heater.

The horsepower of heater per square foot of surface is $\dfrac{1}{S}$.

The value of S given by the formula is for copper tubes.

TABLE VI

AREA OF HEATING SURFACE REQUIRED IN FEEDWATER HEATERS PER BOILER HORSEPOWER

Initial Temperature of Feedwater Degrees Fahrenheit	Temperature of Water When Fed Into the Boiler Degrees Fahrenheit											
	180			190			200			210		
	Square Feet of Heating Surface Required for											
	Copper	Brass	Iron	Copper	Brass	Iron	Copper	Brass	Iron	Copper	Brass	Iron
50	.17	.19	.28	.21	.23	.34	.27	.29	.43	.39	.42	.63
60	.16	.18	.27	.20	.22	.33	.26	.28	.42	.38	.41	.62
70	.15	.17	.26	.19	.21	.32	.25	.27	.41	.37	.40	.61
80	.14	.16	.24	.18	.20	.30	.24	.26	.40	.36	.39	.60
90	.13	.15	.23	.17	.19	.29	.23	.25	.38	.35	.38	.58
100	.12	.14	.21	.16	.18	.27	.22	.24	.36	.34	.37	.56
110	.11	.13	.20	.15	.17	.25	.21	.23	.34	.33	.36	.54
120	.10	.12	.18	.14	.16	.24	.20	.22	.33	.32	.35	.53
130	.09	.10	.16	.13	.15	.22	.19	.20	.31	.30	.34	.51

For brass or iron tubes larger areas will be required; these can be obtained by multiplying the result given by the formula by 1.12 for brass tubes and 1.67 for iron tubes.

EXAMPLE.—What area of copper-tube surface will be required for a 100-horsepower heater that is supplied with steam at 215° F. and is to heat the water to 190° F., the temperature of the feedwater being 60° F.

SOLUTION.—In formula 2, $T_s = 215°$, $T_i = 60°$, $T_a = 190°$; hence

$$S = .248 \log \frac{215 - 60}{215 - 190} = .248 \log \frac{155}{25} = .20 \text{ nearly}$$

hence, for 100 H. P. the tube surface should be $.20 \times 100 = 20$ sq. ft. Ans. If iron tubes were used the area would be $20 \times 1.67 = 33.4$ sq. ft.

72. Table VI shows the area of copper heating surface required per horsepower for initial temperatures of feedwater ranging from 50° to 130° and final temperatures ranging from 180° to 210°. A boiler horsepower is taken as equivalent to an evaporation of 34.5 pounds of water per hour from and at a temperature of 212° F. The temperature of the steam in the heater is taken as 215°, because the pressure in the heater is always slightly in excess of atmospheric pressure.

HEATING BY WASTE FLUE GASES

73. The gases going to the chimney carry off, on an average, according to good authority, from 15 per cent. to 50 per cent. of the heat units contained in the fuel. Some portion of this heat is always available for heating the feedwater, by using what are known as **economizers**, and frequently the feedwater may be carried nearly to the temperature of high-pressure steam, making a saving in some instances of 80 per cent. The heating surface of the economizer intercepts and absorbs a large percentage of the heat from the gases passing to the chimney. The more wasteful the boiler, the greater is the benefit derived from the economizer; but for large plants it is always a valuable adjunct, and particularly where condensing engines are used. In many cases water having been heated by exhaust steam may be raised to a much higher temperature by being passed through an economizer.

Fig. 19 illustrates a Green fuel economizer. It consists of groups of vertical tubes h connected into headers k and so arranged that the water may be forced through the tubes in a direction opposite to the flow of the hot gases circulating around them. Thus the water enters at f and the coolest gases meet the coolest water; as the water becomes heated

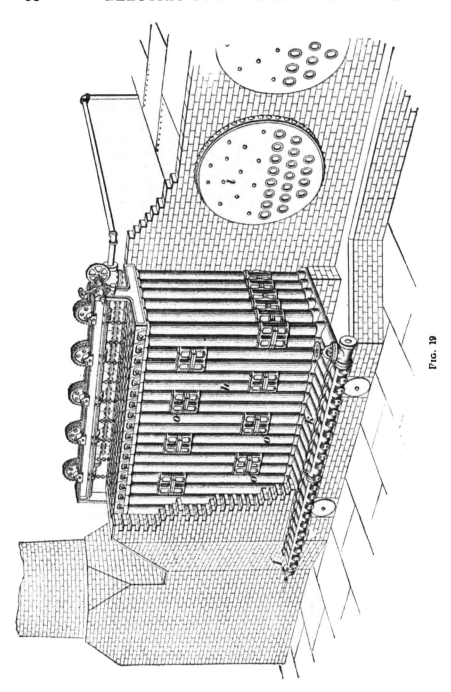

FIG. 19

by passing to the right it comes into contact with the hotter gases nearer the boiler. The economizer is placed in an enclosing chamber, usually of brick, through which the gases pass on their way from the boilers *l* to the chimney. Fig. 20 shows a typical arrangement of economizer with a by-pass flue for use in case it is desired to shut off the gases from the economizer.

No economizer is complete without some device for cleaning the soot and ashes from the tubes, because a coating of soot retards the absorption of heat from the gases by the water. In the Green economizer this cleaning is effected by

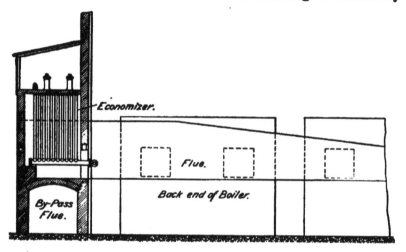

Fig. 20

scrapers *o, o* that are slid slowly up and down the tubes by means of suitable gearing. Cast-iron pipes are used to avoid pitting and corrosion, to which wrought-iron pipes are subject. The temperature to which an economizer can raise the feedwater depends on the temperature of the gases leaving the boilers. Where sufficient boiler heating surface is used to reduce the gases to a low temperature before their discharge to the chimney, economizers will not be of so great advantage as where less boiler heating surface is applied.

74. The economy resulting from the use of an economizer depends on the temperature at which the feedwater is supplied, and also on the temperature of the escaping flue gases from the boiler furnaces, which should be sufficient to increase the temperature of the water 75° to 100°, at least. As the economizer is expected to raise the water to a temperature above 212° F., water containing scale-forming impurities should be treated with a solvent that will prevent the deposit of scale on the inner surface of the economizer tubes. As the tubes of the economizer retard the flow of the hot gases, an intense draft is necessary. If natural draft is to be used an extra height should be estimated for when designing the chimney, but with draft induced by mechanical means, such as fans or steam blowers, the required intensity is readily controlled.

75. If the economizer raises the temperature of the water from 75° to 100° an increased efficiency of from 7½ to 10 per cent. may be expected. It is not practicable to reduce the temperature of the gases below 250°, and if they do not reach the economizer higher than 350°, the temperature of the feedwater will be raised but 50°, giving an increase of 5 per cent. in efficiency.

The following calculation illustrates the value of an economizer. Assuming that it requires 3½ pounds of coal per hour to generate 1 boiler horsepower, a plant operating 24 hours daily, or 8,760 hours during the year, will require 8,760 × 3.5 = 30,660 pounds, or 13.7 tons of coal per horsepower per year, which at $3 per ton will amount to $41.10. Assuming that the economizer saves 8 per cent. of this the saving will be $3.29. The estimated cost of the economizer erected, including by-pass, flues, and dampers, is $1.75 per square foot. Allowing 4 square feet per horsepower, the cost is $7 per horsepower installed. Charging 12 per cent. of the cost against the economizer for attendance, maintenance, depreciation, and interest, we have 84 cents, which amount deducted from saving in coal as above, leaves $2.45 net saving under the above stated conditions. As boiler heating surface costs about 60 cents and upwards per square foot, and economizer heating surface about $1.75 per square foot, it becomes a nice question to decide wherein lies the more useful and economical investment in heating surface.

76. Instructions Relating to Economizers.—Where economizers are used the following instructions relative to their care and management should be observed: The foundations should be substantial. A by-pass flue with dampers must be provided; the flues should have the fewest possible turns, and these should consist of easy curves. Holes, cracks, and air leaks must not be allowed. The main steam-regulated damper should be placed between the economizer and the chimney, and the safety valve should be blown daily or oftener to free any scum collected in the outlet pipe. The blow-off valve should be opened a few seconds daily, as the condition of water may require. All scrapers should be kept well lubricated, and the soot chamber cleaned out frequently. The economizer should be cleaned more or less frequently, according to the quality of the feed-water, to clear the inside of scale and sediment. Thermometers should be fixed in inlet and outlet pipes to determine temperatures.

HEATERS USED WITH CONDENSING ENGINES

77. Steam engines are generally run condensing whenever it is possible to obtain sufficient water. That is, instead of allowing the exhaust steam from the engine cylinder to escape into the air, it is led into a condenser where, by means of cold water, it is condensed, thereby creating a partial vacuum behind the piston, thus practically adding the equivalent pounds pressure represented by the vacuum to the effective steam pressure and effecting a considerable economy.

Heating feedwater where condensing engines are used is a more complicated problem than with simple non-condensing engines. The intermediate feedwater heater connected in the line of exhaust between the engine and the condenser will be beneficial in obtaining heat from the exhaust steam before the steam passes into the condenser; the intermediate heater must be vacuum-tight. The feedwater passing through this heater will reach from 108° to 110° F. and can then be delivered to an open exhaust-steam heater.

The steam from all pumps, fan engines, and other engines used as auxiliaries is generally not used expansively, and consequently, the temperature of the exhaust steam from these auxiliaries is much higher than from the larger engines. The exhaust from the auxiliary pumps and engines can very profitably be used for further raising the temperature of the feedwater in the open heater from 190° to 210°, thus permitting the delivery of the feedwater to the boilers almost at boiling point. If in addition to this the feedwater is forced' through an economizer, the temperature will be raised considerably above the boiling point, and as it enters the boilers the water will be ready to expand into steam at once.

78. In a plant where condensers are used, the discharge of condensed steam has the advantage that all of the vegetable and mineral impurities have been removed through evaporation; therefore, if this water is used for boiler feed it is free from these impurities. The disadvantage is the impregnation with cylinder oil, which must be removed before the water is fit for boiler feeding. There are many types of grease extractors and oil separators on the market; the most efficient of these are claimed to remove from 95 to 98 per cent. of the oil contained in the water, thus rendering it sufficiently pure for boiler feed.

PUMPING

79. The cost of pumping water for boiler-feed and condensing purposes in electric power stations is no small item of expense. Every legitimate effort should be made at the time of designing the station to do this work in the most economical manner. The steam consumed by the ordinary steam pump is far beyond what is generally realized, and the subject of steam economy for pumping is most important. It is no uncommon thing to work duplex pumps so that they take steam from seven- to eight-tenths of their stroke, resulting in an enormous waste of steam. The ordinary steam pump will use 100 to 300 pounds of steam per horsepower per hour.

Several kinds of pumps and injectors are available for forcing the feedwater into the boilers. They may be steam-driven direct-acting, geared from the main engine, or geared to an auxiliary engine. In electric power stations, the pumps are frequently geared to motors that receive their current supply from the main generators. In an injector, the feed-water is forced into the boiler by virtue of the kinetic energy imparted to a jet of water by a jet of steam acting on it. Injectors may be operated either by live steam or by exhaust steam, but the latter cannot be used for forcing against high pressures. The injector is not as reliable as the pump and is not generally used in the regular operation of electric power stations.

Stated in the order of economy of operation, the boiler-feeding appliances would stand as follows:

Exhaust-steam injectors feed the boiler up to a pressure not exceeding 75 pounds and heat the feedwater to 185° or 190° with exhaust steam. This is only noted as a method of boiler feed, but is not considered practicable for electric power stations.

A *geared pump run from the engine* derives its power from the most economical source in the station.

A *geared pump driven by an electric motor* derives its power through the generator from the second most economical source.

A *direct-acting pump* with compound steam cylinders, and the exhaust used in an auxiliary heater.

A *direct-acting pump* using steam at high pressure with the exhaust used for heating feedwater in an auxiliary heater.

A *live-steam injector* feeding through a heater.

80. Pump Governor.—For best service, the supply of boiler feedwater should be continuous, and regulated by automatic control according to the demands on the boilers. The water-line in the boilers should be maintained at a constant level.   If the level be allowed to fall, the feed will be required more rapidly at intervals, and when so supplied will affect the uniformity of the steam pressure.   There are several automatic regulators on the market for this purpose.

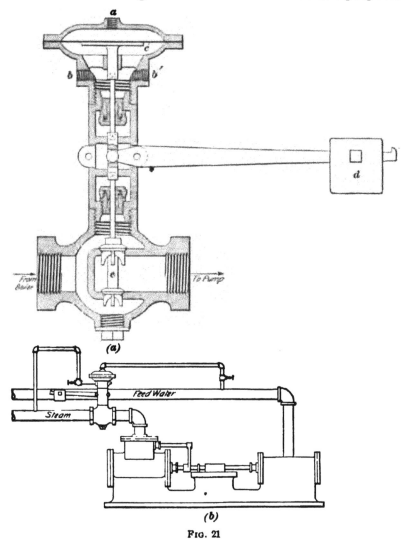

Fig. 21 illustrates a good device wherein the steam pressure in the boilers is balanced against the water pressure in the feed-line and thus automatic control is obtained.   Connection

to the feedwater pipe is made at a, Fig. 21 (a), and to the
steam pipe at b or b'. The steam is admitted through suitable openings, to the underside of the diaphragm c and the
water pressure acts on the upper side. By means of the
weight d, the governor can be adjusted so that when a certain level is maintained the throttle valve e will be closed and
the pump stopped. Both the upper and lower sides of c

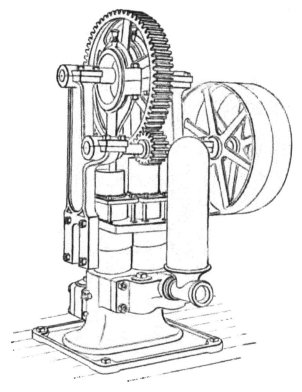

Fig. 22

are connected to the boiler, the upper side connecting to the
water space (through the feedpipe) and the lower side to the
steam space. A change in water level will therefore affect
the resultant pressure on the diaphragm. If the level lowers,
the pressure on the upper side becomes less and the steam
pressure raises valve e, thus starting the pump, which continues to operate until the downward pressure on c overcomes

the upward steam pressure and closes the valve *e*, thus stopping the pump. Fig. 21 (*b*) shows the method of piping the governor. Fig. 22 shows a typical **duplex pump** arranged for belt driving from either an engine or an electric motor. Fig. 23 shows a **triplex pump** geared to a motor. **Turbine**, or **centrifugal**, pumps are well adapted for direct connection to electric motors. Considerable attention has recently been given to the development of this type of pump, and they are now frequently used for power station

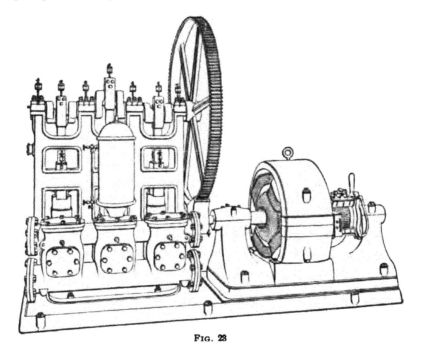

Fig. 23

work. By using two or more pumps in series and thus raising the pressure by two or more stages these pumps can be made to give a good efficiency, and the construction is very simple as compared with a reciprocating pump.

81. Relative Value of Feeding Methods.—The relative value of injectors, direct-acting steam pumps, and pumps driven from the engine, is a question of importance to all steam users. Table VII has been calculated by

D. S. Jacobus, M. E., from data obtained by experiment. It will be noticed that when feeding cold water direct to boilers, the injector has a slightly better economy, but when feeding through a heater, a pump is the most economical.

TABLE VII

RELATIVE ECONOMY OF DIFFERENT METHODS FOR FORCING FEEDWATER INTO BOILERS

Method of Supplying Feedwater to Boilers	Relative Amount of Coal Required per Unit of Time	Saving of Fuel Over Amount Required When Boiler Is Fed by Direct-Acting Pump Without Heater Per Cent.
Direct-acting pump feeding water at 60°, without a heater	1.000	.0
Injector feeding water at 150°, without a heater. .	.985	1.5
Injector feeding through a heater in which the water is heated from 150° to 200°	.938	6.2
Direct-acting pump feeding water through a heater in which it is heated from 60° to 200°	.879	12.1
Geared pump, run from the engine, feeding water through a heater in which it is heated from 60° to 200°	.868	13.2

NOTE.—The temperature of the feedwater delivered to the pump or injector is 60° F. The evaporation of the boiler is at the rate of 10 pounds of water from and at 212°, per pound of coal. In this table, the amount of coal per unit of time required by a direct-acting pump feeding water at 60° F., without a heater, is taken as unity.

82. The Comparative Economy of Motor-Driven Pumps.—The most desirable and economical type of motor for pump service will probably be a compound-wound double-commutator machine. This combined with series-field commutation can be regulated for at least four speeds, thus giving a practical range of feedwater supply equal to four ranges of piston speed on the usual type of steam pumps. If multi-voltage control is available it also constitutes an excellent method for the control of a pump motor. Other methods of control can also be used, as described in connection with direct-current motors. The efficiency of a small size motor will be 70 to 75 per cent., which is readily obtained in every-day practice. If we assume the average economy of the engine-generator unit to be only .025 electrical horsepower per pound of steam (40 pounds of steam per electrical horsepower per hour), the corresponding economy of the steam pump at 100 pounds of steam per horsepower-hour would be $\frac{40}{100} \times .025 = .01$ horsepower per pound of steam and the motor-driven pump at 70 per cent. efficiency would show an economy of $.025 \times .7 = .0175$ horsepower per pound of steam delivered to the engine, or on the full horsepower-hour basis the steam pump would require 100 pounds and the motor-driven pump $\frac{1}{.0175} = 57$ pounds, the saving by the motor-driven pump being 43 pounds of steam per horsepower-hour.

Assuming the pump to require 5 horsepower on 24 hours daily service, we have $5 \times 24 \times 43 = 5,160$ pounds per day, or for 365 days a saving of 1,883,400 pounds of steam. Assuming a good grade of coal and evaporating 6 pounds of water per pound of coal, we have $1,883,400 \div 6 = 313,900$ pounds of coal, or about 140 gross tons, which at $3 per ton would cost $420. This capitalized at 10 per cent. would be $4,200, showing the extra value of the motor-driven pump. In spite, however, of the greater economy of motor-driven pumps, there is considerable difference of opinion as to the advisability of their use. A source of current must always be available to run the pumps and this means that the generators must either be running or else the station must be equipped with a storage battery. The steam pump is thus more independent than the electrically driven pump and for this reason is often preferred.

ELECTRIC POWER STATIONS
(PART 2)

CHIMNEYS AND MECHANICAL DRAFT

THE COMBUSTION OF FUEL

1. In connection with the production of steam, certain salient points regarding the combustion of fuel should be understood. **Fuel** is any substance that, by means of the introduction of natural air, can be burned with economy to generate heat. Each pound of fuel requires a certain quantity of oxygen for its complete combustion, and the amount of air required will vary with the grade of fuel. Each pound of carbon, for its perfect combustion, will, theoretically, require 2.66 pounds of oxygen.

Oxygen is classed as a supporter of combustion, and a **combustible** is a substance capable of rapidly combining with oxygen to produce heat or light.

2. Pure air is composed of oxygen, .213 part, and nitrogen, .787 part, or approximately 1 part of oxygen to 4 parts of nitrogen, per unit volume; but the atmosphere is more or less affected by carbonic acid, moisture, and other impurities.

The economic value of fuel is measured by its heating power, which is principally determined by the elements of its composition—carbon, hydrogen, oxygen, nitrogen, water, and ash in the form of non-combustibles. **Combustion** is the chemical union or combination of the constituents of the fuel with the oxygen of the air. Each atom of carbon unites with two atoms of oxygen and during the combination heat is given out.

3. Table I shows the proportions of carbon, hydrogen, and oxygen contained in several kinds of fuel, and the pounds of air, theoretically, required for their complete combustion.

Since each pound of carbon requires $2\frac{2}{3}$ pounds of oxygen to burn it to CO_2 and air contains 23 per cent. of oxygen, by weight, $2\frac{2}{3} \div .23$ or 11.6 pounds of air is required, theoretically, to burn 1 pound of carbon. The weight of 1 cubic foot of air at sea level is 536 grains, and there are 13.141 cubic feet in 1 pound at atmospheric pressure and 62° F.

TABLE I

AIR REQUIRED FOR COMBUSTION OF FUEL

Kind of Fuel	Proportional Weight of Given Constituents per Unit Weight of Coal			Pounds of Air Required per Pound of Fuel
	Carbon	Hydrogen	Oxygen	
Charcoal	.93			11.60
Coke	.94			11.28
Coal, anthracite	.92	.035	.026	12.13
Coal, bituminous	.87	.05	.04	12.06
Coal, coking . . .	.85	.05	.06	11.73
Coal, cannel . . .	.84	.06	.08	11.88
Coal, lignite . . .	.70	.05	.20	9.30
Wood, dry	.50	.06	.31	7.68
Petroleum oil . .	.85			15.65

4. The amount of air, theoretically, required for perfect combustion of fuel is not attained in practice. Experiments have shown that from 48 to 70 per cent. of air in excess of the theoretical quantity is required with natural draft, while from 20 to 30 per cent. excess is required with forced or mechanical draft. Because of the size of the coal and its arrangement on the fire, the passage of the air through it is more or less restricted, and in practice it usually becomes necessary to supply air in excess of the amount theoretically calculated to insure complete combustion.

5. Coal, in burning, combines with the oxygen of the air, giving up its carbon; first, to form carbonic oxide, or carbon monoxide CO, and further combining with oxygen to form CO_2, carbonic acid or carbon dioxide, the presence of which indicates complete combustion. The carbonic oxide in uniting with oxygen gives up one-third more energy than if passed out as carbonic oxide, and the condition of combustion is indicated by the percentage of carbonic oxide that exists in the gas leaving the furnace; a large percentage of carbonic oxide indicates that the combustion is incomplete. Insufficient air supply, or incomplete combustion of the coal will change the ratio of carbonic oxide to carbonic acid in the gas issuing from the boiler.

The average amount of air required for the combustion of 1 pound of coal is 21 pounds, or 276 cubic feet, at 62° F. To deliver this immense quantity of air in a manner such that it can be properly utilized, it becomes necessary to provide either *chimneys* or *mechanical draft*. A very brief calculation will serve to impress strongly on the memory the magnitude of the work required and the vast quantity of air to be moved. Take 50 square feet of grate surface on which 25 pounds of coal is burned per square foot per hour; we then have $50 \times 25 \times 276 = 345,000$ cubic feet of air to be moved per hour to facilitate the burning of 1,250 pounds of coal. To secure the rapid and almost automatic movement of this large volume of air, it is evident that intensity of draft is required.

DRAFT

6. Definition.—As the term *draft pressure* is usually employed, it refers to the difference in weight between the column of hot air or gas in the chimney, and a column of cold air outside, of the same height and area. The intensity of draft pressure, which controls the velocity imparted to the movement of the air, is measured in inches of water

by a **draft gauge.** This instrument, illustrated in Fig. 1, is
a **U**-shaped tube partially filled with water. One end is open
to the atmosphere and the other end is connected with the
flue or chimney of which it is desired to measure the draft

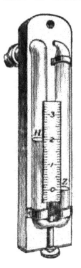

pressure. The liquid rises in one arm and
lowers in the other arm of the gauge; the pressure of the external air pushes the water up in
the arm connected to the chimney until the difference between the levels H and Z is sufficient
to balance the difference in pressure between
the flue and the atmosphere, and represents the
height of a column of the liquid that will be
sustained by the excess of pressure. The difference between the levels H and Z can be read
off by means of the scale and the intensity of
the draft is expressed in inches of water.

Draft must be expended in two ways: First,
to impart the necessary velocity to a sufficient
volume of air for the direct purpose of combustion, and second, to overcome all resist-

Fig. 1

ance offered to the air in its passage through the grate, fuel,
flue passages, and chimney. The velocity increases as the
square root of the pressure, or intensity.

CHIMNEYS

7. The purpose of the chimney is, first, to produce a
draft of sufficient intensity to facilitate combustion, and
second, to carry the obnoxious products of combustion to
such an elevation as will make them unobjectionable. In
designing a chimney, sufficient height must be given to
obtain the required intensity of draft, and sufficient area to
pass freely the products of combustion.

8. The numerous rules that have been formulated to
determine the area and height of chimneys differ more or
less, and while such rules may serve somewhat as a
guide, the proportioning becomes very largely a matter of
judgment gained from experience. The best practice and

judgment, backed by experience, will always decide in favor of the taller chimney, particularly keeping in mind the inevitable future increase of boiler capacity. As the number of boilers served by the chimney increases, the height will increase in greater ratio than the area of the chimney, because greater intensity will be required to overcome the friction in the flue connections between the boilers and the stack.

The horsepower capacity of the boiler plant will also be a guide to the height of the chimney. On this basis the height should be about as follows: Up to 250 horsepower, not less than 100 feet high; from 250 to 500 horsepower, 110 to 130 feet; from 500 to 2,000 horsepower, 130 to 150 feet. These heights are irrespective of the kind of coal to be used.

TABLE II
SIZE OF CHIMNEYS WITH APPROXIMATE HORSEPOWER OF BOILERS

Diameter Inches	Height of Chimneys and Commercial Horsepower						Effective Area Square Feet	Actual Area Square Feet
	100	110	125	150	175	200		
36	182						5.47	7.07
39	219						6.57	8.30
42	258	271					7.76	9.62
48	348	365	389				10.44	12.57
54	449	472	503	551			13.51	15.90
60	565	593	632	692	748		16.98	19.64
66	694	728	776	849	918	981	20.83	23.76
72	835	876	934	1,023	1,105	1,181	25.08	28.27
78		1,038	1,107	1,212	1,310	1,400	29.73	33.18
84		1,214	1,294	1,418	1,531	1,637	34.76	38.48
90			1,496	1,639	1,770	1,893	40.19	44.18
96				1,876	2,027	2,167	46.01	50.27

9. Table II gives the relation between the diameter and height of chimneys and the boiler capacity, in commercial horsepower. The figures at the head of the columns represent different heights of chimneys, in feet, while those in the left-hand column represent the diameter, in inches, the flue in the chimney being circular in cross-section. The two right-hand columns give the effective area of flue, in square feet, and the actual area, in square feet. The effective area is always less than the actual area because the air next the wall of the flue is retarded by friction and the whole area of cross-section is not, therefore, equally effective in allowing the draft to pass up the chimney. In Table II, a chimney 42 inches in diameter and 110 feet high will be capable of supplying draft for 271 horsepower of boiler capacity. The actual area of cross-section of the flue will be 9.62 square feet, but, allowing for friction, the effective area is reduced to 7.76 square feet.

FUEL BURNED PER SQUARE FOOT OF GRATE SURFACE

10. Modern practice has increased the amount of fuel burned per square foot of grate surface per hour until 25 pounds per square foot is quite common, and 30 to 35 is attained at times under maximum overload. The kind of coal to be burned must be considered, as well as the possibility of change at some future time from one kind to another for commercial reasons. The following heights of chimneys are recommended as the least to be used with the coals mentioned: 100 feet for burning bituminous slack, 120 feet for slow-burning bituminous, 130 feet for anthracite pea, 150 feet for anthracite buckwheat. The rate of combustion that can be obtained in any given case will depend on the draft pressure, and this in turn depends on the height of the chimney. Table III shows the combustion that is possible per square foot of grate surface with various heights of chimneys and corresponding draft pressures.

The designer should anticipate the future increase in the cost of coal, and the necessity of burning lower-grade coal,

TABLE III

RATE OF COMBUSTION FOR DIFFERENT TOTAL DRAFT PRESSURES

Height of Chimney Above Grate Feet	Total Draft Pressure Inches of Water	Rate of Combustion per Square Foot of Grate Pounds per Hour
100	.729	22
110	.802	24
120	.875	27
130	.948	30
140	1.029	34
150	1.095	40
180	1.313	50
200	1.459	60
225	1.641	70
250	1.825	80
300	2.189	90

and will therefore be justified in providing a strong, intense draft; he must also provide for increase of boiler capacity.

Where economizers are used in connection with natural draft, as illustrated later in connection with Fig. 5, the importance of sufficient intensity of draft to overcome the resistance through the economizers becomes evident. This must be taken into account when determining the height of chimney.

RULES RELATING TO CHIMNEYS

11. The capacity of the chimney will vary as the square root of the height; and the capacity also varies directly as the area. For example, in a chimney 160 feet high, the velocity will be, theoretically, twice that in a 40-foot chimney and should discharge twice the amount of gas in a given time, or if the cross-section is twice as great, a similar result will be obtained. In calculating the area of a chimney the volume of the escaping gas must be duly considered according to the composition of the coal.

12. Rule for Finding the Cross-Sectional Area of Chimneys.—On the basis that the ordinary horsepower will require the burning of an average of 5 pounds of coal per hour, the following rule, due to Kent, will be of service for calculating the cross-sectional area of chimneys. This gives a liberal allowance because with first-class installations the coal consumption would not be as great as 5 pounds per horsepower per hour. The large allowance is made to cover the use of poor coal or the forcing of the boilers on account of overloads.

Rule.—*Divide the horsepower by 3.33 times the square root of the height. The quotient will be the required effective area, in square feet. To the diameter so found add 4 inches to compensate for friction.*

Example.—What should be the diameter of a chimney 150 feet high to carry off the gases from boilers of 300 horsepower capacity?

Solution.—From the above rule, effective area, in square feet,
$$= \frac{300}{3.33 \times \sqrt{150}} = \frac{300}{3.33 \times 12.25} = 7.35.$$ This corresponds to an effective diameter of 3.06 ft. (or about 3 ft. $\frac{3}{4}$ in.), nearly. To this must be added 4 in. to allow for friction so that the actual inside diameter will be about 3 ft. 4$\frac{3}{4}$ in. Ans.

13. Relation Between Draft and Temperature of Gases.—In the diagram, Fig. 2, curve *B* shows the draft (inches of water) for a chimney 100 feet high where the

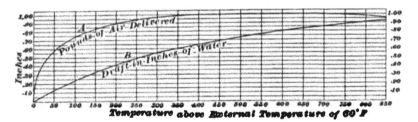

Fig. 2

temperature of the products of combustion varies from 0° to 850° above the assumed external atmospheric temperature of 60° F. Each division of the vertical scale represents $\frac{1}{20}$ inch. Curve *A* shows the relative quantity, pounds of air, that would be delivered at the various temperatures. If, for example, the quantity of air at 100° were represented by .75, the quantity at 200° would be represented by .90.

EXAMPLE 1.—If a stack is capable of discharging 12,000 pounds of air per hour when the difference between the temperature of the gases supplied to the stack and the outside temperature at 60° is 250°, what amount can it discharge if the difference is 375°?

SOLUTION.—Fig. 2 shows that for a difference in temperature of 250° the ordinate of the curve A showing the relative weights of air delivered is .95. For a difference in temperature of 375° the ordinate is 1, hence the weight delivered in the second case will be $12,000 \times \dfrac{1}{.95}$ = 12,630 lb., approximately. Ans.

The importance of keeping the escaping gases down to 350° to 450° F., as indicated in this diagram, will be readily appreciated, as no practical gain in draft is effected under the specified conditions by carrying the temperature in the chimney more than 400° F. above the atmospheric temperature of 60° F. It will be noted that practically no increase in the pounds of air is carried through the chimney, at any temperature above 410° F. This fact should be carefully fixed in mind in connection with proportions and setting of boilers, and the use of fuel economizers.

Rule.—*To determine the quantity of air, in pounds, that a given chimney will carry per hour, multiply the ordinate for curve A at a given temperature on the diagram, Fig. 2, by 1,000 times the effective area, in square feet, and by the square root of the height, in feet.*

EXAMPLE 2.—How many pounds of air will be carried per hour by a chimney 150 feet high and having an effective area of 20 square feet? The temperature of the gases is 250° above the outside temperature of 60°.

SOLUTION.—From Fig. 2, the ordinate of curve A corresponding to 250 is .95 inch, hence, applying the rule, we have

air carried $= .95 \times 1,000 \times 20 \times \sqrt{150} = 232,693$ lb. Ans.

This rule gives the maximum capacity of the chimney, and allowance must be made for friction through all flue and draft connections between the furnace and the chimney.

FORMS OF CHIMNEY

14. Round Versus Square Chimneys.—Practice has shown that the **round chimney** has several advantages over the **square chimney.** Externally, a round chimney offers least resistance to wind pressure, the **octagonal** shape is next desirable on this account, and the square chimney offers greatest resistance. Internally, the round flue offers less resistance and gives a more effective draft than the square flue, the corners of which are filled with eddy currents and are not effective for carrying away the products of combustion.

15. Types of Construction.—Four types of chimney construction may be considered briefly as follows: The chimney of least cost can be constructed of steel plates riveted together, not lined with brick; it can be supported on a breeching, or flue connections of the boilers, or on an independent foundation, and must be secured in true perpendicular position by guy rods suitably anchored. The next advance beyond this type of cheap chimney would be the self-supporting steel-plate chimney secured with anchor bolts to a foundation of sufficient weight and area to afford the necessary stability to withstand all strains due to wind pressure. The next type of construction is a steel-plate chimney similar to this and lined with brick for a portion of the height. The durability of steel-plate chimneys depends on the thickness of the metal, the kind of fuel burned in the furnaces, the atmospheric conditions surrounding the location, and the care taken to maintain the chimney in good condition. This last requires that the chimney shall be frequently painted with a good quality of paint. The annual cost of maintenance can here be considered as offsetting the interest charge on the higher cost of a brick chimney. The next in order is the chimney constructed of selected hard brick, having an external wall and an internal wall forming the core of the chimney, and so constructed that the core is free to expand or contract without strain on the external wall. Square chimneys, being undesirable, will not be considered.

16. The octagonal chimney can be built at as low a cost as the square chimney, is more attractive in design, and offers less resistance to wind pressure. The round core may be easily constructed in the octagonal chimney. Fig. 3 shows a section of a round chimney with core, and Table IV, in combination with Fig. 3, gives data on chimney construction. This table shows the dimensions of various chimneys varying from 34 inches flue diameter and 70 feet in height to 92 inches diameter and 125 feet in height. The vertical dimensions in Fig. 3 are for a 125-foot chimney.

Another type of chimney largely used in European countries, and also becoming popular in the United States, is the *Custodis chimney*, of especially molded and perforated radial brick. These chimneys are neat in design, effective in service, very substantial in their methods of construction, and do not require an internal core or flue. They have

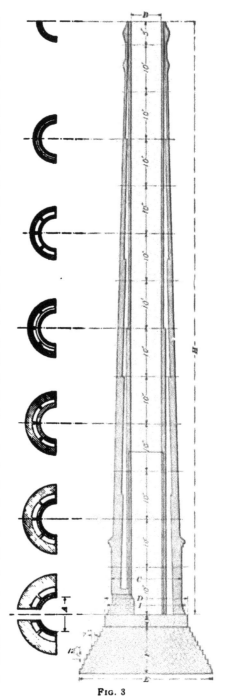

FIG. 3

TABLE IV
DIMENSIONS OF CHIMNEYS

Horsepower	A (In.)	B (In.)	C (Ft. In.)	D (Ft. In.)	H (Ft.)	Chimney 20-Inch Wall (Ft.)	Chimney 16-Inch Wall (Ft.)	Chimney 12-Inch Wall (Ft.)	Chimney 8-Inch Wall (Ft.)	Chimney 4-Inch Wall (Ft.)	Chimney Half Batter (In.)	Flue 8-Inch Wall (Ft.)	Flue 4-Inch Wall (Ft.)	Flue Height of Firebrick (Ft.)	Flue Half Batter (In.)	Total Number of Bricks	Number of Firebricks	Approximate Depth of Foundation (Ft. In.)
100	34	41	7 8	8 8	70			25	30	40	2.5		70	20	.5	17,000	1,150	8
132	38	45	8 8	9 8	70			25	30	40	2.5		70	20	.5	18,000	1,275	8
168	40	48	8 5	10 5	80			30	40	40	2.5		80	20	.5	23,000	1,350	8
200	44	52	9 9	10 9	80			30	40	40	2.5		60	25	.5	24,000	1,475	8
264	46	54.5	9 4	10 4	85		25	30	25	35	2.5	25	60	25	.5	32,000	1,900	8 6
336	50	58.5	9 8	11 8	85		25	30	25	35	2.5	25	65	25	.5	33,000	2,075	8 6
400	54	63	10 1	11 1	90		25	30	30	30	2.5	25	65	25	.5	39,000	2,225	8 6
450	58	67	10 3	12 3	90		25	30	30	30	2.5	25	70	25	.5	41,000	2,400	8 6
500	60	69.5	11 5	12 5	95		25	30	30	35	2.5	30	70	25	.5	45,800	2,475	9
550	62	71.5	11 7	12 7	95		25	30	30	35	2.5	30	70	25	.5	47,200	2,550	9
600	64	73.5	11 5.5	12 7	95		25	30	30	35	2.5	30	70	25	.5	48,600	2,625	9
650	66	75.5	12 7.5	13 9	100		25	25	25	25	2.2	35	70	30	.5	50,000	2,700	9 6
700	68	78	12 9.5	13 11	100		25	25	25	25	2.2	35	70	30	.5	63,000	3,475	9 6
750	70	80	12 11.5	13 2	100		25	25	25	25	2.2	35	75	30	.5	64,500	3,550	10
800	72	82	13 4	14 6	110	25	25	25	25	35	2.2	45	75	30	.5	66,000	3,625	10 6
850	74	84	13 6	14 8	110	25	25	25	25	35	2.2	60	75	30	.5	67,500	3,700	10 6
900	76	87	13 8	14 10	110	25	25	25	25	35	2.2	60	65	30	.5	72,000	3,800	11
950	78	89	14 6	14 2	125	25	25	25	25	35	2.2	60	65	30	.5	74,000	3,900	11
1,000	80	91	14 8	15 2	125	25	25	25	25	25	2.2	60	65	35	.5	76,000	4,000	12
1,100	84	95	14 10	16 4	125	25	25	25	25	25	2.2		65	35	.5	79,000	4,200	12
1,200	86	98.5	15	16 6	125	25	25	25	25	25	2.2			35	.5	108,000	5,000	12
1,300	88	100.5		16							2.2				.5	111,000	5,100	12
1,400	90	102.5		16							2.2				.5	114,000	5,200	12
1,500	92	104.5		16							2.2				.5	117,000	5,300	12

been used for many of the most modern power stations of large size.

Fig. 4 shows the method of construction used in the Custodis chimneys. A number of different sizes of hard-burned radial brick are used, these bricks being perforated as shown. The perforations allow the bricks to be burned hard and uniform and also insure tight and well-locked joints, since the mortar is worked into the perforations about ⅛ inch. The perforated bricks are considerably larger than ordinary bricks and the number of mortar joints in a chimney is thus reduced.

All brick chimneys should be finished with a cast-iron cap, well anchored in place, and fitted with a lightning rod substantially erected and thoroughly grounded. An iron ladder should be built in the brickwork to give access to the top.

Fig. 4

17. Fig. 5 illustrates the construction of a brick-lined steel stack supported on an octagonal brick foundation. The outer steel-plate shell a is anchored to the foundation by means of eight foundation bolts b. The stack is lined with both common brick and firebrick, as shown. In many cases the firebrick is not carried more than one-half or two-thirds the height of the chimney, the balance being lined with a good quality of hard-burned brick. Fig. 5 also shows a method of arranging the economizer. The gases on their way from the boiler to the stack can be made to take the path through the economizer by opening the cast-iron doors c shown in the figure, or these doors can be closed and the gases allowed to pass through the central smoke flue, as indicated by the full-line arrows.

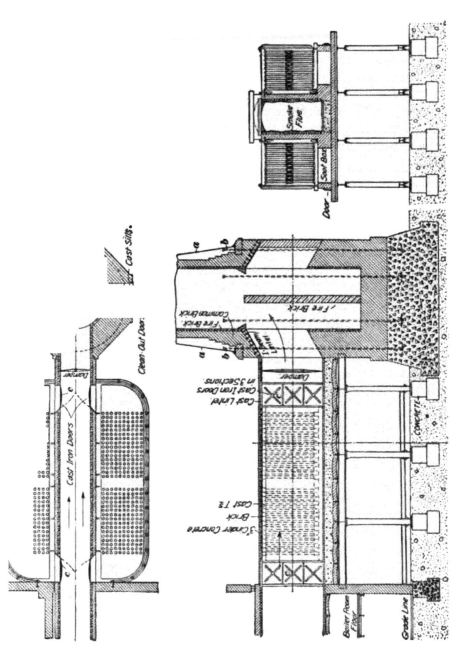

FIG. 5

STRENGTH OF CHIMNEYS

18. As chimneys must successfully withstand pressure in a heavy storm, the material employed and the construction adopted must give solidity and strength to a sufficient extent to withstand a strong pressure per square foot concentrated entirely on one side. A summary of the records and reports from the various observatories shows that the wind pressure may vary from 1 pound per square foot in a gentle breeze, to 50 pounds per square foot in a strong hurricane blowing at the rate of 100 miles per hour. The stability and power to withstand the pressure of high wind requires a proportionate relation between the weight, height, breadth of base, and exposed area of the chimney. For a square chimney the total weight should be fifty-six times the breadth of base multiplied by the exposed area; for an octagonal chimney it should be thirty-five times, and for a circular chimney twenty-eight times. For example, suppose that a square chimney is 100 feet high and has an average breadth of 8 feet; the area of a side exposed to the wind is then 100×8. If the breadth of base is 10 feet, the total weight, in order to secure stability, should be $56 \times 10 \times 100 \times 8 = 448,000$ pounds. Brickwork weighs from 100 to 130 pounds per cubic foot, hence such a chimney must average 13 inches in thickness in order to be safe. A round stack may weigh half the above amount or have less base. The breadth of base of chimneys is ordinarily made one-tenth the height, and great care should be taken to see that a substantial foundation is provided.

MECHANICAL DRAFT

19. Mechanical draft is any system whereby artificial draft is created, and may be either *forced* or *induced*. In laying out an electric power station, the question comes up as to whether natural or mechanical draft will most effectively and efficiently secure the desired result. All valuable points of each should receive due consideration, it being understood that when desired, economizers can be used with either system. The essential feature is sufficient intensity of draft; this is a most important factor, and must be under good control.

_____

20. Fans.—**Forced draft** may be produced by several methods. One method is to close the ash-pits tightly and use fans for forcing the air blast under the grate, as shown in Fig. 6. The air from the blower a, driven by the small steam engine b, is introduced by leading a duct into the bridge wall c under the grates. The distribution of air is controlled by means of dampers or deflecting plates, so

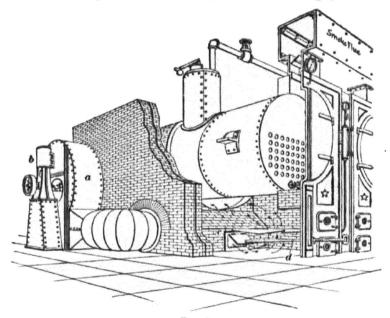

Fig 6

arranged as to send the air to all parts of the ash-pit, and regulated by means of a rod d reaching through the front of the boiler.

21. Another method is to build an air duct under the floor in front of the boilers and introduce the air through suitable passages to the ash-pit in the front, controlling with dampers regulated by means of hand rods, as above mentioned. This method will be understood by referring to Figs. 7 and 8. This air-blast method is not desirable for continuous service, but in connection with chimney draft is very advantageous as an auxiliary to increase the steaming

. Fig. 7

capacity of the boilers and to aid in carrying heavy peak loads for a few hours.

22. McClave Argand Steam Blower.—Another method of producing a forced draft is by means of the McClave

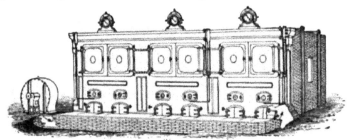

Fig. 8

Argand steam blower, illustrated in Fig. 9. This is an effective device for producing a forced draft for burning low grades of anthracite or bituminous coal. It consists of an air tube *t* discharging from the end *s* below the grate. In the other end of the tube is placed a ring-shaped tube *r* perforated on the right with small holes. Steam from the boiler is led into the ring by the pipe *l* and escapes in jet through the perforations, carrying air with it into the ash-pit. A small amount of steam is thus thoroughly mixed with a large volume of air and delivered under the grates at such a pressure as to augment materially the effect of the natural draft. Each blower has the advantage of independent regulation instead of relying on a single fan delivering air through a conduit to a number of boilers. The combination of steam and air is

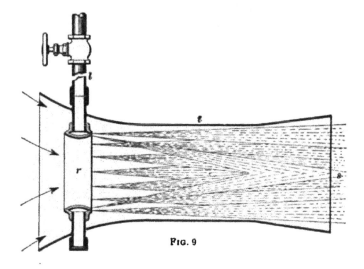

Fig. 9

beneficial in securing more perfect combustion of the low-grade fuels and reducing the hardness of the clinkers. This method is good for burning low-grade anthracite, but cannot be recommended for general use.

Steam jets may be used directly under the grates as an auxiliary in conjunction with natural draft. This plan has nothing to recommend it for adoption in a new plant, and is simply a poor makeshift to tide over an emergency in helping out poor chimney draft.

23. Enclosed Fireroom Method.—The enclosed fireroom method, as applied in some steamships, is so arranged that the air in the fireroom is maintained under pressure by powerful blowers, and the boilers can be worked with ash-pit doors open; an air pressure of 1 inch will afford an intense draft. This method is not practicable for power stations.

ADVANTAGES AND DISADVANTAGES OF FORCED DRAFT

24. The several methods of forced draft give good results under favorable conditions, but the following disadvantages are to a greater or less degree always present. Soot and ashes accumulate rapidly on the heating surface of the boilers; there is a tendency for the fires to burn unevenly; when firing, the soot, ashes, and gas blow into the face of the fireman, unless the draft is cut off or the gases pass out through the chimney; ashes cannot conveniently be removed without shutting off the draft.

Forced-draft methods may frequently be applied to boilers already in use, and thus increase their evaporative capacity, also making it possible to burn a cheaper grade of fuel than was originally intended, and thereby compensating in a measure for these disadvantages. .

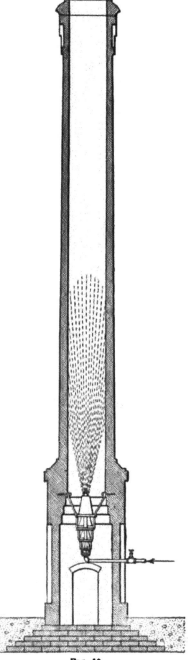

Fig. 10

INDUCED DRAFT

25. **Induced** or **suction draft** may be described as an artificial means of inducing intensity of draft by the introduction of the appropriate appliances in the flue passages beyond the boilers.

26. **Steam Jet in Stack.**—The introduction of a steam jet at the base of the stack increases the intensity of the draft

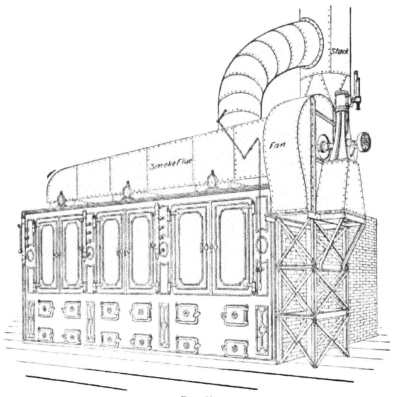

FIG. 11

by reason of the velocity of the steam issuing at high pressure. Fig. 10 illustrates a jet especially designed for this purpose. This method is not economical; it may serve to hasten a slow fire, but it is often found that the combined volume of steam added to the volume of gas from the furnaces is more than the chimney can quickly discharge.

27. Fans.—In this method slow-moving fans are arranged in the flues that carry off the hot gases. The suction created draws fresh air through the fire by reason of the reduced air pressure above the grates. The whole suction of the fans is through the fire, so that the intensity of the draft can be readily controlled by varying the speed of the fans. The fans may be driven by a direct-connected engine, an electric motor, or a belt from any convenient source of power. A simple application of this method, as installed by the American Blower Company, is illustrated in Fig. 11, and the arrangement may be modified to accommodate boiler equipments of any desired capacity. Greater reliability is obtained by installing two sets of fans, one of which may be used if the other is out of order.

ADVANTAGES AND DISADVANTAGES OF INDUCED DRAFT

28. The principal advantages claimed for the induced-draft system are as follows: (a) The ability to closely regulate the amount of air required for perfect economy of combustion; (b) the providing of an efficient means for producing any desired intensity of draft at less first cost than the amount required to build a good chimney; (c) the ability to absolutely control a uniform draft regardless of atmospheric conditions, to burn low-grade coals, and to economically increase the intensity of the draft and thereby increase the evaporative capacity of the boilers.

These combinations of control render the method of induced draft well adapted for use in connection with economizers. Fig. 12 (a) and (b) shows plan and sectional views of an economizer installation used with induced draft. Two fans, located at a and b and driven by independent engines, draw the products of combustion through the economizer and discharge them into the stack. Fig. 12 (b) is a sectional view along the line x x and shows the by-pass flue under the economizer and the various dampers and sliding doors by means of which the movement of the air is controlled.

It is claimed that the steam required to drive the fans for induced draft is from 1 to 1½ per cent. of the steaming capacity of the boilers. The disadvantage of induced draft lies in this constant daily cost of operating and maintenance.

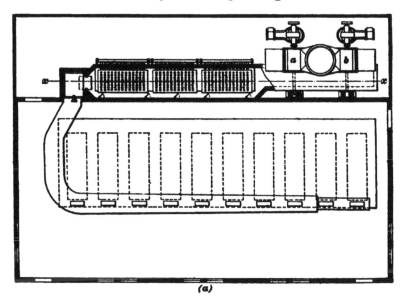

(a)

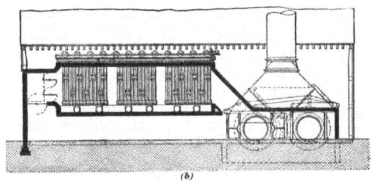

(b)

FIG. 12

The fan and enclosing iron casing are exposed to the action of the waste gases and are particularly liable to deterioration from this cause; the bearings must be cooled by water circulation, and are difficult to keep successfully lubricated. It is not considered wise to estimate less than 10 per cent.

per annum for depreciation and maintenance, on the induced-draft system, and to this must be added the cost of operation and interest on investment. It must not be understood that mechanical draft is under all circumstances equally economical or desirable. The economy of mechanical draft must depend on the conditions under which the plant is to be installed. The true measure of its value in each individual case must be estimated on a commercial basis with reference to the net saving in first cost of construction and in operating expenses, and a possible reduction in operating expense because of the ability to burn a low grade of coal. As before stated, the quantity of air required by mechanical methods may prove to be somewhat less than that needed with natural draft, but as air is obtained free of cost it is a question whether it pays to operate an engine and fan to cause its movement, or to induce such movement by natural means.

The advantages of the chimney are that after the first cost of construction has been met it requires no daily attention, the cost of repairs is trifling, and the products of combustion are delivered at such an elevation as not to cause a nuisance in the vicinity.

STEAM BOILERS

29. The **boiler** must be considered one of the most important parts of the steam plant because within it is generated the steam that supplies the energy and sustains the movements of the whole plant. On the selection of a proper type of boiler and on its successful operation largely depend the economy of the entire station. If the boilers are of poor material and workmanship, deficient in number or capacity, or not properly proportioned, the reliability of the station will be imperiled and abnormal amounts paid for fuel and repairs. Boiler horsepower should not be understood as the equivalent of engine horsepower, but as already explained, a boiler horsepower is equivalent to the work done in evaporating 34½ pounds of water per hour, at the pressure and temperature of unconfined boiling water.

TYPES OF BOILERS

30. For electric power station purposes boilers of two classes are worthy of consideration. One class comprises **water-tube boilers,** in which the water travels through the tubes and the heated products of combustion circulate around the tubes. Boilers of this type are often called *safety boilers.* The *Babcock and Wilcox, Heine, Stirling,* and *Climax* are representatives of this class. The other class comprises **fire-tube boilers,** in which the water is contained in the shell and surrounds the tubes, and the products of combustion pass through the tubes. The *horizontal tubular,* the *vertical tubular,* and *Sederholm boiler* are representatives of this class.

WATER-TUBE BOILERS

31. Babcock and Wilcox Boilers.—The Babcock and Wilcox boiler, shown in Fig. 13, is a type that is used very largely in electric power stations. It consists essentially of a main horizontal drum *a a* and a series of inclined tubes *b, b.* Only a single vertical row of tubes is shown in the figure, but it will be understood that each nest of tubes is composed of several vertical rows; there are usually seven or eight vertical rows to each horizontal drum. The front and rear ends of the tubes *b, b* are expanded into hollow *headers d, d.* The front and rear headers are connected to the main drum by the tubes or risers *c, c.* In front of each tube, a handhole is placed in the header for the purpose of cleaning, inspecting, and renewing the tubes.

The boiler is supported from the I beams *e, e* by means of straps passing around the drum. These I beams are supported by cast-iron columns, the brickwork setting not being depended on as a means of support. This make of boiler, in common with all others of the water-tube type, requires a brickwork setting to confine the furnace gases. The furnace is placed under the front end of the nest of tubes. The bridge wall *f* is built up to the bottom row of tubes and another firebrick wall *f* is built between the top row of tubes and the drum. The walls and baffle plates g force the hot furnace gases to follow a zigzag path back and forth between the tubes. The gases finally pass through an opening at h in the rear wall and from there to the chimney flue. The feedwater is introduced at the front of the boiler through the pipe k and the main steam pipe is attached at l. At the bottom of the rear row of headers is placed the mud-drum m; since this drum is at the lowest point of the water space, most of the sediment naturally collects there and can be blown off from time to time through the blow-off pipe n or

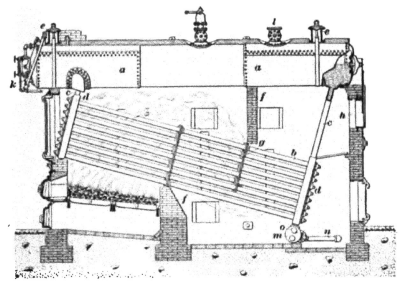

Fig. 13

removed through the handhole o. A manhole is provided in each end of drum a for access to the interior.

The water fed into the boiler through pipe k, passes to the back of the drum and descends through the back risers c. The water becomes heated in the tubes and rises through the front risers into the drum. There is thus a continuous circulation of water from front to back through drum $a\,a$ and from back to front through tubes b, b.

32. Heine Boiler.—The Heine boiler, shown in Fig. 14, is another prominent example of the water-tube type. It consists of a large main drum A that is above **and** parallel with the nest of tubes T, T. Both drums and tubes are inclined at an angle with the horizontal that brings the water level to about one-third the height of the drum in front and about two-thirds in the rear. The ends of the tubes are expanded into the large wrought-iron water legs B, B. These water legs form the natural support of the boiler, the front legs resting on a pair of cast-iron columns E that form part of the boiler front, while the rear leg rests on rollers at F.

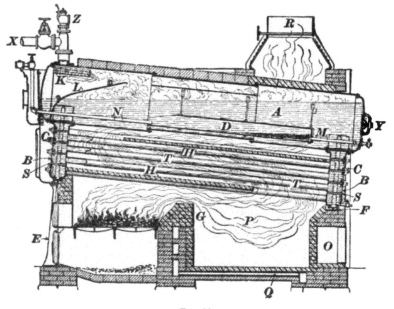

Fig. 14

These rollers allow the boiler to expand freely when heated. The bridge wall G is made largely of firebrick and has openings in the rear to allow air to pass into the chamber P and mix with the furnace gases. In the rear wall is an arched opening O that is closed by a door and further protected by a thin wall of firebrick. When it is necessary to enter the chamber P, the wall may be removed and afterwards replaced.

The feedwater is brought in through the feedpipe N that passes through the front head. As the water enters it flows into the mud-drum D, which is suspended in the main drum below the water-line, and is thus completely submerged in the hottest part of the water. This high temperature is useful in precipitating the impurities contained in the feed-water and they may then be blown through the blow-off pipe M. Layers of firebrick H, H are laid at intervals along the rows of tubes and act as baffle plates, forcing the hot gases to pass

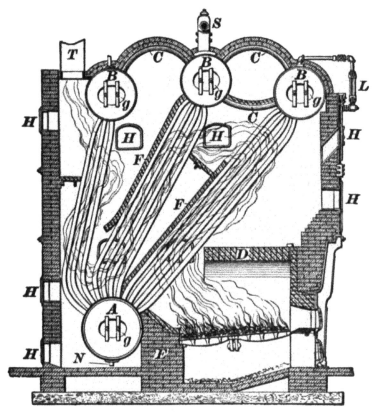

Fig. 15

back and forth along the tubes until they escape through the chimney R. The steam is collected and freed from water by the perforated dry pipe K. The main steam pipe, with its stop-valve, is shown at X, the safety valve at Z. The deflecting plate L is placed in the front end of the drum to prevent a spray of mixed water and steam from the front header from entering the dry pipe K.

33. Stirling Boiler.—The Stirling water-tube boiler, shown in Fig. 15, is a departure from the regular type of water-tube boilers. It consists of a lower drum A connected with three upper drums B, B, B by three sets of nearly vertical tubes. These upper drums are connected by the curved tubes C, C, C. The curved forms of the tubes allow the different parts of the boiler to expand and contract freely without strain. The brickwork setting is provided with various openings H, H so that the interior may be inspected or repaired. The bridge E is lined with firebrick and is built in contact with the lower drum A and the front nest of vertical tubes. The arch D built over the furnace, in connection with the bafflers F, F, directs the course of the heated gases.

The cold feedwater enters the rear upper drum and descends through the rear nest of tubes to the drum A, which acts as a mud-drum and collects the sediment brought in by the water; a blow-off pipe N permits the removal of the sediment. The steam collects in the upper drums, and the steam pipe 'and safety valve are attached to the middle drum. The chimney T is located behind the rear upper drum and the water column L, with its fittings, is placed in communication with the first upper drum. All drums are provided with manholes g.

34. Morrin Climax Boiler.—Fig. 16 shows a vertical boiler of the water-tube type known as the **Climax boiler**. It consists of a stand pipe, or main vertical drum, a fitted with a large number of looplike tubes b, b, the ends of which are expanded into the shell. The furnace is circular, and in order to give free access to the fire, four furnace doors are provided. A deflector plate d is fitted to the shell a little above the water level, which tends to throw back any water carried up by the steam. The upper portion of the central shell is divided by a series of diaphragms e, e into a number of superheating chambers, through which the steam is compelled to circulate successively by the connecting looplike tubes. The steam thus becomes thoroughly dried and

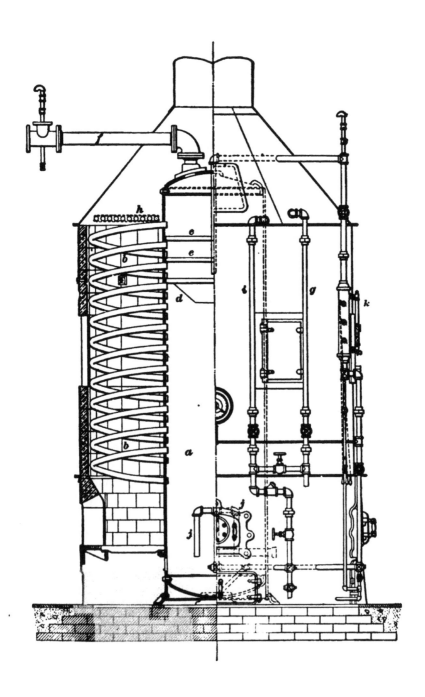

somewhat superheated before it enters the main steam pipe *f*.
The feedwater is discharged through the delivery pipe *g* into
a spiral feed-coil *h* resting on top of the tubes, where it is
heated to a high temperature. It leaves the coil through the
pipe *i* and passes downwards, finally being discharged into
the bottom through the internal feedpipes *j, j*. The water
column *k* is connected to the top and bottom of the central
shell. These examples give a general idea of the construction of water-tube boilers. There are so many types that it
is impossible to describe them all. The advantages of the
water-tube boilers are: economy of floor space, thin heating
surface, joints not exposed to fire, large draft area, quick
steaming, generally effective circulation, safety from explo-
sions, ease of transportation, durability, and accessibility for
cleaning. Water-tube boilers having restricted circulation
are liable to accident in the event of pushing the steam production much beyond the rated capacity.

FIRE-TUBE BOILERS

35. Horizontal Return Tubular Boiler.—This boiler
is probably used more for general purposes in the United
States than any other type. It is simple, effective, inexpensive, and easy to install. Fig. 17 shows a side view of it

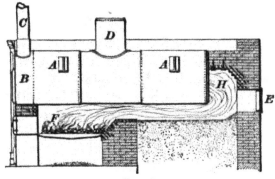

Fig. 17

and Fig. 18 a cross-section through the tubes. The boiler
consists of a main outer cylindrical shell supported on the
brickwork by the brackets *A, A* riveted to the shell. The
tubes extend the whole length of the boiler and are expanded

into holes in the heads of the boiler. The front end of the shell projects beyond the head, forming the smokebox B into which opens the stack C. The steam pipe and safety valve are attached to the top of the steam drum D and a door E gives access to the rear of the boiler. The tubes form a series of flues through which the hot gases pass from the back of the boiler to the smokebox B and thence out of the stack. The gases pass to the back of the boiler under the shell and then return to the front through the tubes, hence the name given to the boiler. It is thus seen that the fire, or rather hot gases, pass through the tubes of the boiler and the tubes are surrounded by water contained in the outer shell, whereas in the water-tube boiler the reverse is the case.

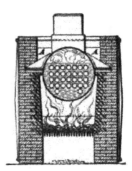

Fig. 18

36. Vertical Tubular Boiler.—Fig. 19 shows a vertical fire-tube boiler. The gases from the furnace F pass through the tubes t, t and out of the stack K. The steam is taken from the steam space S by means of the steam pipe G.

37. Sederholm Boiler.—This boiler is a special type of horizontal tubular boiler. It consists of a large main shell a fitted to tubes and connected to cylindrical drums b placed underneath, as shown in Figs. 20 and 21. These form a roof over the furnace and protect the main shell from the action of the flames, removing the danger of burning out the shell and rendering it possible to make the main shell as thick as required. It can therefore be made of large diameter, even when built for high pressure, and large units become possible. In Fig. 20 it is seen that each furnace drum b is connected to the main shell by means of three circulating pipes c, d, e, of which the central one extends almost to the bottom of the drum. The arrangement of these pipes insures a very effective circulation, and their curved shape gives considerable elasticity to the construction. This makes it possible to get up steam in a very short time, without giving rise to any local strains, because the perfect circulation entirely prevents unequal heating. The main shell is fitted with tubes placed in two nests, leaving ample space in the center for a steady downward current of water. The boiler heads above the tubes are reenforced by doubling plates riveted on, and braced by strong stayrods f, f, Fig. 21, running the full

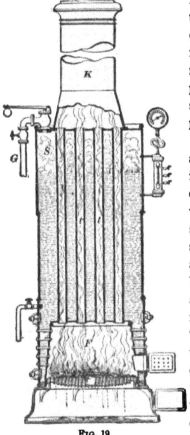

FIG. 19

length of the boiler. Access to the inside is obtained by means of one manhole on top for each furnace drum. The boiler is fitted with a perforated dry pipe g so arranged that the steam is taken from the highest part of the boiler.

A special feature is the unusually high furnace, which makes it possible to secure perfect combustion with the soft coals used in the Middle States; the special construction of the boiler setting still further favors this, and practically smokeless combustion is the result. Of course, any kind of special grate or furnace may be fitted to the boiler. The gases of combustion heat the lower side of the furnace drums, and then pass through the tubes h, Fig. 21, after which they are conducted either directly to the smokestack or around the main shell, according to circumstances. This secures a very uniform rate of evaporation throughout the whole boiler, resulting in the whole of the large liberating surface being effective, and therefore giving unusually dry steam, even when the boilers are forced to their limit. The exposed parts of the boiler are arranged to be covered with non-conducting material,

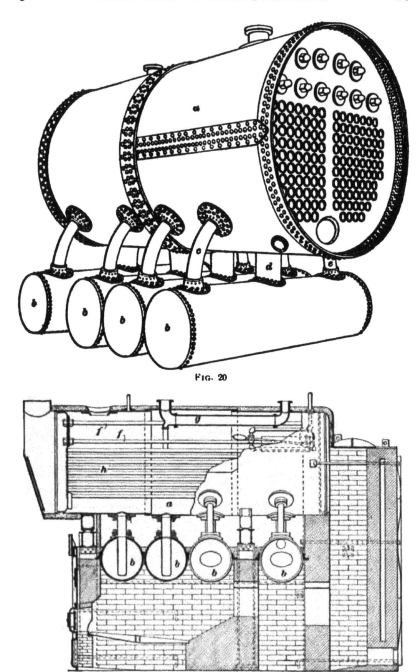

FIG. 20

FIG. 21

and the boiler rests on an iron framework, so that no weight comes on the boiler setting. The brickwork is also very low, and the quantity of bricks required for the setting is therefore small. These boilers are built in sizes ranging from 150 to 600 horsepower and for pressures up to 250 pounds and higher.

<hr>

SELECTION OF BOILERS

38. When about to select boilers, the following details should receive careful consideration: (*a*) The area of the heating surface; (*b*) the area of the grate surface; (*c*) the area and route of the gas passages from the combustion chamber; (*d*) the free circulation of water in the boiler; (*e*) facilities offered for inspecting, cleaning, repairing, and keeping the boiler free from soot and ashes without, and free from scale and sediment within; (*f*) the kind of fuel to be used.

Experience has proved that boilers of the water-tube and the fire-tube type, when equally well designed and proportioned, give equally good economy, provided they are operated with the same skill and quality of fuel; therefore, no definite rule can be adopted as to which type should be selected. Where only moderately high pressures are required, and where there is ample floor space, the horizontal tubular boiler will give excellent results. Where pressures above 120 pounds are to be maintained, and more power is required per foot of floor space, the water-tube boilers will be most desirable.

The standard specifications of the Association of American Steam-Boiler Manufacturers may be accepted as embodying the best experience in the description of tubes, steel plates, riveting, etc. When inviting proposals for boilers, it is good practice to specify the number of square feet of heating surface required per rated horsepower. For water-tube boilers, this should not be less than ten, and for fire-tube boilers not less than twelve, and all competitive proposals should be figured out to the established basis, regardless of claims made by selling agents as to the advantages of their special design.

39.   Heating Surface.—There has been a difference of opinion as to what constitutes the available heating surface of a boiler tube—the surface exposed to the flame, or that in contact with the water.   A committee appointed by the American Society of Mechanical Engineers has decided in favor of the outside surface of the tube for both water-tube and fire-tube boilers.   The Hartford Steam-Boiler Inspection and Insurance Company has for some time maintained the correctness of this as the true heating surface of a tube, and it is accepted by most boiler manufacturers.

The heating surfaces of value in water-tube boilers are the tubes and usually half the area of the steam drums; and for horizontal tubular boilers, half the area of the shell and all the surfaces of the tubes.   The calculation of heating surface at the ends is an unnecessary refinement, as such surfaces are of but little real value.

40.   Influence of Type of Engines.—When selecting boilers for an electric power station, the type of engine to be used must be taken into consideration, as the capacity of the engine to consume steam will create a relative demand on the boiler for the evaporation of more or less water into steam; therefore, the actual heating surface of the boilers selected should be considered in connection with the steam required by the type of engines to be used.   The extent of heating surface in tubular boilers usually allowed per horse-power when the type of engine is known, is as follows:

	SQUARE FEET
Corliss compound condensing engine	7½ to 8
Simple Corliss condensing engine	9
Simple Corliss non-condensing engine	10 to 11
Automatic cut-off medium-speed engine	11½ to 12
High-speed automatic cut-off engine	13 to 15

41.   Size of Boiler Units.—For electric power station service, the units of boiler capacity should be selected in proportion to the unit of engine and generator capacity; this can be determined by allowing one, two, or three boilers per engine generator unit, according to the steam requirements of the latter, and the steam-producing ability of the boilers. Water-tube boilers can be obtained in larger units of capacity than fire-tube boilers.

. The load on the usual power station increases or diminishes at regular hours, excepting on the occasion of heavy storms or some unusual demand for current, and to successfully meet such emergencies boilers of quick-steaming and large reserve capacity are needed. Every electric power station should have sufficient extra boiler capacity to allow time for cleaning and repairs without interfering with the regular service demanded from the station. The boilers should be regularly inspected, their condition noted, and advantage taken of the season of small loads to put them in perfect condition for the next season of heavy loads.

42. Evaporation.—The maximum results that may be obtained for the evaporation from and at 212° F., from boiler tests, should not be taken as a basis for every-day working conditions of boilers in service. Tested under favorable conditions the best boilers will show from 8.5 to 11.5 pounds of water evaporated from and at 212° F., per pound of combustible, but in daily practice the average rate of evaporation will be from 6 to 9.5 pounds of water. This difference may be accounted for because of variable loads, careless firing, difference in quality of coal, condition of boilers, and other incidental causes that occur in daily service.

The coal required per horsepower-hour will be dependent on the economy of the combined equipment of boilers, engines, pumps, etc. of the station. Table V shows the average evaporation, in pounds per hour, from a feedwater temperature of 150° F. into steam of 100 pounds pressure, per each square foot of grate surface for different surface ratios and weight of fuel. The figures in this table are very conservative.

The *surface ratio* of a boiler is the ratio of the heating surface to the grate surface; or, in other words, the number of square feet of heating surface per square foot of grate surface. For example, from Table V it is seen that if a boiler has a

TABLE V

AVERAGE EVAPORATION, IN POUNDS PER HOUR, FROM A FEED TEMPERATURE OF 150° F. INTO STEAM AT 100 POUNDS PRESSURE, PER SQUARE FOOT OF GRATE SURFACE, AND FOR DIFFERENT SURFACE RATIOS AND WEIGHT OF FUEL

Coal per Square Foot of Grate per Hour (Pounds)	Surface Ratio (Heating Surface ÷ Grate Surface)															
	30	32	34	36	38	40	42	44	46	48	50	52	54	56	58	60
	Pounds of Water Evaporated per Hour															
10	77	79	82	84	87	90	93	96	99	103	106	110	114	118	123	127
12	89	91	94	96	99	102	105	108	111	115	118	122	126	130	133	137
13	95	97	100	102	105	108	111	114	117	121	124	128	132	136	141	145
14	101	103	106	108	111	114	117	120	123	127	130	134	138	142	147	151
15	107	109	112	114	117	120	123	126	129	133	136	140	144	148	153	157
16	113	115	118	120	123	126	129	132	135	139	142	146	150	154	159	163
17	119	121	123	125	128	131	134	137	140	144	147	151	155	159	164	168
18	125	127	130	132	135	138	141	144	147	151	154	158	162	166	171	175
19	131	133	136	138	141	144	147	150	153	157	160	164	168	172	177	181
20	137	139	142	144	147	150	153	156	159	163	166	170	174	178	183	187
21	143	145	148	150	153	156	159	162	165	169	172	176	180	184	189	193
22	149	151	154	156	159	162	165	168	171	175	178	182	186	190	195	199
23	155	157	160	162	165	168	171	174	177	181	184	188	192	196	201	205
24	161	163	166	168	171	174	177	180	183	187	190	194	198	202	207	211
25	167	169	172	174	177	180	183	186	189	193	196	200	204	208	213	217
26	173	175	178	180	183	186	189	192	195	199	202	206	210	214	219	223
27	179	181	184	186	189	192	195	198	201	205	208	212	216	220	225	229
28	185	187	190	192	195	198	201	204	207	211	214	218	222	226	231	235
29	191	193	196	198	201	204	207	210	213	215	218	222	226	230	235	239
30	197	199	202	204	207	210	213	216	219	223	226	230	234	238	243	247

surface ratio of 40, and if 15 pounds of coal is burned per square foot of grate per hour, the boiler should, for each square foot of grate surface, be able to convert 120 pounds of water, supplied at 150° F., into steam at 100 pounds pressure per hour. The numbers at the tops of the vertical columns in Table V are surface ratios and the numbers` in the body of the table are pounds of water evaporated per hour.

43. Grate Surface.—The allowance of grate surface, as compared with heating surface, varies with the type of boiler and the general custom of the builders, and should be carefully investigated. It varies from ¼ square foot to ⅓ or ½ square foot of grate surface to each 15 feet of heating surface; this is equivalent to from 60 to 30 square feet of heating surface per square foot of grate surface. The object of the grates is to furnish a surface to support the fuel in process of combustion, and offer the least obstruction to the passage of the air.

The proper proportion of grate surface of the boiler to the heating surface is important, to the end that the fuel may be burned most advantageously. Table V shows the ratio of heating surface to grate surface, but does not take any account of any special type of boiler, as with any well-designed and proportioned water-tube or tubular boiler the results will be practically the same.

44. Grate Bars.—The kind of fuel to be used should be determined, if practicable, because it will to some extent influence the style of grate bars to be employed. The rate of combustion, intensity of draft, and quality of fuel will determine the extent of grate surface. The smaller sizes of coal, because of packing together more closely, offer more resistance to the passage of the air through the fresh fuel than the coarse coal, and the coal burns more slowly; also a greater intensity of draft is necessary to maintain the circulation of air through the fire. When a coarse coal is used the air spaces are wider, the air goes through the fire more readily, and great care is required to prevent holes being burned out in the bed of the fire, resulting in loss of economy.

The sizes of air space required for grate bars for different kinds of fuel are indicated in Table VI.

TABLE VI

SIZE OF GRATE BARS AND AIR SPACE FOR DIFFERENT, KINDS OF FUEL

Kind of Fuel	Size of Bar Inch	Width of Air Space Inch	Kind of Fuel	Size of Bar Inch	Width of Air Space Inch
Anthracite, lump. . . .	$\frac{1}{2}$	$\frac{3}{4}$	Bituminous, run of mine	$\frac{3}{8}$	$\frac{1}{2}$
Anthracite, egg	$\frac{1}{2}$	$\frac{3}{4}$	Bituminous, slack . . .	$\frac{5}{16}$	$\frac{3}{8}$
Anthracite, nut.	$\frac{3}{8}$	$\frac{1}{2}$	Bituminous, lump . . .	$\frac{1}{2}$	$\frac{5}{8}$
Anthracite, pea.	$\frac{3}{8}$	$\frac{3}{8}$	Wood	$\frac{3}{8}$	$\frac{3}{4}$
Anthracite, buckwheat	$\frac{3}{8}$	$\frac{5}{16}$	Sawdust. . .	$\frac{5}{16}$	$\frac{1}{4}$

45. Coal Burned per Square Foot of Grate Surface.—With anthracite of quick combustion, high furnace temperature, and little flame, a limit is reached to the value of extra tube surface. From 5 to 28 pounds of anthracite can be burned successfully per hour per square foot of grate surface with natural draft, and the heating surface may reach as high as 45 square feet per square foot of grate surface. With bituminous coal, which is gaseous and requires more combustion space, larger tube surface becomes more valuable, and may be from 45 to 55 square feet per square foot of grate surface. Ordinarily, from 8 to 15 pounds of bituminous coal may be burned per hour per square foot of grate surface. The heating surface and combustion space should be so proportioned that the escaping gases

TABLE VII

COAL CONSUMPTION PER SQUARE FOOT OF GRATE SURFACE FOR DIFFERENT AREAS OF GRATES AND VALUES OF TOTAL FUEL CONSUMED PER HOUR

Square Feet of Grate Surface	Pounds of Fuel Burned per Hour										
	50	100	200	300	400	500	600	700	800	900	1,000
4	12.50	25.00	50.00	75.00	100.00	125.00	150.00	175.00	200.00	225.00	250.00
6	8.33	16.60	32.30	50.00	66.60	83.30	100.00	116.00	133.30	150.00	166.60
8	6.25	12.50	25.00	37.50	50.00	62.50	75.00	87.50	100.00	112.50	125.00
10	5.00	10.00	20.60	30.00	40.00	50.00	60.00	70.00	80.00	90.00	100.00
12	4.16	8.33	16.60	25.00	33.30	41.66	50.00	58.33	66.60	75.00	83.32
14	3.57	7.14	14.28	21.40	28.57	35.71	42.80	50.00	57.14	60.90	71.42
16	3.12	6.25	12.50	18.70	25.10	31.25	37.40	43.75	50.20	56.25	62.50
18	2.77	5.55	11.11	16.60	22.22	27.70	33.20	38.80	44.44	50.00	55.40
20	2.50	5.00	10.00	15.00	20.00	25.00	30.00	35.00	40.00	45.00	50.00
22	2.27	4.55	9.11	13.68	18.22	22.72	27.36	31.81	944	40.90	45.44
24	2.08	46	8.33	12.50	16.66	20.83	25.00	28.33	33.32	37.50	41.66
26	1.92	3.84	7.69	11.53	15.38	19.23	23.06	26.92	30.76	34.61	38.46
28	1.78	3.57	7.14	10.71	14.28	17.85	21.42	25.00	28.56	32.14	35.60
30	1.66	3.33	6.66	10.00	13.33	6.66	20.00	23.33	26.66	30.00	33.32
32	1.56	3.12	6.25	9.37	12.50	15.62	8.74	21.87	25.00	28.12	31.25
34	1.47	2.94	5.88	8.82	11.77	14.70	17.64	20.58	23.54	2647	29.40
36	1.38	2.77	5.55	8.33	11.11	13.88	16.66	19.44	22.22	25.00	27.76
38	1.31	2.63	5.25	7.89	10.50	13.15	15.78	8.42	21.00	23.68	26.30
40	1.25	2.50	5.00	7.50	10.00	12.50	15.00	17.50	20.00	22.50	25.00
42	1.19	2.38	4.73	7.14	9.52	11.90	14.28	6.66	19.04	21.42	23.80
44	1.13	2.27	4.54	6.81	9.09	136	13.62	15.99	18.18	20.45	22.72
46	1.08	2.17	4.34	6.52	8.69	10.86	13.04	15.21	17.38	196	21.72
48	1.04	2.08	4.16	6.29	8.33	10.41	12.58	14.66	16.66	18.75	20.82
50	1.00	2.00	4.00	6.00	8.00	00	12.00	14.00	16.00	18.00	20.00

from anthracite do not exceed 380° to 400° F., and from bituminous coal 415° to 460° F.

46. Desirability of Ample Flue Areas.—The flue
openings and other passages from the furnace must be of
sufficient area to permit the volume of the products of combustion a free and unobstructed exit after leaving the boiler.
The volume of the products of combustion will, of course,
depend largely on the weight of fuel burned, and the area
of the openings for the escape of the hot gases has considerable to do with the successful operation of the boiler.
Special emphasis is placed on the importance of an intense
draft.   An electric power station with a slow draft may fail
to get up steam sufficiently quick to meet the demands of a
sudden load.   When the draft is slow and the coal is of a
poor quality, and burns slowly, a larger grate area will be
necessary to give the same results that would be obtained
with a better coal, stronger draft, and higher rates of combustion.   Table VII shows the coal consumption per square
foot of grate surface per hour for different areas of grates
and values of the total pounds of fuel consumed per hour.
The first column at the left-hand side indicates the area of
grate surface, the columns parallel thereto indicate the
pounds of coal burned per hour per square foot of grate
surface, and the row of figures at the top of columns indicates the total pounds of coal from 50 pounds to 1,000 pounds
per hour burned at the several ratios.   The heavy zigzag
line shows about the maximum limit for economical consumption of fuel under favorable conditions.   For example,
in order to burn 400 pounds of fuel per hour with 12 square
feet of grate area, the consumption per square foot of grate
surface would be $\frac{400}{12} = 33.3$, as indicated in the column
headed 400 and opposite the grate area 12.   This consumption is at the extreme limit for economical combustion, as
the number 33.3 is just under the zigzag line.

LOCATION OF BOILERS

47. The convenient and proper placing of boilers is an important matter. Special care should be taken to see that sufficient space is allowed to withdraw and renew tubes, and also that the boilers are accessible for examination, cleaning, and repairs. Fig. 22 shows an improper location for a boiler setting because the external wall is exposed to the weather

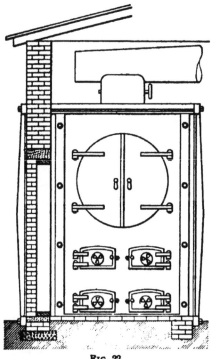

FIG. 22

and also carries the wall of the building; Fig. 23 shows a proper location for the same boiler setting. In a plant of comparatively small capacity, the boilers should be so placed that there is convenient access between the fireroom and the engine room, as there will be certain times during hours of light load when one man can look after both; in larger plants, it may not be possible to carry out this arrangement. If water-tube boilers are used they should be set in batteries of two, and each group of boilers should be easily accessible on all sides. Where horizontal tubular boilers are used several may be set side by side in the same battery.

48. Boiler Foundations.—The foundation walls should be built to the level of the floor line and if possible should be started on rock or solid earth. The excavation should not be less than 3 feet deep; the foundation walls should be laid of good concrete or broad flat stones in cement mortar. If soft earth should be encountered, the entire area should be excavated; if the earth is constantly wet, lay two courses of $3'' \times 12''$ plank and fill with concrete to within 1 foot of the floor line, from which level the foundation walls may be started. Where boilers are set above a basement or on upper floors, it is of course presumed that the columns, girders, and beams of the building have been substantially designed to support the load, which will consist of the weights of the brick-work, boilers, and water combined.

49. Boiler Settings. The plans for boiler set-tings vary with the type and size of the boiler, but certain features of good practice apply to all cases. The proper design and execution of the work is a matter of great im-portance, as the boiler setting is subjected to greater strains in pro-portion to the weight sup-

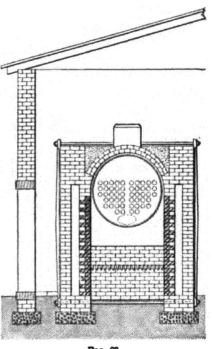

Fig. 23

ported, than is the case with ordinary walls because of the extremes of temperature within and without the enclosing walls. The best standard of work will include the follow-ing points:

(a) The rear and side walls should be double with an air space between, to avoid the leakage of cold air through the walls.

(b) The double walls need not be tied together, but at every fifth or sixth course the bricks can project from the outer wall and touch the inner; a 1-inch air space is quite sufficient. Fig. 23 illustrates this method but emphasizes the air space to make it clear.

(*c*) For a durable setting the outer wall should be one brick thick, or a 9-inch wall, and the inner wall should be one and one-half bricks thick, or a 13-inch wall.

(*d*) The furnace walls should be lined with the best fire-brick having six courses laid with the length of the bricks or ends exposed to the fire, and the seventh course a row of stretchers. When laid in this way the furnace wall will stand twice as long without repairing; the ends of the bricks will burn off or fuse away, but the wall will not fall down as is frequently the case when five courses of stretchers are laid with the sixth course headers. The remaining part of the combustion chamber can be lined with the usual single course of firebrick stretchers tied into the inner wall.

(*e*) The joints of the red brickwork should not exceed $\frac{3}{16}$ inch, and for all red brick a good mortar can be made of

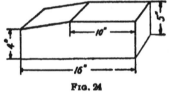

Fig. 24

two parts lime, one part hydraulic cement, and four parts clean, sharp sand. Each brick should be dipped in water before laying. The fire-brick should be laid in a mortar made of firebrick ground to a fine powder, and mixed with water to a proper consistency. The joints should be very close, not exceeding $\frac{1}{8}$ inch.

(*f*) The bridge wall, if laid of the usual firebrick according to standard methods, will soon be knocked down. The form of firebrick blocks shown in Fig. 24 laid side by side with fireclay joints and backed up with the usual inclined flame bed will stand many years of service.

This setting will save in repairs many times its cost. Ample time, from 30 to 60 days, should be allowed for the brickwork to dry out, and a slow, easy fire should be used for several days before getting up steam pressure.

50. Ash-Pits.—Where ashes are removed from a basement below the boiler room, a suspended hopper-shaped ash pocket built of $\frac{5}{16}$-inch boiler plates may be used; the hopper should be lined with slabs of firebrick, and the opening closed with a vizor gate and lever. Another plan is a pocket built of brickwork with a front delivery to car or conveyer; this arrangement is shown at *a*, Fig. 25, which shows a cross-sectional view of the boiler plant for a large station. Where neither of these can be adopted, the sides of the ash-pit should be sloped at an angle of 45° from the walls and the floor should slope from the front and rear to the center. The floor of ash-pit should be of vitrified brick laid on edge in cement mortar, well grouted, and made water-tight. The floor of the boiler room in front of the boilers will be very durable if. made of vitrified brick laid on edge and grouted; the bed should be 6 inches of concrete with 1 inch of sand to level the brick. All water and blow-off pipes that can be laid below the floor line, are preferably so located in channels of ample size. The walls of pipe channels are easily formed of concrete, and iron plates will make most satisfactory covers.

AUTOMATIC STOKERS

51. The **automatic stoker** is a device for reducing the labor of hand firing, maintaining a uniform rate of fuel supply, and more uniform and perfect combustion and furnace temperatures. Automatic stokers may be divided into two classes: *overfeed* and *underfeed*.

OVERFEED STOKERS

52. Roney Mechanical Stoker.—This stoker, which is shown in Figs. 26 and 27, is one of the most widely used stokers of the **overfeed type.** Fig. 26 is a perspective view of it, as applied to a horizontal tubular boiler, and Fig. 27 a sectional view, showing the relation of the different parts. Like parts have been lettered the same in the two figures. The coal is fed into the hopper *a*, from which it is pushed by the pusher plate *b* on to the dead plate *c*, where it is heated and coked. From *c*, the coke passes to the grate *ddd*, which consists of cast-iron bars that form a series of steps; each bar is supported at its ends by trunnions and is connected by an arm to a rocker bar *i*, Fig. 27, which is

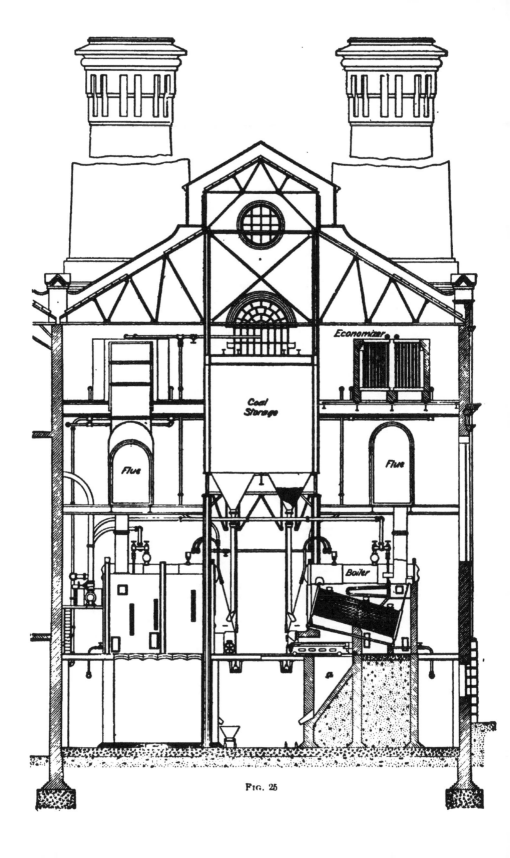

FIG. 25

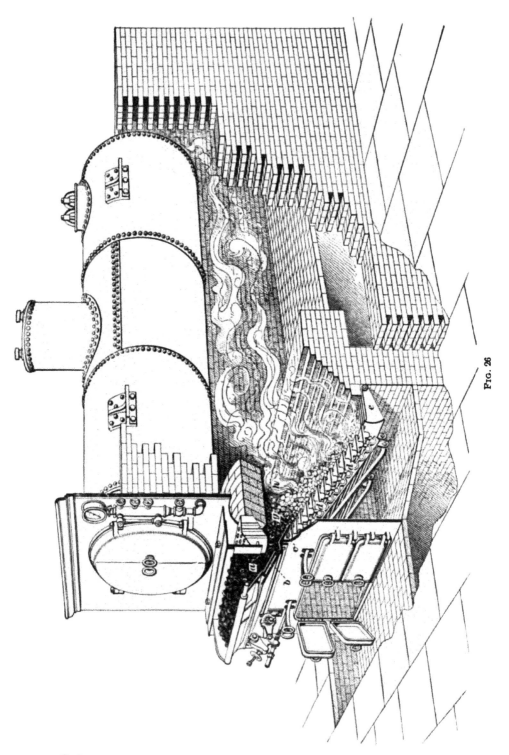

FIG. 26

slowly moved to and fro by an eccentric on the shaft s, so as to rock the grates back and forth between the stepped position shown and an inclination toward the back of the furnace; the grates thus gradually move the burning coke downwards. The ashes and clinkers are discharged from the lower grate bar on to the dumping grate e, which can be lowered so as to drop them into the ash-pit below. A guard f, Fig. 27, may be raised, as shown by the dotted lines, so as to prevent coke or coal from the grate bars from falling into the ash-pit when the dumping grate is lowered. Air for burning the gases is admitted in small jets through holes in the hot air tile g, and the mixture of gas and air is burned in the hot chamber between the firebrick arch h and the bed of burning coke below.

The **Roney stoker** is designed especially for burning all grades of bituminous coal, but may be successfully used for burning fine anthracite.

ENDLESS-CHAIN, OR TRAVELING, GRATES

53. Another class of stokers that belongs to the overfeed type is the **traveling grate;** in this stoker, the coal is carried into the furnace by means of a slowly moving grate made in the form of an endless chain. Among the many types of chain-grate stokers on the market are the Duluth stoker, the Green traveling-link grate, the McKenzie furnace, and the Playford chain-grate stoker. All traveling-grate stokers are similar in general character; the fuel is supplied from a hopper to the front end of a moving grate, it ignites and burns as the grate travels to the rear, and the refuse products of combustion are dumped at the rear end of the furnace near the bridge wall to an ash-pit below. These stokers are generally moved by a small engine or by an electric motor. The average power required to move the grate varies with the size and speed; it ranges from $\frac{1}{8}$ to $\frac{3}{4}$ horsepower. The traveling grates vary in details of construction, in the style of the bars, and methods of coupling the sections together. Extending across the front of the furnace is a

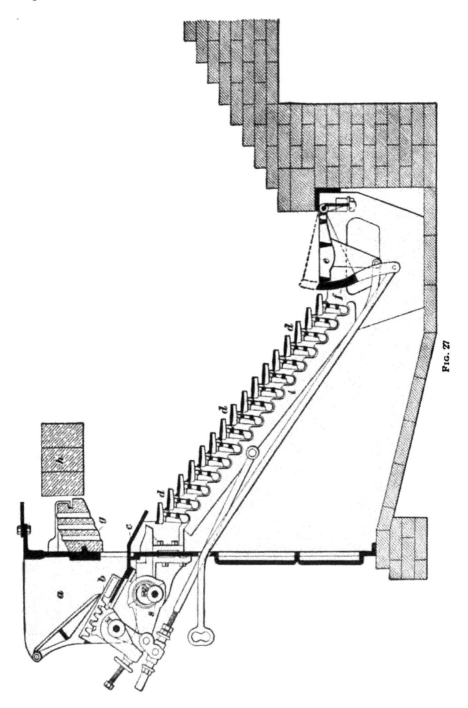

FIG. 27

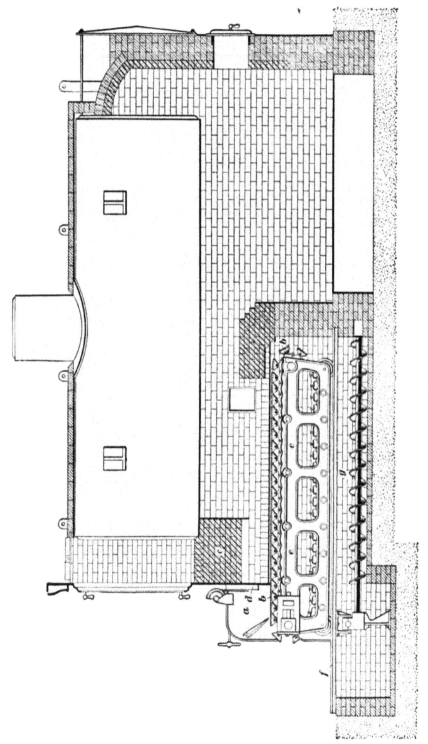

Fig. 28

hopper into which coal may be shoveled or automatically distributed by tubes from overhead storage. The feed of coal is adjustable and may be graded to whatever thickness of fire may be desired, even to an overload on the boilers.

Traveling grates are set up on substantial frames, and mounted on wheels and a track so that they can be run out from under the furnace for inspection or repairs. It is usual to build a firebrick arch at the front end of the furnace, where the heat is most intense; this facilitates the coking of .the fresh fuel as it is first supplied, and further aids the combustion of the volatile gases evolved and thereby tends to prevent smoke.

54. Playford Chain-Grate Stoker.—The Playford stoker, which is here shown as a typical example of the chain-grate class, is illustrated in Fig. 28. It consists of a heavy cast-iron frame *e* which is provided with suitable sprocket wheels and rollers on which travels a grate *b b* made up of sections attached to endless chains. The top of the grate is driven slowly toward the rear of the furnace, taking with it coal from the hopper *a*. The amount of coal fed to the furnace is regulated by the speed of the grate and by the

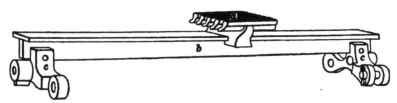

FIG. 29

opening of a gate *d*, which is water-cooled to prevent the heat of the fire from igniting the coal in the hopper. The gas is distilled from the coal in the front of the furnace under the firebrick arch *c* and burns as it rises and passes toward the back. The motion of the grate carries the coke backwards at a rate that permits the carbon to be completely burned before the rear end of the furnace is reached. The ashes and clinkers are dumped into the ash-pit at the back. A spiral conveyer *g* conveys the ashes from the rear of the furnace to a point near the front or to any convenient point from which they can be removed. The frame *e* rests on rollers that run on rails *f* and make it possible to withdraw the stoker from the furnace when repairs are needed.

In order to make the removal of burned-out grates easy and inexpensive, the grates are made in small sections, as *a*, Fig. 29, which slide over steel **T** bars *b*. The latter are, in turn, easily removed from the chain links *c* by taking out the pins at the ends.

_____

UNDERFEED STOKERS

55. In order to secure a high temperature of the gas and air, a number of systems of firing have been devised, in which the gas liberated from the freshly fired coal, together with most of the air required for its combustion, are drawn through the bed of burning coke. Such systems, if properly managed, bring the mixture of gas and air into the closest possible contact with the incandescent coke, and, consequently, secure practically perfect and smokeless combustion. The mechanical stokers to which this principle has been applied are known as **underfeed stokers**; the coal is forced by some mechanical device into a magazine or chamber and then through an opening at the top into the bed of burning coke. In this magazine distillation takes place; the coke that is formed in the magazine is forced upwards by the fresh supplies of coal and burns above and at the sides of the magazine. The gas produced meets a supply of air from openings in the sides of the chamber and the mixture arises through the bed of burning coke.

56. The **American stoker** illustrates the principles of construction of the underfeed stoker. Fig. 30 shows sectional views of this stoker as applied to a return tubular boiler. Coal is fed into the hopper *a*, from which it is drawn by the spiral conveyer *b* and forced into the magazine *d*, in which it is coked. The incoming supply of fresh fuel forces the coke to the surface and over the sides of the magazine on to the grates *i, i*, where it is burned. A blower forces air

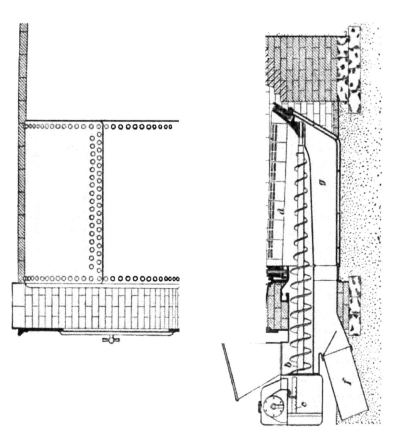

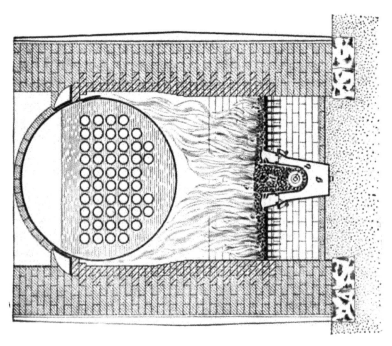

FIG. 30

through a pipe *f* into the chamber *g* surrounding the magazine. From *g* the air passes through the hollow cast-iron tuyère blocks and out through the openings, or tuyères, *e,e,e*. The gas formed in the magazine, mixed with the jets of air from the tuyères, rises through the burning coke above, where it is subjected to a sufficiently high temperature to secure the combustion. Nearly all the air for burning the coke is supplied through the tuyères, only a very small portion of the supply coming through the grate.

The ashes and clinkers are gradually forced to the sides of the grate against the side walls of the furnace, from which they are removed from time to time through doors in the furnace front similar to the fire-doors of an ordinary furnace.

57. The construction of this stoker is such that the fire must be cleaned and the ashes removed by hand. This has the disadvantage of a somewhat greater expenditure of labor than is required with those furnaces that discharge their ashes into the ash-pit, especially where it is desired to use ash-handling machinery; it also subjects the boiler to the deleterious influences of inrushes of cold air when the cleaning doors are opened. In this connection it may be stated that it is claimed by the makers that the fires do not need cleaning oftener than once in 8 or 10 hours with the poorer grades of coal, and that once in 12 hours is sufficient with the better grades; it is also a fact that all furnaces require occasional hand stirring and cleaning in order to secure a thoroughly satisfactory distribution of the fire on the grates and to prevent the formation of masses of clinkers that will occasionally stick to the grates, no matter how carefully the stoker is designed and operated.

ADVANTAGES OF AUTOMATIC STOKERS

58. The advantages claimed for automatic stokers over hand firing are: (*a*) The ability to burn a low-grade fuel; (*b*) the prevention of a large amount of smoke from bituminous coal; (*c*) no waste of fuel from cleaning fires; (*d*) an increase in evaporative capacity of the boilers; (e) a uniform supply of fuel and constant high furnace temperature and uniformity of steam pressure; (f) a material saving in fuel and labor where large steam boiler plants are used.

There can be no question of these advantages where the conditions are favorable for the use of stokers; these conditions become more apparent with the increasing capacity of the boiler equipment. Reliable tests have shown a saving of from 10 to 20 per cent. in fuel from the use of stokers as compared with hand firing. The lower percentage would be better taken as the basis of estimate for the value of the stoker, and after deducting the cost of repairs it can easily be estimated how long a period would be required for the saving in fuel to pay the cost of the stoker. No saving in labor can be made in a boiler plant of 500 horsepower or less, but in excess of this it may be possible to economize in the number of firemen or helpers.

The whole combination of automatic coal delivery, mechanical stokers, and ash removal by conveyer or cars from a large receptacle below the boiler-room floor, cannot fail to economize in labor and fuel and result in a clean and neat boiler room.

ELECTRIC POWER STATIONS

(PART 3)

STEAM ENGINES

1. Electric power station service, because of very stringent requirements, has been a most important factor in the recent development of types of engines that are far superior to engines of the older classes of construction. The successful engine, be it of any size, type, or speed, must combine the following features:

All parts subjected to strains must be proportioned to withstand higher initial pressures and speeds than engines for commercial manufacturing service. All working parts and wearing surfaces must be very liberal, and also be fitted with the most improved devices for automatic lubrication for continuous service. All bedplates and frames must be exceedingly strong and so proportioned as to maintain perfect alinement. Ample provision must be made for quick and accurate adjustment. All steam passages connecting with the cylinder must be clean cut and so arranged that the entrance of steam into, and its discharge from, the cylinder shall take place with the minimum of loss. The clearance space in the cylinder should be reduced to the smallest possible percentage. The valve gear should be simple, neat, noiseless in operation, having small angles of travel and the moving parts and rubbing surfaces of such metal as will give the maximum amount of durability. The governor must be durable in its construction, highly sensitive in operation, and possess all the regulating qualities described later.

TYPES OF ENGINES

2. The selection of a suitable type of engine is altogether a question of location, service required, and the cost of fuel and water. Engines may be considered under *vertical* and *horizontal types*, and these again subdivided as to speed, and again as *simple* or *compound, condensing* or *non-condensing.*

VERTICAL ENGINES

3. Engines of the vertical type may be used where real estate is costly and floor space limited. For similar classes of engines the economy is nearly the same, whether the horizontal or vertical type is used; but in attendance and adjustment, the vertical engine will require closer attention. The weight of the piston, piston rod, crosshead, connecting-rod, and boxes will be added to the steam pressure on each descending stroke, and must be deducted therefrom on each upward stroke. Therefore, the work cannot be so uniform as that of the horizontal engine.

HORIZONTAL ENGINES

4. Horizontal engines have the advantage of being wholly under the eye of the engineer from the floor level, the parts are more accessible for adjustment, and in many details the horizontal type is more readily examined, and can be operated with more comfort and less anxiety.

SIMPLE ENGINES

5. This class includes all those engines in which the expansion of the steam is effected in a single cylinder. The simple engine is very largely used for small high-speed units or in places where the cost of fuel is of secondary importance. In modern stations of large output, it has been replaced by the compound engine because the increased economy of steam more than offsets the increased first cost of engine. Fig. 1 shows a cross-section of a typical, high-speed, simple engine of the self-oiling type. These engines are popular for small plants, such as isolated plants in office buildings or hotels, because they have few parts to get out of order and can be run with a minimum amount of attention. The details of self-oiling engines have now been brought to such a high state of efficiency that this feature may be considered reliable when combined with the high standard of work always required in the open engine. High-speed engines of good design have, as individual engines, a wide range of power, which is obtained by the combination of increased or reduced speed, low- or high-steam pressure, and early or late cut-off.

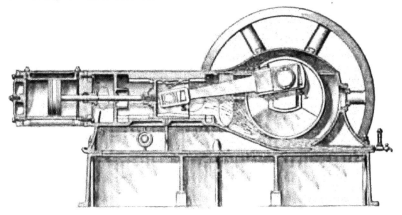

FIG. 1

It has been demonstrated by frequent trials that the favorable point of cut-off in a simple non-condensing engine is at about one-fourth stroke, when using steam at 90 to 100 pounds initial pressure. Cutting off earlier produces a greater percentage of loss through cylinder condensation; cutting off later increases the loss because the steam is exhausted or thrown away while at a considerable pressure.

COMPOUND ENGINES

6. Compound engines, as distinguished from simple engines, are those in which the steam is expanded in two or more stages. The steam is first passed into the *high-pressure cylinder* and there expanded down from boiler pressure to an amount depending on the design of the engine and the conditions under which it is operated. The steam is then passed into the *low-pressure cylinder* and further expanded, after which it is exhausted into the atmosphere or a condenser, as the case may be. The number of stages in which the expansion of the steam is carried on is denoted by calling the engine a *compound, triple-expansion,* or *quadruple-expansion engine.*

7. Cylinder Condensation.—An important advantage of compounding lies in the reduced range of temperature occurring in the cylinders, as compared with the range of temperature in the cylinder of a simple engine, using steam between the same limits of pressure as the compound. In a simple condensing engine using steam at an initial pressure of 100 pounds, the entering steam has a temperature of 340° F. If the exhaust pressure is 11 pounds below atmospheric pressure, the escaping steam has a temperature of 150°. The range of temperature in the cylinder is the difference of these figures, or 190°, and this change occurs at every stroke of the engine. Of course, it cannot be said that the iron of the cylinder responds to the above changes to the extent that the figures indicate, but the inner surface of the iron no doubt changes its temperature during each stroke, and at the time of admission its temperature is below the average of the temperatures given. The entering steam, therefore, comes in contact with surfaces whose temperature is far below its own; this causes condensation at those surfaces during admission. During exhaust, heat is transferred to the exhaust steam from the surface of the cylinder. These two actions cause a loss of effect, which is usually stated as that due to **cylinder condensation.** It is greater in amount the greater the range of temperature, and is materially reduced by dividing the expansion into two stages, each being performed in a separate cylinder. In a well-arranged compound engine using steam at the pressures given, the range of temperature in each cylinder will be about one-half as great as in the case of the simple engine, or about 95°.

8. Arrangement of Cylinders for Compound Engines.—In considering compound engines, only those engines in which the steam is expanded in two stages will be described. In power-station practice, the triple-expansion engine does not usually give results of sufficiently high economy over a well-designed compound engine to justify the increased expense and complication. Tests made in some cases have actually shown better economy with the intermediate cylinder cut out of service, than with the three cylinders in use. For best economy, the triple-expansion condensing engine must be of large size and operated under a constant load at its best efficiency. This remark applies with greater force to quadruple-expansion engines. Compound engines are usually classed as *tandem compound* or *cross-compound;* either of these types may be horizontal or vertical. A third type which may be classed as *duplex vertical and horizontal compound* has recently been used for larger power-station engines.

9. Tandem Compound Engines.—Fig. 2 (*a*) and (*b*) shows two arrangements of the cylinders for tandem compound engines. In this type, the two cylinders are placed in line and there is but one crosshead and connecting-rod. Fig. 3 shows a high-speed tandem compound engine arranged for direct connection to a dynamo. The cylinders are arranged, as in Fig. 2 (*a*), with the low-pressure cylinder *L. P.* next the crank-shaft.

The advantage of the tandem compound engine is its lower cost than the cross-compound engine. Its principal dis-advantage is the inaccessibility of the low-pressure cylinder. The cross-compound engine permits a higher rated speed and easier access to both cylinders; it also gives a more uniform turning effort on the crank-shaft. It occupies less floor space in length and greater floor space in width than the tandem compound.

10. Cross-Compound Engines.—In these engines the cylinders are arranged side by side, as shown in Fig. 2 (*c*) and (*d*). The steam first passes into the high-pressure

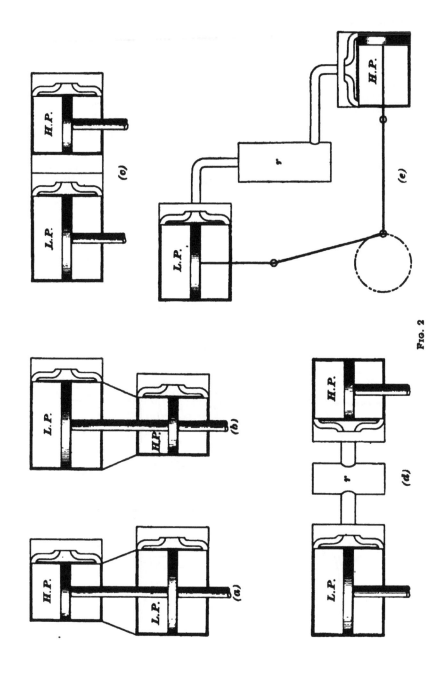

FIG. 2

cylinder *H. P.*, then exhausts into the receiver *r*, if one is provided, and from thence passes into the low-pressure

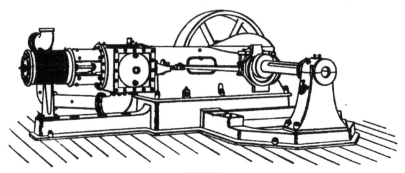

FIG. 8

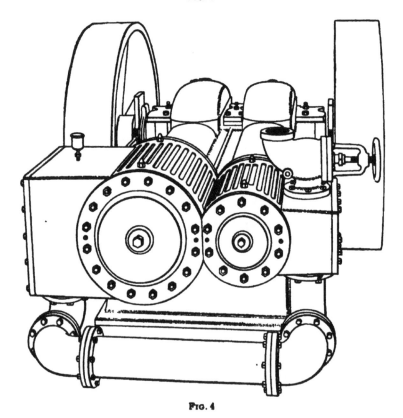

FIG. 4

cylinder *L. P.* The cylinders may be so close together that the receiver *r* forms a jacket connecting the cylinders, or

47—10

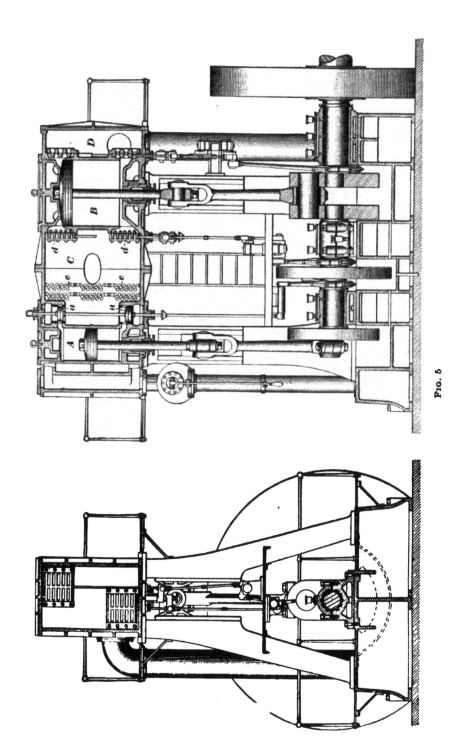

FIG. 5.

two distinct engines may be used, as in Fig. 2 (*d*). Fig. 4 shows a rear view of a cross-compound, high-speed engine. Fig. 5 shows a sectional view of a vertical cross-compound engine of moderate size. Steam is admitted to the high-pressure cylinder *A* by means of the piston valves *a*, *a*, which also allow the exhaust to pass into the receiver *C*. From *C*, the steam is admitted into the low-pressure cylinder *B* by the gridiron slide valves *d*, *d*, which give a large port opening with a small range of movement. From *B*, the steam passes into the exhaust chamber *D*, the exhaust being controlled by gridiron valves that are driven independently of the inlet valves *d*, *d*. Each set of valves is driven by an

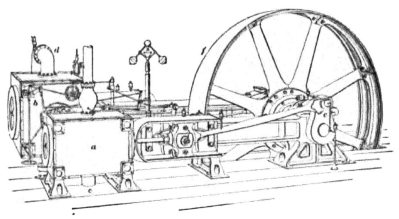

Fig. 6

eccentric on the main shaft. The cross-compound engine gives a more uniform turning effort on the crank-shaft than the tandem compound or simple engine, because the two crankpins used with the cross-compound can be placed at approximately right angles and thereby avoid the dead centers that are always present where only a single crank is used. Fig. 6 shows a horizontal cross-compound engine where the cylinders are separated, thus making in effect two separate engines; *a* is the high-pressure cylinder, *b* the low-pressure, *c* the exhaust from *a* leading to the receiver, and *d* the pipe supplying steam from the receiver to the low-pressure cylinder.

11. Fig. 7 shows a type of vertical cross-compound engine that has been used in a number of large electric power stations. It has a Corliss valve gear and is direct-connected to a dynamo, as shown. The steam is led to the engine through the main steam pipe *a* and, before passing into the high-pressure steam chest, flows through a separa-

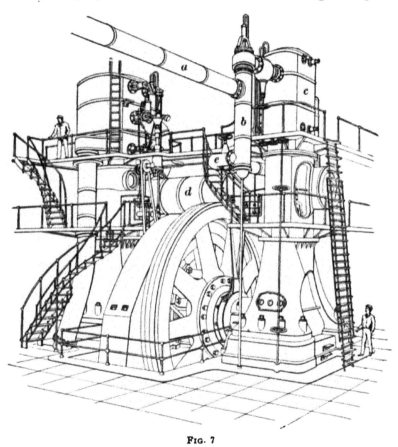

Fig. 7

tor *b*, which removes the entrained water. The exhaust steam from *c* passes into the receiver *d* where it is reheated, as explained later, by live steam taken from the bottom of the separator through the pipe *e*. This particular engine is about 4,500 horsepower, has cylinders 46 and 86 inches in diameter, a stroke of 60 inches, and runs at 75 revolutions per minute.

12. Duplex Vertical and Horizontal Compound.
Fig. 2 (*e*) shows the arrangement of high- and low-pressure
cylinders for this type of engine and Fig. 8 shows one of
the engines as built by the Allis-Chalmers Company and
used in the power station of the Manhattan Elevated Railway, New York. It consists of four engines, two high-
pressure horizontal and two low-pressure vertical, arranged
at right angles to each other as shown, *a, a* being the high-
pressure cylinders and *b, b* the low-pressure. Each pair of

FIG. 8

engines connects to a common crankpin *c*. The cranks of
this engine are placed 135° apart, so that the crank-shaft
receives eight impulses during each revolution, which gives
such a uniform turning effect that the flywheel is dispensed
with, its place being taken by the revolving field *d* of the
dynamo. The steam from the high-pressure cylinders passes
into the receivers *e, e* and thence into the low-pressure
cylinders. These engines are of 8,000 horsepower, the

low-pressure cylinders are 88 inches in diameter, the stroke is 60 inches, and the speed 75 revolutions per minute.

13. Compound Engines Used With Condenser. The compound condensing engine, when used under favorable conditions of steam pressure and load, is highly desirable, because of its economy of steam. It is not good practice to operate a compound engine without a condenser, for regular service, neither is it good practice to work a compound condensing engine constantly below its rated capacity; because when so operated the amount of steam exhausted from the high-pressure into the low-pressure cylinder is insufficient to do the full work intended for the latter and the engine must drag the low-pressure piston.

Compounding in connection with the condenser renders possible a higher ratio of expansion, either by means of a lower final terminal or a higher boiler pressure, or both, and as the steam is first received in one cylinder and the vacuum formed in another, neither cylinder is exposed to such extreme changes of temperature as when the highest and lowest pressures alternate in the same cylinder. The condensation from internal changes is therefore less, but to realize the most gain, very good external protection from radiation is necessary, as there is increased surface to protect. Compound condensing engines require correspondingly less boiler capacity than simple engines; and this effects a saving in the initial cost of boiler installation as well as maintenance.

14. Reheaters for Compound Engines.—Reheaters are frequently used between the high-pressure and low-pressure cylinders of compound engines, to heat the exhaust steam from the high-pressure cylinder before it enters the low-pressure cylinder. Such appliances should be used with caution. It is frequently claimed that there is a liberal percentage of economy gained by the use of reheaters, but it often happens that the reheater is a source of actual loss. This can be readily determined by analysis of the indicator cards showing the work done by the low-pressure cylinder with and without the reheaters in service. Instances are known where, by actual test of the amount of water of condensation derived from the live steam used to heat the exhaust passing through the reheater, it has been found that the live steam used in the reheater was as much as from 40 to 60 pounds per hour for each horsepower-hour gained in the low-pressure cylinder, whereas the engine itself was operating at a consumption of 17 pounds per horsepower-hour. The location of a reheater *e* is shown in Fig. 5. It consists of a large number of small tubes through which live steam is circulated, thus reheating the steam in the receiver.

SPEED CLASSIFICATION

15. For electric power station service, engines have been classified as *high-speed*, *medium-speed*, and *low-speed*. It is a difficult matter to state where the dividing line between these classes lies, though, roughly speaking, engines running at 200 revolutions or over would be classed as high-speed; from 100 to 200, as medium-speed; and below 100 as low-speed. Electric generators for driving by direct connection to the engine shaft are designed to conform with the rotative speed of the engines. With any generator of given capacity, reduction of rotative speed increases the weight and cost almost in direct ratio, and increase of rotative speed reduces weight and cost. There can be no question, from the practical operative standpoint, that the high-speed engine requires closer daily attention to keep it properly adjusted for good work, and that the cost of maintenance will average a trifle higher than for engines of moderate speed; yet the fact must not be overlooked that electric lighting is indebted to the high-speed engine for its successful commercial development. The low cost of the high-speed engine, its good regulation, and ready adaptability made success possible.

16. Piston Speeds. — The piston speed limits the length of stroke for a given rotative speed, and is itself limited by the practical mechanical difficulties of lubrication and of successfully reversing the motion at the end of each stroke, of a large mass of metal moving at a very high velocity. The limits of piston speed for engines of long stroke have been found to be from 700 to 800 feet per minute, and for short-stroke engines, 600 feet per minute. With high-speed engines, the clearance is of a larger proportionate percentage to the volume of piston displacement; this is a disadvantage that pertains to short-stroke engines, but the short stroke permits high rotative speed, thereby reducing the cost per horsepower of the engine and generator, making the unit more compact and requiring the minimum of floor space.

STEAM CONSUMPTION

17. The amount of steam per horsepower-hour required for engines of the different classes, whether vertical or horizontal, is given in tabulated form in connection with Condensers and Condensing Appliances. Regarding steam consumption it is only fair that the steam engine should receive credit for its ability to show the development of an indicated horsepower at a specified economy according to its design and the employment of the necessary auxiliary appliances that make for its economy, but devices extraneous to the engine, which are applied to obtain better economy by recovering heat units that would otherwise be wasted, belong to the plant in the aggregate, and not to the engine as the prime mover. It is not correct to state the economy of steam engines in pounds of coal per horsepower-hour, as this involves the kind and quality of the fuel, the efficiency of the boilers and piping, and the skill of the firemen. The correct expression of engine economy is in terms of the pounds of steam per indicated horsepower-hour per hour, which with single-cylinder non-condensing engines will be 24 to 35 pounds when operated under favorable conditions. The same engines operated condensing will take from 19 to 28 pounds of steam, and compound condensing engines will take 13 to 16 pounds. Finally, to get at the real value of the developed economy, the steam consumption must be calculated down to the pounds of steam per kilowatt-hour developed at the generator. In expressing the economy of steam engines the terms *steam consumption* and *water consumption* are synonymous. For example, if an engine is said to require 30 pounds of water per horsepower-hour, it is equivalent to saying that it uses 30 pounds of steam per horsepower-hour, because the weight is the same whether the steam is condensed into the form of water or not.

18. Engine Friction.—Remembering that the indicated horsepower is greater than the actual horsepower obtained from the engine, it is most important to closely analyze the

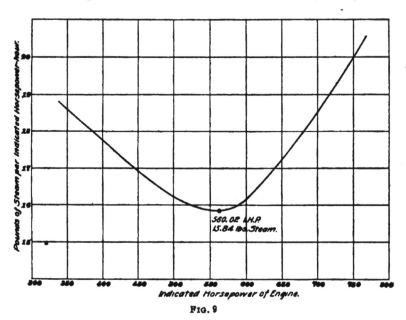

Fig. 9

working parts and to know what the friction indicator card shows, as the engine friction may vary from 6 per cent. to 15 per cent., or possibly more, of the maximum indicated horsepower.

As excessive friction will tend to reduce the commercial efficiency, it is unwise to select engines having complicated valve motions or cumbersome and useless moving parts.

19. Combined Efficiency of Generating Units. A point often overlooked is that the engine may be selected too large or too small for the generator. Each engine has a point of maximum efficiency, which is attained by the combination of the most favorable point of cut-off at a given initial pressure, at which point the indicated horsepower is developed with the minimum of steam consumption. This is illustrated by the efficiency curve shown in Fig. 9, which is derived from a series of tests of a compound condensing engine rated at 600 horsepower, and shows the relation between the indicated horsepower and the steam consumption per indicated horsepower. It is seen that for a load of 560.02 indicated horsepower a minimum consumption of 15.84 pounds of steam per horsepower is reached.

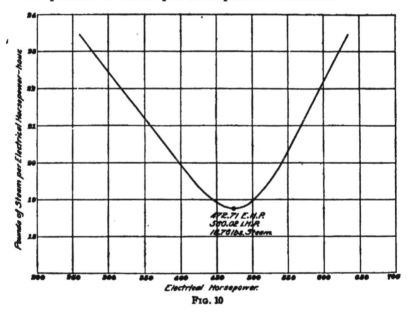

Fig. 10

An electric generator also has its point of maximum efficiency. Now, if the engine and generator as a combined unit are so selected and operated that they reach their points of maximum efficiency simultaneously, it is evident that the unit will show the highest results of its class, indicated by the minimum pounds of steam per electrical horsepower developed.

Fig. 10 shows the relation between the electrical horsepower, delivered at the terminals of the generator, and the steam consumption per electrical horsepower. Both of these curves show that any considerable departure from the load giving maximum economy results in quite a large increase in the steam consumption per horsepower-hour, and that in order to secure most efficient operation it is necessary to keep the load in the region represented by the lower part of the curves.

STEAM PRESSURES

20. Starting out with what was formerly the customary standard of 75 to 90 pounds boiler pressure, it has been theoretically and practically demonstrated that a higher pressure and earlier cut-off leads to greater economy of steam, so that the pressures now commonly used are 100 to 120 pounds for single-expansion engines, 140 to 160 pounds for compound condensing engines, and 180 to 190 pounds for triple-expansion condensing engines. These higher pressures permit the use of smaller pipes, valves, and fittings than would be required for a similar aggregate horsepower at the low-pressure basis. Thus far it has been found that high-steam pressure should be less than 200 pounds for successful service. The losses, annoyances due to unexpected leakage, difficulties of engine lubrication, wear and tear, and repairs, thus far offset the gain theoretically to be obtained from any higher pressures. Improvements are constantly being made to perfect the satisfactory use of still higher pressures, which may be particularly adapted for very large stations.

DESIRABILITY OF SEVERAL ENGINES

21. In a power station having a variable load, such as incandescent lighting, the best engine economy is attained when the engine units are selected of such capacities and duplicated, so that as the load increases or decreases the engine units may be put in or taken out of service; and each engine while in service shall be operated at its highest efficiency during the greatest possible number of hours.

The capacity, type, and speed of engines to be selected

for a given station can best be determined by careful consideration of the following points:

1.   The character of the load, whether for incandescent lighting, arc lighting, motive power, electric railways, or a combination of part or all.

2.   The probable area of the load diagram, and the fluctuations in the load line.

3.   The cost of fuel for steam production, the cost of real estate, etc.

4.   The location of the station as regards water supply for condensing purposes, coal delivery, ash haulage, etc.

As to whether the units of engines and generators shall be direct-connected or belt-driven, the selection for the smaller station is frequently and non-scientifically decided in favor of the belt-driven, because of a few hundred dollars extra investment required for the direct-connected unit, often losing sight of the cost of the extra floor space required, belt maintenance, loss by friction, and similar expense occasioned by the use of belt-driven apparatus.

STANDARD DIMENSIONS AND SPEEDS FOR DIRECT-CONNECTED ENGINES AND GENERATORS

22.   In order that, as far as practicable, standard dimensions may be adopted by builders of engines and generators for combination as direct-connected units, a committee of the American Society of Mechanical Engineers have, after full investigation and consultation, recommended the capacities, speeds, and dimensions shown in Table I and Fig. 11.   These recommendations refer to generating units made up of a high-speed engine coupled to a dynamo mounted on a sub-base that forms an extension of the engine bed, as in Fig. 3. The size of units does not exceed 200 kilowatts, and the schedule provides for a variation of 5 per cent. above or below the mean speed.   The dimensions of shafts apply only to engines of usual proportions with the generators attached at the side of the engine.   Two styles of generators are

TABLE I

SIZES, SPEEDS, AND STANDARDIZED DIMENSIONS OF DIRECT-CONNECTED GENERATING SETS

Capacity of Unit Kilowatts	Revolutions per Minute	Armature Bore Inches		Diameter of Engine Shaft at Armature Fit Inches		Space on Shaft Between Limit Lines Inches		*B*, Length of Extension Pieces Inches	*C*, Axis of Shaft Above Top of Base Inches	*R* Inches	*D*, Width of Top of Subbase Inches	Key (a Feather) Inches				Holding-Down Bolts	
		Center-Crank Engines	Side-Crank Engines	Center-Crank Engines	Side-Crank Engines	Long Class *A*	Short Class *A'*					Width	Thickness	Depth in Shaft at Edge	Projection Above Shaft at Edge	Diameter Inches	Number
25	310	4	4½	4 + 0.001	4½ + 0.001	30	25	5	23½	Flat	48	1	¾	⅜	⅜	1	4
35	300	4	5½	4 + 0.001	5½ + 0.001	33	28	5	25	Flat	54	1	¾	⅜	⅜	1	4
50	290	4½	6½	4½ + 0.001	6½ + 0.002	37	31	6	28	Flat	60	1¼	¾	⅜	⅜	1	4
75	275	5½	7½	5½ + 0.001	7½ + 0.002	43	37	6	31	Flat	66	1½	1	½	½	1¼	4
100	260	6	8½	6 + 0.001	8½ + 0.002	48	42	6	34	Flat	72	1½	1	½	½	1¼	4
150	225	7	10	7 + 0.002	10 + 0.002	51	45	6	37¾	41¼	84	1¾	1¼	⅝	⅝	1¼	4
200	200	8	11	8 + 0.002	11 + 0.002	54	48	6	42¾	47½	96	2	1¼	⅝	⅝	1½	4

NOTE 1.—Five per cent. variation of speed permissible above and below speeds in table.
NOTE 2.—Distance from center of shaft to top of base of outboard bearing may be less than *C* (to suit engine builder), though not less than possible outside radius of armature.

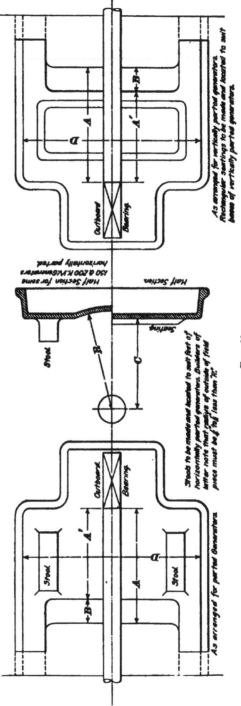

FIG. 11

provided for—those having the fields divided vertically and those having the fields divided horizontally. Seatings for each style of machine are provided by two patterns of extension subbase for the generator. The overload capacity should not exceed 25 per cent. of the rated capacity for 2 hours, or 50 per cent. for 30 minutes.

23. Armature Fit.—The bore of the armature is to be the exact size stated in the table; the allowance for a pressed fit is to be made by a slight increase in the diameter of the engine shaft. The allowance of .001 inch for shafts of 4 to 6 inches, inclusive, and of .002 inch for shafts $6\frac{1}{2}$ to 11 inches, inclusive, represent the best practice. The best results will be attained by working to a definite gauge; and the generator builder should be required to furnish one of the exact diameter of armature bore, and the engine builder will make due allowance for a pressed fit.

Holding-down bolts, keys, and outboard bearings should be furnished by the engine builder. The length of key is to be adjusted between the builders of engine and generator in each individual case, and it should be specified in the contract which builder shall fit and press the armature on the shaft. For engine and generator units of other types and larger capacity, the specifications should clearly set forth the particular details required for each manufacturer on all of the above points, to the end that there shall be no clash of opinion or division of responsibility, and also that the purchasing company shall not be put to unexpected extra expense by the failure to specify clearly these matters in advance.

ENGINE REGULATION

24. Flywheels.—In reciprocating engines, the elements that disturb regulation are variations of load and the engine impulses. The following factors have their influence on the uniformity of rotative effect during each revolution of the shaft:

1. The weight and momentum of the reciprocating parts,

which are the piston and piston rod, the crosshead, and connecting-rod.

2.   The mean effective steam pressure on the piston.

3.   The inertia of the reciprocating masses that must be overcome at each end of the stroke.

In vertical engines, the conditions are different from horizontal engines, as in the vertical engine the weight of the reciprocating parts acts on the downward stroke the same as an equal amount of steam pressure, adding to the work done; and, correspondingly, the same force must be subtracted from that of the up stroke, as these weights must then be lifted. This makes a difference in the power of each stroke. Allowance must also be made for the influence of the different positions of the connecting-rod, which influence is different in vertical and horizontal engines.

To overcome these disturbing influences, the flywheel is mounted on the engine shaft with the purpose that the heavy rim shall aid in maintaining a certain degree of uniformity in angular velocity during each engine revolution. For electric power station service, it is necessary, not only that the number of revolutions per minute should be practically constant within narrow limits, but, particularly for driving alternating-current generators in parallel, it becomes vitally essential that the variation of angular velocity during each revolution shall be confined within exceedingly narrow limits. It is for this reason that the rim of the flywheel, which is the only revolving mass having any influence on the regulation, must be of sufficient weight to provide the momentum requisite to overcome not only the influences of the reciprocating parts that are operating against uniformity of rotation, but also variations in the retarding effects of the work done by the revolving element of the generator.

25.   The commercial requirements for the governing of engines for electric service, are now sufficiently well known for first-class engine builders to guarantee the required regulation. From the nature of the power-station service, the load is continually changing. Any variation in engine

speed causes disturbances on the system, and it is therefore essential that the speed shall be uniform during all changes of load; and even under variations of boiler pressure, say for 5 pounds above or below normal, there should not be any disturbance on the system.

The governor must regulate to a high degree of perfection with very rapid adjustment, and without the slightest instability or tendency to race. The modern inertia governor is well adapted for this service. In a well-designed governor of this type, the resisting strains will be so well equalized as to balance each other and not cause any excessive strain on the studs or bearings of the governor. The engines must not race under any conditions of service and must be so governed as to permit the generators to be operated individually or in parallel with each other.

REGULATION FOR ALTERNATING CURRENT

26. Engines for driving alternating-current generators require closer regulation than those driving direct-current generators. Definite conclusions have not yet been reached concerning the exact changes that take place in operating alternating-current machines in parallel. It is essential, however, that they run with sufficient regularity not to get out of phase with each other. In other words, they must remain in step. In addition to the steadying effect of the flywheel, the control of the steam admitted to the engine must be so well regulated that the flow shall not be in excess of that needed in combination with the stored energy in the flywheel; therefore, as power cannot be obtained from the engine without the requisite quantity of steam, the admission of steam must be controlled by a governor sufficiently sluggish in its action to prevent periodic pulsations; this is best accomplished by a proper dashpot arrangement attached to the governor.

27. The parallel operation of polyphase alternators driven by well-regulated turbine waterwheels has been uniformly successful, undoubtedly because the turbine wheel is not forced to overcome the tangential effort of the crank-movement and the reversal of heavy reciprocating parts twice during each revolution, as is the reciprocating engine. The governors of engines used to drive alternators should be so designed and constructed that there will be no tendency to cause a periodic transfer or surging of the load between one engine and another, as surging is similar to throwing the load quickly off or on a single engine at short intervals. The natural function of the governor is to regulate the supply of steam in proportion to the load, and for the satisfactory driving of alternators the natural tendency of the governor to perform its task quickly must be overcome by attaching a dashpot, or some form of friction brake, to prevent the governor from responding too quickly to changes in load, or with small changes give no response at all. The use of governors that are adjustable while running is desirable, and a motor adjustment controllable from the switchboard is commonly provided.

The variation of the rotative speed of the generator during any single revolution at constant load not exceeding 25 per cent. overload should not exceed one-sixtieth of the pitch angle between two consecutive poles from the position it would have if the rotative motion were absolutely uniform at the same mean velocity. The maximum allowable variation, which is the amount the rotating part forges ahead plus the amount that it lags behind the position of uniform rotation, is therefore one-thirtieth of the pitch angle between two poles. According as the number of poles on the alternator increases the permissible angular variation decreases, as shown by Table II.

Where several engines are located in one station to drive alternators in parallel, they should have the same characteristics of speed regulation to the end that the power delivered to their respective generators may be proportional to the load.

28. In some cases where parallel operation is possible, the angular variation may be such that the operation of rotary converters or synchronous motors is unsatisfactory. If the engines in the station are set in such perfect alinement that it is possible to look through the spokes of the flywheels, it is easy to see the change in relative angular position when the several engines are in operation.   It is important that the governors shall be fully competent to control the engines under all changes of load, from no load to 25 per cent. or 50 per cent. overload, when operating the generators in parallel; and that as the load rises to the maximum or falls to

TABLE II

MAXIMUM PERMISSIBLE ANGULAR VARIATION OF ENGINES DIRECT-CONNECTED TO ALTERNATORS

Poles	Angular Degrees	Per Cent. of Circumference	Poles	Angular Degrees	Per Cent. of Circumference
2	3.000	.833	40	.150	.042
4	1.500	.417	44	.136	.038
6	1.000	.278	48	.125	.035
8	.750	.208	52	.115	.032
12	.500	.139	56	.107	.030
16	.375	.104	60	.100	.028
20	.300	.083	64	.094	.026
24	.250	.069	68	.088	.025
28	.214	.059	72	.083	.023
32	.187	.052	76	.079	.022
36	.167	.046	80	.075	.021

the minimum, the governor shall permit the load to be shifted from engine to engine, to the end that the engines may be taken out of or put in service and be loaded to the most economical point of operation.

Specifications of requirements of engine regulation and guarantees of performance cannot be drawn too carefully, and the engine builder must be one who understands from a practical standpoint the detrimental results of defective regulation and from experience knows how to furnish what is required.   The object of the peculiar specification as to engine regulation for driving alternators in parallel is to limit the amount of cross-currents that can flow between the generators and prevent troubles with synchronous apparatus, such as synchronous motors and rotary converters. This whole matter is one in which no risk can be taken with engine builders of doubtful experience.

The weight of the flywheel should be such that its momentum will not permit one machine to drag behind the other, and the amount of variation either way in the angular velocity in a single revolution should not exceed one-sixtieth of the pitch angle between two consecutive poles. In a 24-pole alternator, for example, the angle between poles would be $\frac{360}{24}°$ and the allowable variation either way would be $\frac{1}{60} \times \frac{360}{24}° = \frac{1}{4}°$, or .0693 per cent. of a revolution. Under any change of load, therefore, the internal variation of speed will not exceed one quarter of one geometrical degree from the position of uniform rotation during any single revolution.

STEAM TURBINES

29. In principle, the **steam turbine** is not new, as its type is the first heat motor of which we have record. It dates from as far back as 120 B. C. The apparatus of Hero of Alexandria, was a reaction turbine and is described as a spherical vessel mounted on trunnions through which steam was admitted to issue finally from openings tangential to the sphere; the reaction of the steam jets causing the sphere to revolve. The steam turbine, in recent years, has been modified and improved to such a degree that in future power-station equipments, it will be a strong rival of the reciprocating engine. It is in many ways particularly well suited for the driving of dynamos and is now being installed in many new power stations.

30. In the steam turbine, the steam is made to produce rotary motion by causing jets to act on a suitably arranged wheel, or series of wheels, carrying vanes or buckets. The energy stored in the steam is thus made to produce a uniform

rotary motion of the turbine shaft and the intermediate reciprocating motion of the ordinary engine is eliminated. Early
types of steam turbine were very wasteful of steam and thus
were not able to compete with the reciprocating engine.
Improvements, however, have been made until now the turbine is at least as economical, if indeed not more economical
than the reciprocating engine. The pressure or expansive
force of the steam delivered to the steam turbine is caused
first to act on the steam itself, thus generating velocity in
the jet, which is directed on the blades of the turbine, and the
energy of the steam thus transferred thereto.

In a waterwheel it is clear that the wheel is turned by
the impact or pressure of a heavy mass of water having
density and substance to cause the revolutions. The steam
makes up in velocity what it lacks in mass. A jet of
water 1 inch in diameter issuing at a pressure of 100 pounds
per square inch will discharge 41 pounds per second at a
velocity of 121 feet per second, representing, approximately,
9,300 foot-pounds of work per second.* This amount of
work expended per second on a turbine wheel (if one of 100
per cent. efficiency) will generate about 17 horsepower. A
jet of steam issuing through an orifice 1 inch in diameter,
under 100 pounds gauge pressure, will deliver 1.293 pounds
of steam per second at a velocity of 1,466 feet per second.
If this steam is allowed to strike against a wheel so that the
whole of the kinetic energy of the jet is converted into power,
the jet can do, approximately, 43,200 foot-pounds of work per
second. Since 1 horsepower is equivalent to 550 foot-pounds
per second, the jet will be able to deliver $\frac{43200}{550}$ = 78.5 horsepower. The amount of steam delivered per hour will be
1.293 × 3,600 = 4,654.8 pounds, equivalent to nearly 59.3
pounds of steam per horsepower-hour.

In the above example relating to a steam jet it should
be noted that the steam is supposed to blow through a
plain orifice and impinge on the wheel to which it imparts

---

* Kinetic energy = ½ mass × velocity² = ½ $m v^2$ = ½ $\frac{W}{g} v^2$ = ½ × $\frac{41}{32.16}$
× (121)² = 9,332 foot-pounds.

energy due to the velocity of the steam as it passes through the orifice. By suitably arranging the nozzles or turbine wheels so that the steam is allowed to expand, a considerable part of the heat energy stored in it can be made available, thus materially reducing the steam consumption per horsepower-hour. Thus, in the nozzle of the De Laval turbine the steam expands until, when it strikes the wheel, it is at practically the same pressure as the steam in the chamber in which the wheel revolves. A large part of the heat energy stored in the steam is thus given up during the process of expansion and goes to increase the velocity. A velocity of over 4,000 feet per second can be obtained in this manner; and since the energy increases with the square of the velocity, the steam consumption per horsepower can thus be greatly reduced from the amount given in the above example, which merely compares a steam jet with a water jet.

All successful types of steam turbine make provision for this expansion of the steam to increase its velocity. In the De Laval turbine, the expansion takes place wholly in the nozzles and the steam is delivered to the buckets at a low temperature and pressure but at a very high velocity. In the Curtis turbine, the steam is expanded partly in the nozzles and partly during its progress through the turbine wheels, the continuous expansion and consequent abstraction of heat energy from the steam keeping up the velocity until the steam is finally discharged from the last wheel at the temperature and pressure corresponding to the degree of vacuum in the condenser. In the Parsons turbine, the steam is expanded during its passage through the turbine and is not expanded in the inlet nozzles.

TYPES OF STEAM TURBINES

31. The three most important types of steam turbines at present manufactured in the United States in competition with reciprocating engines, are the De Laval, manufactured by the De Laval Steam Turbine Company; the Parsons, manufactured by the Westinghouse Company; and the Curtis,

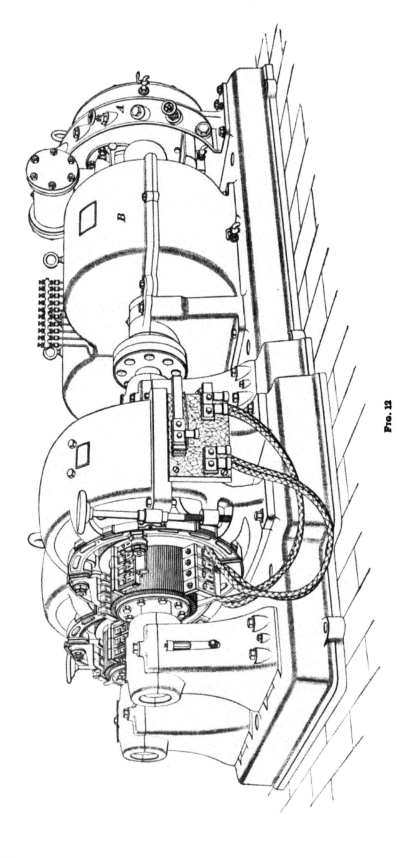

Fig. 12

manufactured by the General Electric Company. The principles of operation of these turbines have been described elsewhere. The Parsons and Curtis turbines run at a comparatively low rotative speed and can therefore be direct-connected to dynamos; while the De Laval is essentially a high-speed machine and is connected to the dynamo through special gearing. Turbines are particularly well suited to the driving of alternators, since the latter can be readily designed for high rotative speeds, especially if the revolving field construction is used. It is more difficult to construct direct-current machines of large output for high rotative speeds, because the extremely rapid reversal of the current in the armature coils during the time they are passing through the neutral region makes it difficult to secure sparkless commutation. Moreover, the mechanical difficulties at the commutator are liable to increase with the speed.

32. De Laval Turbine-Dynamo Set.—Fig. 12 shows the general arrangement of a De Laval turbine as applied to the driving of two direct-current dynamos. The turbine wheel is in case A while case B contains spiral gears running in oil and driven in similar directions by a spiral pinion mounted between them. Each dynamo is of 100 kilowatts capacity, so that the turbine develops, approximately, 300 horsepower. The whole outfit occupies a floor space only 6 feet 3 inches by 15 feet.

The loss of efficiency caused by friction in this type of turbine is claimed to be much below that of a compound reciprocating engine, and the steam consumption per horsepower-hour is almost uniform from one-quarter load to 25 per cent. overload. Under test, it is claimed that this turbine showed the use of 15.43 to 15.99 pounds saturated steam per brake horsepower-hour, and with steam superheated 84° F. it required from 13.55 to 14.21 pounds of steam per brake horsepower-hour, the saving by the use of superheated steam being about 8¼ per cent.

33. Parsons Turbo-Alternator.—Fig. 13 shows a 1,000-kilowatt set consisting of a turbine coupled to a polyphase alternator. The steam enters at *a* and passes through one or both of the governing valves *b*, *b′* and from thence between the sets of stationary and revolving blades contained in casing *c*. The steam passes through the turbine in an axial direction and exhausts from *c* into a condenser. The flyball governor is located at *d* and is connected to pilot valves which control governing valves *b*, *b′*. Governing valve *b* controls the speed from no load to about full load. On overloads the secondary governing valve *b′* is brought into action and admits steam directly to the intermediate drum of the rotor where the working steam areas are greater. The governor acting through valves *b* and *b′* thus automatically controls the speed from no load to a very

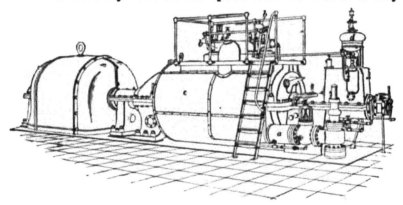

Fig. 13

considerable overload. Means are provided for forced lubrication of the bearings, oil being circulated by means of a pump driven by the turbine. A turbine of this size runs at a speed of about 1,500 revolutions per minute; one of 400 kilowatts capacity runs at about 3,500 revolutions per minute. Fig. 14, curve *a a*, shows the pounds of steam (water) used per electrical horsepower-hour by a Parsons turbine. At full load of 300 kilowatts the consumption is about 16.5 pounds. Line *b b* shows the total steam consumption per hour.

34. Curtis Turbo-Alternator.—Fig. 15 is a perspective view of a 500-kilowatt Curtis turbine set consisting of a turbine direct-connected to a three-phase revolving-field alternator designed for a speed of 1,800 revolutions per minute. Fig. 16 is a sectional view of the same set showing the arrangement of the various parts. The shaft is vertical, the alternator A being mounted on top of the turbine B, and the turbine is in two stages—an upper a and a lower b. Each stage contains three steel turbine wheels c with buckets, or vanes, d cut on their peripheries. The stationary vanes e between the wheels are supported by the turbine casing. Steam enters at f and passes through the nozzles

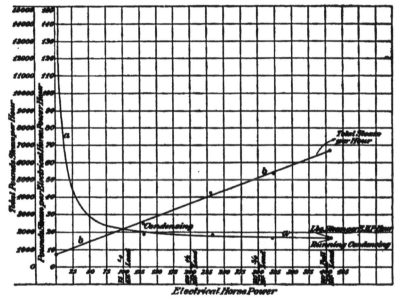

Fig. 14

at g, thus impinging on the first wheel. The flow is then reversed by the first set of stationary blades, the steam strikes the second wheel, and so on until it passes through the three wheels of the first stage. It then passes into the wheels of the second stage through openings at h, and finally passes, through k, into the condenser. In case a condenser is not used, the steam is exhausted through the exhaust connection l, the first stage only being used.

In order to obtain a high economy, it is necessary to operate turbines with a high vacuum, hence they are not

operated non-condensing if it is possible to avoid it.   In the
later types, a surface condenser is placed in the base of the

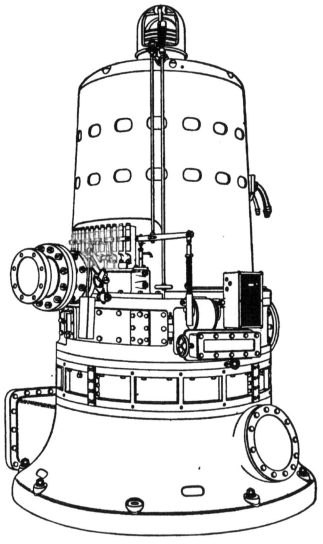

Fig. 15

turbine instead of being a separate device.   By this arrangement the liability to leaks between turbine and condenser is
reduced and a high vacuum secured, provided a plentiful

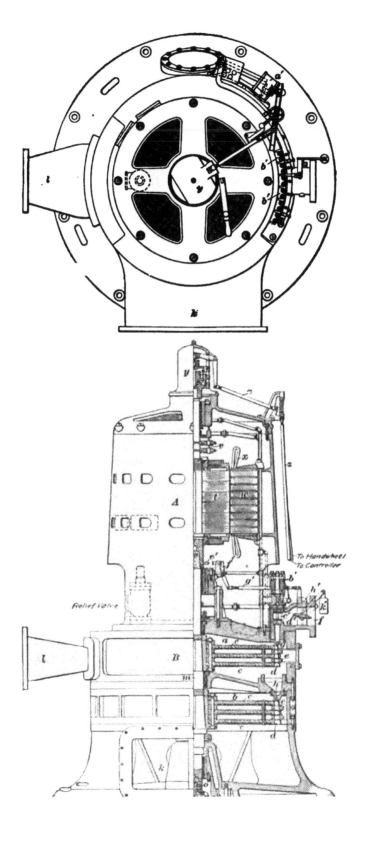

Relief Valve

To Handwheel
To Controller

supply of condensing water is available and a brisk circulation maintained through the condenser.

In Fig. 16 the turbine shaft *m* is supported on a thrust bearing *o* and oil is forced in under the shaft through pipe *p*. The oil flows up and out through the bearing and returns through pipe *r*. The end of the shaft, therefore, rests on a thin film of oil and the downward pressure on the bearing is taken up by the oil pressure. The alternator shaft is coupled to the turbine shaft at *s* and carries a four-pole revolving field *t*, which is built up of sheet-steel stampings. The magnetizing coils are made of copper strip wound on edge and are supplied with exciting current through the collector rings *v*. The stationary armature is shown at *w*; it is constructed in the usual manner and consists of toothed sheet-steel stampings which, when assembled, provide grooves for holding the armature coils *x*.

The governor *y* is located on top of the machine and consists of centrifugal weights acting against a heavy spring. The movement of the weights operates rods *z*, *z* connected to an electric controller located at *a'*. The controller regulates the flow of current (furnished from the exciter, which is driven independently of the turbine set) through a series of small valves operated by electromagnets *b'*. These small pilot valves operate valves *c'* that admit the steam to the several nozzles. Any decrease in speed, caused by an increase in load, causes a corresponding movement of the governor. This moves the controller that operates the electromagnets so as to bring more nozzles into action and thereby supply more power to carry the load. In addition to the regular throttle valve, the turbine is provided with a valve *d'* that closes automatically if, for any reason, the speed should become excessive. A centrifugal device arranged on the shaft at *e'*, flies out when the speed rises above the predetermined amount, thus moving lever *f'*, pulling on rod *g'*, and releasing catch *h'*. This allows weight *k'* to drop and close the valve. Small adjustments in speed, as required, for example, in synchronizing the alternators, can be made by turning the small hand wheel *l'*, thereby

changing the action of the governor slightly. In some of
the turbines, particularly in the larger sizes, this adjustment
is made by means of a small electric motor controlled from
the switchboard.

Fig. 17 shows the steam consumption of a 600-kilowatt
Curtis turbine at various loads. The turbine ran at a speed of
1,500 revolutions per minute and was used with a condenser.
Steam turbines, like ordinary engines, can be used either

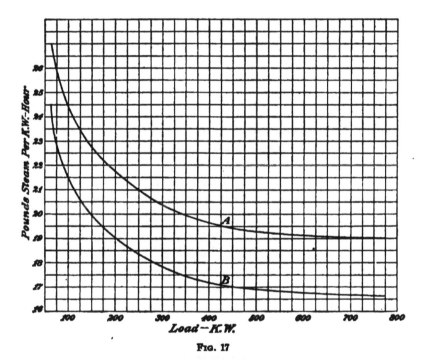

Fig. 17

with or without condensers, but their economy is much
better when used with condensers, because if a high vacuum
can be maintained it is possible to work with much higher
ratios of expansion than are practicable with ordinary
engines. In Fig. 17, curve _A_, it will be noted that the
steam consumption at full load (600 kilowatts) is about 19.1
pounds per kilowatt-hour. This is equivalent to about
14.2 pounds of steam per horsepower-hour. This, however,
is the steam consumption measured at the terminals of the

generator. If the efficiency of the generator is assumed to be 95 per cent. and the mechanical efficiency of the turbine 95 per cent. also, the steam consumption per horsepower, which would be comparable with that per indicated horsepower of an ordinary engine, would be about 12.8 pounds of steam per indicated horsepower-hour. This is much better than would ordinarily be obtained with a reciprocating engine of this size. The steam consumption of the turbine can be considerably reduced by using superheated steam, as shown by the lower curve *B*. This curve was calculated on the assumption that the steam was superheated 150°. The use of superheated steam in turbines is not attended by the difficulties cf lubrication that are met with in reciprocating engines.

35. Advantages of Steam Turbines.—The important advantages that the steam turbine has over the reciprocating engine may be summed up, briefly, as follows: (*a*) The ability to use highly superheated steam, resulting in greater economy; (*b*) reduced cost per kilowatt capacity of the generating unit, because of increased speed and less weight per kilowatt; (*c*) reduced floor space, resulting in less cost for land and power-station building; (*d*) reduced cost of lubrication, as no cylinder oil is required and less oil is needed for bearings; (*e*) saving in labor, as engine oilers are not required and one engineer can attend to more output than five engineers on reciprocating engines; (*f*) reduced cylinder condensation, because all parts of the turbine are maintained at practically constant temperature; (*g*) reduced cost of foundations, as the turbine is perfectly balanced and has no reciprocating parts; (*h*) a commercial efficiency higher than the reciprocating engine, because there is much less friction; (*i*) increased steam economy and high economy at light loads. The fact that the turbine gives a good steam economy over a wide range of load, whereas the economical load of a reciprocating engine is quite sharply defined, is an important argument in favor of the turbine, particularly in stations handling a variable load. If it becomes necessary to operate a turbine unit at a comparatively light load, say one-fourth or one-half load, the increase in steam consumption per horsepower-hour or kilowatt-hour output is not nearly as great as it would be with a reciprocating engine. Also, a turbine unit will work more efficiently on overloads. These points will be seen at once by comparing the steam-consumption curve, Fig. 10, with the curves shown in Figs. 14 and 17. The forces acting on the turbine wheels are continuous, hence a uniform rotary motion is secured without the necessity of heavy flywheels. This feature is of particular advantage when power is to be furnished for the parallel operation of alternators, all the difficulties met with on this account in reciprocating engines being absent.

CONDENSERS AND CONDENSING APPLIANCES

36. In non-condensing engines (that is, engines that are not supplied with a condenser) the steam is exhausted into the atmosphere, and therefore must have, at least, the pressure of the atmosphere acting against it; in practice, the back pressure of steam in a non-condensing engine is scarcely ever less than 16 pounds above vacuum, and is oftener 17 pounds or more. In good condensing engines the back pressure is often as low as 2 pounds above vacuum.

37. If a cubic inch of water is converted into steam at atmospheric pressure, it will occupy 1,646 cubic inches of space, and, conversely, if 1,646 cubic inches of steam at atmospheric pressure is condensed into water, it will occupy but 1 cubic inch of space; hence, if a closed vessel is filled with steam at the atmospheric pressure and that steam is condensed to a cubic inch of water, $\frac{1645}{1646}$ of the space will be, theoretically, devoid of air and a perfect vacuum would be the result. This is not strictly true in practice, because the feedwater of the boilers always contains a small quantity of air, which passes into the condenser with the exhaust steam and is released there when the steam is condensed; more or less air also finds its way into the condenser through leaks around the piston rod and valve stems, and in the case of the jet and the induction condensers, the air contained in the condensing water is also released in the condenser under the influence of the partial vacuum. Moreover, water in a vacuum emits a certain amount of vapor, and if the condenser were successively filled with steam and the steam condensed at each filling, air and vapor would, unless they were removed, accumulate from these various sources until the vacuum was entirely destroyed.

38. The object of the **condenser** is to remove a large part of the back pressure on the exhaust side of the piston. By making the engine exhaust into a condenser, the back pressure will be lowered to the pressure existing in the condenser, and consequently, with the pressure on the steam side of the piston remaining the same as before, the net pressure on the piston will be increased by the use of a condenser.

To get rid of the air and vapor that would otherwise accumulate, the condenser is fitted with an **air pump,** or is provided with other means by which the air and vapor are removed from the condenser along with the condensed steam and condensing water. Sometimes a pump is so arranged that it removes air and vapor only and does not pump out the water and condensed steam; a pump of this kind is usually referred to as a *dry-air pump.*

In electric power stations, condensers are used wherever possible because of the reduction in the pounds of steam required per horsepower-hour and a corresponding reduction in the amount of fuel from what would be required by an engine operating without a condenser.

39. Economy of Condensers.—The advantages and economies pertaining to the use of condensers may be enumerated as follows:

1. The increase in power gained by reason of the vacuum on the exhaust side of the piston, which, for 26 inches of mercury, is approximately equivalent to a net gain of 12 pounds mean effective pressure per square inch of piston area.

2. A reduction in fuel is obtained in ratio to the gain in mean effective pressure and a smaller amount of steam is required; or more power may be obtained with the same fuel consumption.

3. There may also be secured a saving in the cost of boiler feedwater, when the water from the condenser discharge is used over and over again,' provided effective methods are applied to extract the cylinder oil. When the condensed steam is used, over and over, for boiler feed, the scaling impurities will, of course, have been completely removed.

4. Because the condensing engine requires less steam per horsepower, it becomes possible to use less boiler-heating surface for a given horsepower of engine.

5. With condensation, in combination with other economic appliances, it becomes possible to obtain the highest economic results of modern practice by using compound engines.

Extra attendance, if any, will be required with a large plant, but not with a small one, when condensers are used.

40. Comparative Economies.—The important point to be considered as influencing the use of the condensers is the greater economy that will be shown in power produced from compound condensing engines as compared with a station operated with ordinary high-speed, non-condensing engines. In actual practice, the ordinary high-speed engine frequently uses 35 to 40 pounds of steam per hour per indicated horsepower, and the medium high-speed or simple Corliss engines seldom reach as low as 25 pounds in constant daily service. Therefore, the estimated saving as between the most uneconomical (a station with high-speed engines) and the most economical (a station with compound condensing engines) is from $12\frac{1}{2}$ to 20 pounds of steam per indicated horsepower-hour, according to the capacity and environment of the station.

It must not be forgotten that the steam used by the engine only, is different in quantity from the steam per horsepower used for operating the entire station, and the engine should not be charged with the steam it does not use. In this connection, especial attention should be given to selecting economical auxiliary appliances for the station. The higher the cost of fuel and water, the greater will be the inducement to use condensing engines and appliances; but when the whole cost of fuel is as low as $1 to $1.50 per ton, the calculations must be made with exceeding care, as it is possible that the interest on investment for condenser equipment, cost of pumping, maintenance, and attendance will nearly equal the value of fuel and water saved. The real gain by the use of condensers will vary according to the type of engines used, and can be estimated by comparing the pounds of steam per hour required to produce a horsepower with and without condensation.

41. Table III shows the pounds of steam required by engines used in power stations under actual conditions. This table was originally prepared by the late Charles E. Emery, Ph. D., and has been modified according· to some recent tests.

TABLE III

STEAM CONSUMPTION OF VARIOUS TYPES OF ENGINES

Type of Engine Name	Steam per Indicated Horse-power-Hour				Per Cent. Gained by Condensing
	Non-Condensing		Condensing		
	Probable Limits Pounds	Assumed for Comparison Pounds	Probable Limits Pounds	Assumed for Comparison Pounds	
Simple high-speed . . .	40 to 26	33	25 to 19	22	33
Simple low-speed . . .	32 to 24	29	24 to 18	20	31
Compound high-speed .	30 to 22	26	24 to 16	20	23
Compound low-speed .	25 to 18	25	20 to 12$\frac{1}{2}$	18	25
Triple low-speed . . .	24 to 17	20	18 to 12$\frac{1}{2}$	15	20
Triple high-speed . . .	27 to 21	24	23 to 14	17	29

TYPES OF CONDENSERS

42. Condensers may be considered under two types, or classes, for power-station service, and the kind of water, its abundance, cost, levels for pumping, cost of fuel, etc., will determine the type to be used in any given case. These types are those that discharge the two waters separately, and those that unite the cooling water with the water of condensation.

THE SURFACE CONDENSER

43. The **surface condenser** belongs to the class in which the condensing water and the water resulting from the condensed steam are discharged separately. It consists of a large number of small tubes secured at each end into headers and so arranged with connecting chambers that the cooling water passes through the tubes that are surrounded by the exhaust steam.

Fig. 18 shows a section of a Wheeler surface condenser, which will serve to illustrate the principle of surface condensers in general. The exhaust steam from the engine enters at a and passes between the tubes $c\,d\,e$, where it is condensed. The air pump that clears the condenser of the water of condensation and maintains the vacuum is attached to the outlet b. The cold condensing water enters at f and after passing through the nests of tubes d, e passes out at g. The exhaust steam on entering at a first passes between the group of tubes c through which the boiler feedwater is pumped and thus heated before passing into the boiler. This form of condenser therefore contains the features of a feedwater heater and condenser combined.

44. Cooling Surface.—In a surface condenser the cooling surface of the tubes should not be less than 1 square foot of surface (measured on the outside of the tubes and between the tube heads) to each 10 pounds of exhaust steam to be condensed per hour. The tubes used vary in diameter from $\frac{5}{8}$ inch to 1 inch and the thickness from No. 16 to

No. 18 B. W. G.; a common size is ⅝ inch, outside diameter, spaced 1⅜ inch between centers. They should not exceed 12 feet in length, and should be tested under hydraulic pressure to 300 pounds per square inch. When they exceed 6 feet between heads, a central support is desirable. The tubes in the latest condensers are made of an alloy of copper 70 per cent., zinc 29 per cent., tin 1 per cent., though copper tubes have been largely used also; they should be carefully tinned inside and out. Provision must be made for the free expansion and contraction of the tubes under change of temperature, and Fig. 19 shows a method used in the Wheeler condenser for accomplishing this. One end a of each tube is flanged and rigidly held in the tube by means of a screw follower b; the other end of the tube

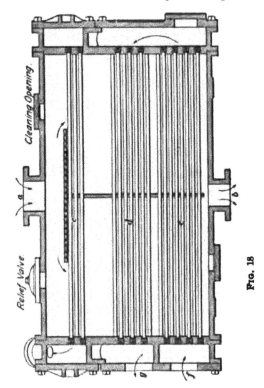

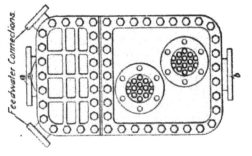

passes through an adjustable gland c that permits free movement of the tube during expansion and contraction. The gland c screws up against the packing d, thus forming a stuffingbox that permits an end movement of the tube without allowing leakage. This method of securing rigidly one end of the tube reduces the number of glands or stuffingboxes just one-half. The glands can be readily removed and the packing replaced if it becomes leaky from long use. The circulating water may flow through the tubes at a speed of from 400 to 700 feet per minute.

45. Air and Circulating Pumps. — To operate the surface condenser, an **air pump** to withdraw the water resulting from the condensed steam, together with the air contained in the water when it was fed to the boilers, and a **circulating pump** to move the water for cooling the condenser are required. These pumps are usually combined on one frame and bedplate. Fig. 20 shows a common

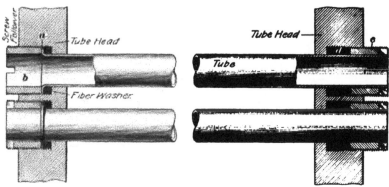

FIG. 19

arrangement of Wheeler surface condenser with air and circulating pumps. The circulating pump is at the right, the air pump at the left, and the steam cylinder for operating the pumps is located between them. The injection water is drawn in at M and passes out at D. The exhaust steam enters at A and after being condensed is discharged, by the air pump, from K to be used again in the boilers.

46. Advantages and Disadvantages of Surface Condensers.—An objection frequently urged against using surface condensers is that grease from the cylinder oil carried over by the exhaust steam accumulates on the steam side of the tubes and reduces the efficiency of the cooling surface;

but this may be prevented by an efficient grease extractor. The surface condenser for a given capacity is the highest in first cost; but where boiler feedwater is an important item of operating expense, the surface condenser will save most of it in the form of distilled water, and this saving in the water will more than equal a large interest on the difference in cost between a surface-condenser and a jet-condenser equipment. Only a trivial amount of water is lost, and fresh water is required to make up such loss; in fact, a little natural water must always be added to the distilled water.

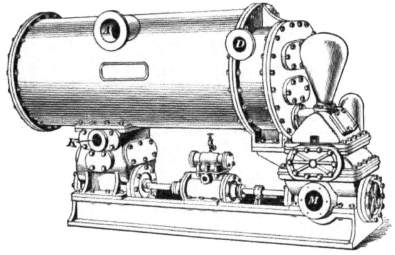

FIG. 20

A valid objection to the use of distilled water in boilers is that it is too pure, it being a well-known fact that pure or distilled water is a solvent of iron. In steam plants using condensing engines, and returning condensed steam only to the boilers, corrosion or pitting is found to take place, unless a portion of undistilled or natural water is mixed with the returns from the condensers.

When testing the exact amount of steam used by an engine, the surface condenser is a satisfactory adjunct, as the water independently discharged by the air pump can be weighed, and if carefully handled, should show the weight of steam used by the engine in performing its work.

47. Operation of Surface Condensers.—Surface condensers require compliance with the following conditions in their care and operation if the best results would be obtained:

To insure good results, it is absolutely necessary to place the air pump below the condenser. In connecting two or more engines to one condenser, with a common steam inlet, use double elbows of large radius, and avoid straight tee connections, otherwise one engine is liable to exhaust against the other.

All suction pipes should be provided with a suitable strainer having openings at least 1½ times the area of the pipe. Wrought iron should not be used for salt water; copper or cast iron will be more satisfactory.

Before starting the main engine, the air and circulating pumps, if independent, should be started with the air pipe on the condenser open to expel the air from the tubes and air chambers. If the condenser is hot, sufficient water should be circulated to cool it, and the engine then started slowly. The automatic open-air exhaust relief valve is closed by atmospheric pressure, and should be examined daily to see if it is properly seated and lifts freely. It is important to remember that a high vacuum requires a tight engine and tight connections. Any vibration of the pointer on the vacuum gauge indicates air leaks. To test the tightness of a condenser, the air-pump connections must be blanked off, and the condenser filled with water, and the water bonnets on the chambers removed. If any tubes that cannot be immediately replaced are found leaking, a pine plug driven in each end of the tube will answer temporarily. If it is necessary to remove a tube that is packed around the end, it should be closed at the expansion end with a proper tool before attempting to pull the tube out.

To avoid cylinder oil in dilution with the condensed water, a vacuum-tight oil separator will be found most efficient, and should be connected between the engine exhaust and the condenser; if this is not done and it is intended to use the condensed water for boiler-feed purposes, it should be purified after leaving the condenser, which will be a more or less troublesome method. To keep the condenser clean, where engines are in constant service, a gallon of kerosene oil or a solution of sal-soda can be occasionally injected with the exhaust steam. Surface condensers should be cleaned once a week. _____

THE JET CONDENSER

48. In the **jet condenser**, which represents the second class of condensers, the exhaust steam from the engine enters at the top of the condenser chamber, and is condensed in an effective manner by meeting at once the spray of injection water. Three classes may be considered—the *air-pump and jet condenser*, the *barometric-column condenser*, and the *induction condenser*.

The valuable features of the jet condenser are that it occupies but little floor space, does not require a costly foundation, can usually be connected close to the engine, and, being independent, can be started in advance of the engine, thus obtaining a vacuum as soon as the engine is in operation. The quantity of cooling water can be varied to meet the demands of the engine according to its load.

49. Examples of Air-Pump and Jet Condensers. The combined outfit of **air-pump and jet condenser** is usually of the horizontal type, all parts being designed for a compact setting on a continuous bedplate. Fig. 21 is a view of a Blake jet condenser. *A* is the condenser, *B* the air pump, and *C* the steam cylinder for driving the pump, which, in this case, is of the direct-acting type.

Fig. 22 shows a section of a Worthington independent jet condenser. The cold water enters the condenser at *b*, passes down the spray pipe *c*, and is broken into a fine spray by the cone *d*. The exhaust steam in the meantime comes in at *a*, and, mingling with the spray of cold water, is rapidly condensed. The velocity of the entering steam is imparted to the water, and the whole mixture of steam, water, uncondensed vapor, and air is carried with a high velocity through the cone *f* into the air-pump cylinder *g*, whence it is forced by the pump through the discharge pipe *j*.

50. This condensing apparatus is operated as follows: The air-pump having been started, a vacuum is formed in the condenser, the exhaust pipe, the engine cylinder, and injection pipe; this causes the injection water to enter through

Fig. 21

the injection pipe attached at *b* and to flow through the spray pipe *c* into the condenser cone *f*. The main engine being then started, the exhaust steam enters through the exhaust pipe attached at *a*, and, coming into contact with the cold

water, is condensed.   The spray pipe *c* has at its lower end a
number of vertical slits through which the injection water
passes and becomes spread out in thin sheets.   The spray
cone *d*, by means of its serrated surface, breaks the water passing over it into fine spray and thus insures a rapid and

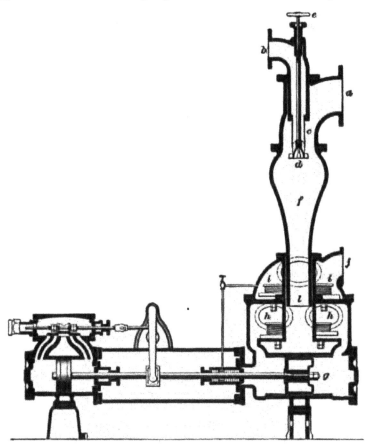

FIG. 22

thorough admixture with the steam.   This spray cone is
adjustable by means of a stem passing through a stuffingbox
at the top of the condenser and is operated by the handle *e*.

51.   In Fig. 23 is shown a jet condenser in connection
with the boiler and engine.   The exhaust pipe *A* leads
directly to the condenser; the injection pipe *B* draws water

from the reservoir *C*. After the steam is condensed, the mixture of exhaust steam and injection water is discharged through *D* into the sewer. A portion of this discharge, however, flows through *E* to the feed-pump *G*, which forces it through the coil in the heater *F* to the pipe *H* leading to the boiler. The exhaust from the two pumps is discharged into the feedwater heater through the pipe *M*. It will be noticed that water from the overflow pipe *D* enters the feed-pump under a slight head. This is because the water is heated by

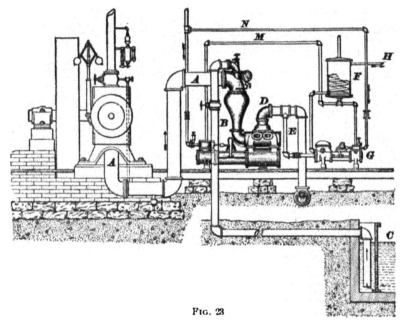

FIG. 23

the exhaust steam, and hot water cannot be raised by a pump like cold water. A pipe *N* leads from the boiler and supplies steam for both pumps.

52. The air pumps shown in Figs. 21 and 22 are driven by single-cylinder, direct-acting engines. With a single steam cylinder, the steam follows nearly the full length of the stroke and thus materially reduces the economy sought for with a condenser. The compound steam cylinder is an improvement on the single-cylinder type, and will more than save, in steam, the extra cost. Fig. 24 shows a large,

Fig. 24

Conover, combined air and circulating pump used in very large stations where a condenser of the highest economy is desired. The air pump *A* is of the single-acting type and the circulating pump *B* is double-acting. The driving engines *C, D*, situated at either end, are of the vertical compound type fitted with Corliss valve gear. This type of outfit, provided with compound steam cylinders and flywheel, possesses many advantages in the way of uniform speed and economy of steam over the direct-acting type; hence, it is suitable for large units where the extra expense is warranted.

53. Barometric Column, or Siphon, Condenser. This differs from the common jet condenser in that no air pump is required to remove the air, uncondensed vapor, and water, but a circulating pump or head of water is needed to supply the injection water when the lift is more than 20 feet. The vacuum is generated and maintained by a column of water flowing downwards through a vertical pipe of not less than 34 feet in length, having its lower end immersed in the water of the hotwell.

54. It will be remembered that a column of water 34 feet in height will just balance the atmosphere at the sea level when the barometer stands at 30 inches, but if an additional amount of water be allowed to enter the upper end of the water pipe, the equilibrium between the column of water and the column of air outside will be disturbed, and an amount of water corresponding to that allowed to enter at the upper end of the pipe or tube will flow out at the lower end.

This is the principle of the siphon condenser. So long as the proper amount of water continues to flow into the upper end of the pipe and a corresponding amount flows out at the lower end, the air and vapor in the condenser will be carried out by the descending water and a vacuum will be formed and maintained. If the area of the pipe is contracted into a neck, or throat, the velocity of the falling water will be accelerated and the action of the condenser will be improved thereby.

It is important that the stream of injection water entering the condenser should have a steady and continuous flow, and there must be no air leaks in the exhaust pipe or condenser. The siphon condenser is often, but wrongly, called the *injector condenser*.

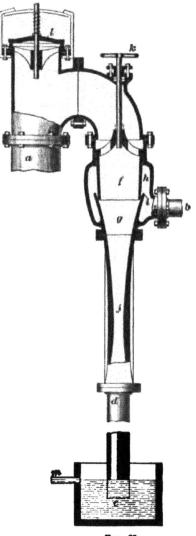

55. An illustration and a description of an example of this type of condenser, known as the *Baragwanath condenser*, is here given.

Fig. 25 represents a sectional view, in which a is the exhaust pipe; b the injection pipe; d the long discharge pipe, or *tail-pipe;* and e the hotwell. The steam enters through exhaust pipe a and flows through the exhaust nozzle f into condensing chamber g. Here it is met and condensed by the injection water that enters from the water-jacket h into the condenser in a thin conical sheet, flowing through the annular opening between the exhaust nozzle f and the prolongation of the shell of the condenser forming the inverted cone i. A vacuum is formed in the condensing chamber g by the condensation of the steam and by the

FIG. 25

air and uncondensed vapor being entrapped and carried out of the chamber by the cylindrical stream of water. The injection water and water of condensation flow from condensing chamber g through the throat j with such velocity as to carry with them the air and vapor that pass over with the steam and the injection water.

The exhaust nozzle f is adjusted by means of the wheel and screw spindle k, and can be set so as to admit just the right quantity of injection water. An automatic atmospheric relief valve l is fitted for the purpose of discharging any excessive accumulation of air, steam, or vapor that may collect in the exhaust pipe into the atmosphere. A hotwell overflow, or discharge, pipe m is always fitted to the hotwell.

56. In the barometric-column form of condenser, the throat of the condenser head is usually elevated at least 34 feet above the water level in the hotwell, so that the descending column of water, combined with the condensed steam, will create and maintain the vacuum. This condenser application combines three features: the exhaust leading into the condenser head at an elevation of about 35 feet above the hotwell; the water supply pipe to furnish the injection water at the condenser head; and the discharge pipe through which is carried off the combination of the condensed steam and the cooling water. This condenser is of low cost, and very economical where the injection water can be obtained from a tank, reservoir, or stream at such an elevation as to render pumping unnecessary, and under such conditions, with a supply pipe of ample size and a continuous flow of water, will show as good economy as can be obtained with a regular surface or jet condenser.

57. Where water is not naturally delivered at a convenient elevation, power must be expended to elevate the water to the height of the condenser head to insure proper supply. When considering the application of this type of a condenser, where the water must be pumped to an elevation of 30 feet, it is important that careful calculation be given to the cost of lifting the water. When the water must be pumped to the condenser head it should also be remembered that the descending columns will produce a siphon action, and will thus partially balance the ascending column, thereby reducing the work on the pump in proportion.

Although the condenser is placed at a height of 34 feet above the hotwell, the vacuum assists the circulating pump

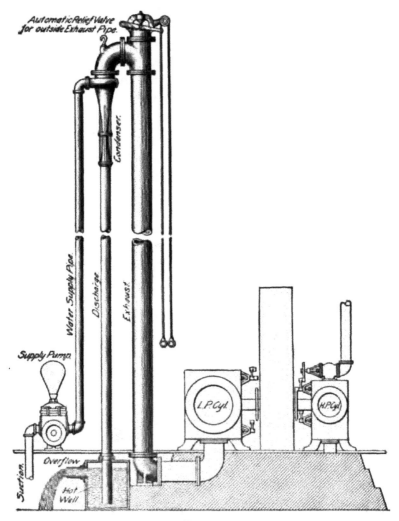

Fig. 26

in a proportionate degree, so that with a vacuum of 24½ inches the actual height that the water is forced by the pump is but 7 feet. Fig. 26 shows the general arrangement of a

47—13

Knowles condenser of this type, the condensing water being supplied from a pump. Fig. 27 shows an arrangement of the same condenser where the exhaust steam is first passed through a feedwater heater and where the condensing water is siphoned from a flume or tank instead of being pumped.

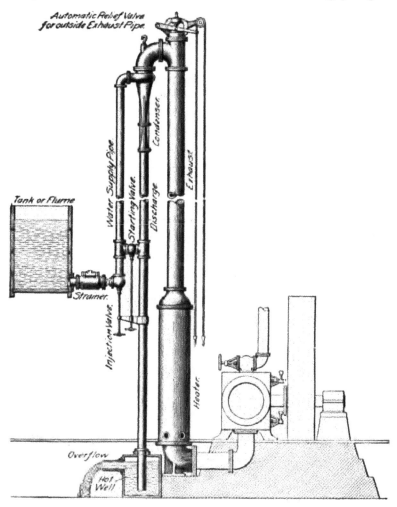

Fig. 27

58. The Induction Condenser.—The operation of the induction condenser is based on the same principle as that of the steam injector, used so largely for boiler feeding, and it may properly be called an **injector condenser,** although it is not given this name by the trade.

Fig. 28 represents a partial sectional view of a condensing apparatus of this type; it is known as the *Korting universal exhaust-steam induction condenser.* Referring to the figure, the exhaust steam enters at *a*, and after passing through the balanced horizontal check-valve *b* enters the water chamber *c*; it then passes through the inclined openings in the tube *d* into the condensing chamber. *e*, where it is met by the injection water and is condensed, forming a partial vacuum. The condenser is started by a supplementary jet of steam or stream of water. The vacuum in the chamber *e* induces the injection water to be siphoned into the condenser from the supply reservoir *f* through the injection pipe *g* and the strainer *h*, from whence it flows into the annular space *i* around the ram *j*, passing into the condensing chamber *e* through the annular opening *k*, where it meets the exhaust steam, which is then condensed. Here the injection water and the water of condensation intermingle and with the air and vapor are carried down the discharge pipe *l* into the hotwell *m*, the surplus water flowing into the sewer *n*.

59. To obtain the best results under the varying quantities of steam it may be called on to handle, this condenser requires that it shall be adjustable. This is accomplished by the ram *j* being made tapering and capable of being raised and lowered at will, which operation varies the size and capacity of the annular opening *k* and controls the volume of water admitted to the condensing chamber *e*. The ram is adjusted by the hand wheel *o* acting through a rack and pinion. The area of opening required by the steam that enters the condenser is regulated by the sleeve *p*, which covers more or less of the openings in the tube *d*, as may be required. This sleeve is raised or lowered by the hand wheel *q*, which also acts through a rack and pinion.

60. Like all condensers, this one requires a valve that opens into the atmosphere to relieve it of any accumulation of steam, air, or vapor that may collect in it. This is

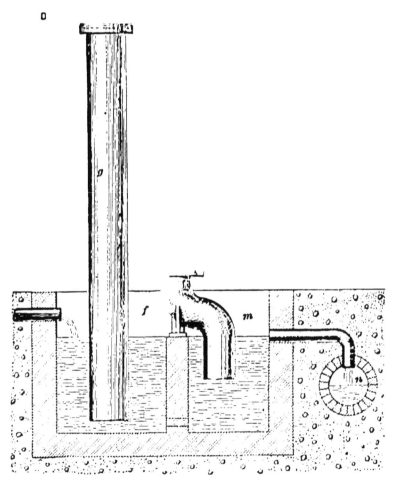

FIG. 28

provided in the *automatic free exhaust valve r*, which valve closes automatically when there is a vacuum in the condenser and opens when the vacuum is destroyed. It is fitted with a piston *s* to prevent the valve hammering. If it should become necessary or desirable to cut the condensing apparatus out of service and run the engine non-condensing, the free exhaust valve may be locked open by turning the hand lever *t* to the left.

The operation of the check-valve *b* is as follows: The inclined suspension bar *u* has a tendency to open the valve, while the inclined supporting bar *v* has a tendency to close it. Thus, the valve is balanced by its own weight, which is so distributed that there is always a slight excess of closing tendency. The object of this valve is to prevent the water in the condenser being siphoned into the steam cylinder.

61. When the injection water is supplied to the condenser under pressure, as from an elevated tank or from the street main, instead of being siphoned up, the openings *w* and *x* are blanked.

When the injection water is siphoned into the condenser and water under pressure is used for starting, the starting water enters at the opening *w* and the opening *x* is blanked.

When the injection water is taken under high suction and steam is used for starting, the steam enters through the opening *w*, a check-valve is attached at the opening *x*, and an overflow pipe is connected with the check-valve to discharge free or into the discharge pipe *l*.

This type of condenser has its limits of operation. If the load on the engine is variable, or the condenser is not of proper proportion, there may be times when the small volume of exhaust steam will be insufficient to impart the necessary velocity to the large amount of water.

The advantages over the ordinary air-pump condenser are: low first cost; absence of moving parts, with consequent certainty of action; small space taken up, thus enabling separate condensers to be applied to each engine, and obviating the use of long exhaust pipes with complicated draining arrangements; and complete avoidance of troublesome air pumps. With this type of condenser, it is possible to use dirty and gritty water, as the condensing jet does not come in contact with the sides of. the water nozzle or condensing tube.

If pumping must be done to maintain the water supply, it will be found that the power required for giving the necessary head for the condenser is less than that absorbed in the working of the combined air pump and jet condenser, which has a low efficiency owing to the fact that it has to discharge both water and air against the atmospheric pressure. Centrifugal pumps direct-coupled to electric motors are largely used, and have the great advantage of simplicity and nonliability to wear and tear. For electric-light units, where the period of peak load and engine overload lasts for a comparatively short time of the engine's run, it is not advisable to adopt a condenser of this type with the expectation of securing the highest vacuum at the time the engine is developing its maximum power, as regulation of the water supply will require exceedingly close attention, and will be very difficult. The usual practice is to adopt such size of condenser as will give 23 to 24 inches at peak load, the average vacuum being 26 inches when the engine is loaded to best efficiency. The induction condenser is suitable under favorable conditions of water supply and where the engine load is practically constant, but under changes in the amount of exhaust caused by a variable load it will not be satisfactory unless a man is stationed at the condenser to manipulate the valves as the load changes.

AUXILIARY FITTINGS AND CONDENSER PRECAUTIONS

62. Air Leaks.—In connection with a condenser equipment, certain auxiliary fittings are necessary; also the observance of special precautions in the execution of the work. Great care must be taken to make all joints and pipe connections perfectly tight, because if a vacuum exists within the system and under atmospheric pressure the external air will leak in at unsuspected places. It is therefore wise to specify that the system shall test a certain vacuum at the engine cylinder, and shall remain air-tight for a specified time, retaining this vacuum after the engine and condenser are closed down. Unsuspected air leaks may be discovered by testing over the entire system at every joint with the flame of a lighted candle; if the air is leaking in, the flame of the candle will be drawn in at the defective point. Close all leaks by taking up all bolts and nuts, and coating the joints with asphaltum varnish.

63. Position of Vacuum Gauge.—In the application of any type of condenser it is important that the vacuum gauge showing the resulting vacuum obtained by using the condenser, should be connected directly, or as close as possible, to the exhaust pipe of the low-pressure cylinder of the engine, in order that the actual vacuum in the engine cylinder may be determined. If the exhaust pipe is a long one and the vacuum gauge is connected near the condenser, it is quite possible that, because of friction or air leaks in pipes or valves, a difference of 2 or 3 inches better vacuum will show at the condenser than at the low-pressure cylinder, and the true results will not be known. The vacuum gauge should be standardized to absolute accuracy.

64. Heating Feedwater.—This can be accomplished by using a closed heater located in the exhaust line near the engine, where the feedwater passing through it will get the benefit of the heat in the exhaust steam before the latter reaches the condenser.

65. Automatic Relief Valve.—Each engine should be protected by an automatic exhaust relief valve that will act promptly and allow the engine to exhaust into the atmosphere in case of any fault in the working of the condensing apparatus. Any condenser is liable to lose its vacuum by the failure of the air pump or its attachments, or by the partial or entire stoppage of the supply of cooling water. Under such conditions, if a relief valve is not provided, the exhaust will accumulate pressure and slow-down or stop the engine. Fig. 29 illustrates one type of automatic exhaust relief valve. In case the back pressure becomes excessive, valve *a* is lifted from its seat and the steam allowed to exhaust into the atmosphere. The dashpot at *b* steadies the movements of the valve.

It is desirable that air pumps and jet condensers be fitted with composition-lined cylinders and composition piston rods, valve seats, valve bolts, springs, etc. The difference between the cost of composition fittings and iron or steel

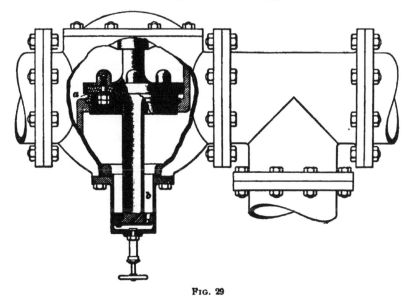

FIG. 29

fittings is more than made up within a short time by the extra cost of repairs and maintenance if an iron-fitted pump is used.

66. Grease Extractors.—There are several effective grease extractors on the market, and every plant should be fitted with an efficient appliance for this purpose. The saving in cylinder oil and the greater purity of the water from the exhaust for boiler feed will well repay the cost of the grease extractors.

WATER USED FOR CONDENSING

67. When considering the application of any condenser, the character of the injection water (whether fresh, salt, acidulous, or otherwise impure), the maximum temperature, the amount of available supply, and the distance horizontally and vertically that the water has to be raised, must be investigated. Instances have occurred where engineers, overlooking the character of the cooling water, have installed surface condensers to use circulating waters so largely impregnated with sewage or vegetable matters that the heat of the exhaust steam caused the impurities contained in the water to be deposited on the tubes of the surface condensers in a jelly-like formation, thus quickly cutting down the efficiency of the condenser, and ultimately making it necessary to change to another type of condenser.

68. Determination of Amount of Injection Water. The amount of water required to condense thoroughly the steam from an engine is dependent on two conditions: the total heat and weight of the steam, which represents the work to be done, and the temperature of the injection water, which represents the value of the cooling agency by which condensation of the steam is to be accomplished. Generally stated, with 26 inches vacuum, and injection water at ordinary temperature (not exceeding 70° F.), from 20 to 30 times the quantity of water evaporated in the boilers (feedwater) will be required for the complete condensation of the exhaust steam from an engine.

The effectiveness of the injection water decreases very rapidly as its temperature increases, and at 80° F. 35 times, and at 90° F. 52 times the feedwater has to be employed; on the other hand, if the temperature be lower, less water is required. The approximate amount of condensing water required per pound of steam condensed can be obtained by the following rule:

Rule.—*Subtract the temperature of the air-pump discharge from 1,190 and divide the remainder by the rise in temperature of the condensing water.*

TABLE IV

POUNDS OF INJECTION WATER REQUIRED PER POUND OF STEAM CONDENSED

Entering Temperature of Injection Water in Degrees F.

Pounds of Condensing Water Required per Pound of Steam

Temperature of Air Pump Discharge From Degrees F.	35	40	45	50	55	60	65	70	75	80	85	90	95	100
90	20.0	22.0	24.4	27.5	31.4	36.7	44.0	55.0	73.3	110.0	220.0			
92	19.2	21.1	23.4	26.1	29.7	34.3	40.7	49.9	64.6	91.5	156.8	549.0		
94	18.6	20.3	22.4	24.9	28.1	32.2	37.8	45.7	57.7	78.1	121.8	274.0		
96	17.9	19.5	21.4	23.6	26.7	30.4	35.3	42.1	52.1	68.4	99.4	182.3		
98	17.3	18.8	20.6	22.7	25.4	28.7	33.1	39.0	47.5	60.7	84.0	136.5	364.0	
100	16.8	18.2	19.8	21.8	24.2	27.2	31.1	36.3	43.6	54.5	72.7	109.0	218.0	
102	16.2	17.5	19.1	20.9	23.1	25.9	29.4	34.0	40.3	49.5	64.0	90.7	155.4	544.0
104	15.7	17.0	18.4	20.1	22.2	24.7	27.8	31.9	37.4	45.2	57.2	77.6	120.7	271.5
106	15.3	16.4	17.8	19.4	21.3	23.6	26.4	30.1	35.0	41.7	51.6	67.7	98.5	180.7
108	14.8	15.9	17.2	18.7	20.4	22.5	25.2	28.5	32.8	38.6	47.0	60.1	83.2	135.2
110	14.4	15.4	16.6	18.0	19.6	21.6	24.0	27.0	30.9	36.0	43.2	54.0	70.0	108.0
112	14.0	15.0	16.1	17.4	18.9	20.7	22.9	25.7	29.1	33.6	39.9	49.0	63.4	89.5
114	13.6	14.5	15.6	16.8	18.2	19.9	22.0	24.5	27.6	31.6	37.1	44.8	56.6	76.9
116	13.3	14.1	15.1	16.3	17.6	19.2	21.1	23.3	26.2	29.8	34.6	41.3	51.1	61.0
118	12.9	13.7	14.7	15.8	17.0	18.5	20.2	22.3	24.9	28.2	32.5	38.3	46.6	59.6
120	12.6	13.4	14.3	15.3	16.5	17.8	19.5	21.4	23.8	26.7	30.6	35.7	42.8	53.5
122	12.3	13.0	13.9	14.8	15.9	17.2	18.7	20.5	22.7	25.4	28.9	33.4	39.6	48.5
124	12.0	12.7	13.5	14.4	15.4	16.7	18.1	19.7	21.8	24.2	27.3	31.4	36.8	44.4
126	11.7	12.4	13.1	14.0	15.0	16.1	17.4	19.0	20.9	23.1	26.0	29.6	34.3	41.9
128	11.4	12.1	12.8	13.6	14.5	15.6	16.9	18.3	20.0	22.1	24.7	27.9	32.2	37.9
130	11.2	11.8	12.5	13.2	14.1	15.1	16.3	17.7	19.3	21.2	23.6	26.5	30.3	35.3
132	10.9	11.5	12.2	12.9	13.7	14.7	15.7	17.1	18.6	20.3	22.5	25.2	28.6	33.1
134	10.7	11.2	11.9	12.6	13.4	14.3	15.3	16.5	17.9	19.6	21.6	24.0	27.1	31.0
136	10.4	11.0	11.6	12.3	13.0	13.9	14.8	16.0	17.3	18.8	20.7	22.9	25.7	29.2

EXAMPLE.—If the temperature of the injection water supplied to a condenser is 70° F. and the temperature of the discharge 110°, how many units weight of injection water will be required per unit weight of steam condensed?

SOLUTION.—In this case, by applying the rule, $\dfrac{1,190 - 110}{110 - 70} = 27$, that is, the weight of the injection water required will be 27 times the weight of the steam exhausted into the condenser. Ans.

69. Table IV shows the quantity of cooling water required under specific temperature conditions. The values in this table are based on the rule given in Art. **68.**

COMPARATIVE COST OF OPERATING CONDENSERS

70. From the saving estimated to be derived by the use of condensers should be deducted the interest on the extra investment, the cost of maintenance, and the cost of operating the air and circulating pumps; therefore, if the power for this pumping can be derived direct from the condensing engine, the cost is the least, as the prime mover is the most economical source of power. If the pumps are operated by an electric motor, the cost is slightly increased because the efficiency is less, on account of the intermediary of the motor, than from the engine direct. A centrifugal pump will be found quite satisfactory only with certain types of condensers, and may be driven by a motor or small engine. If the pumps are independently steam driven, the cost is greater, but some heat may be saved by using their exhaust steam for heating feedwater.

71. Methods for Operating Pumps.—The air and circulating pumps are usually combined, and the methods of drive are stated in the order of their economy.

The following is the estimated steam in pounds per horse-power-hour for pumping:

a, by belt from engine shaft 16
b, in a large station by a high-economy pumping engine . . 20
c, by an electric motor . . . . 30 to 40
d, by compound steam cylinders on pumps . . . . 40 to 70
e, by a direct steam cylinder . . . . 100 to 150

Fig. 30 shows the arrangement of a Knowles jet condenser with a triplex motor-driven pump.

72. Limit of Condensation.—The theoretical limit of condensation would be that of absolute vacuum, which is equivalent to $29\frac{1}{4}$ inches of mercury or $14\frac{3}{4}$ pounds per square inch at sea level. In actual practice, with condensers, this cannot be attained, but the vacuum may range from 23 to 27 inches, or $11\frac{1}{4}$ to $13\frac{1}{4}$ pounds.

73. Relative Vacuum.—A perfect vacuum cannot be attained for the reason that the cooling water entering the condenser immediately absorbs the heat from the exhaust steam and a vapor is formed, thus preventing a perfect vacuum. The ratio between the temperature of condenser discharge and the vacuum maintained under good conditions is as follows:

00 inches vacuum 212° F.   25 inches vacuum 135° F.
11 inches vacuum 190° F. $27\frac{1}{4}$ inches vacuum 112° F.
18 inches vacuum 170° F. $28\frac{1}{4}$ inches vacuum 92° F.
$22\frac{1}{4}$ inches vacuum 150° F.

Twenty-five inches vacuum is considered a point of good efficiency.

_____

COOLING TOWERS

74. The **cooling tower** is a device whereby the water discharged from a condenser may be reduced in temperature by exposure, in the form of spray, drops, or minute streams, to strong currents of air, and thereby sufficiently cooled to be again used in the condenser; the repeated cooling by passing through the tower and circulation through the condenser renders possible the continuous use of the same body of water. The mechanical subdivision and distribution of the water for cooling is obtained by different methods; such as passing it over suspended galvanized-wire mats, over many courses of thin vertical boards laid up like cribwork, or over many series of thin pipes. As early as 1676, the rudiments of our modern cooling tower are known to have been used in Hindustan. An early English traveler describes

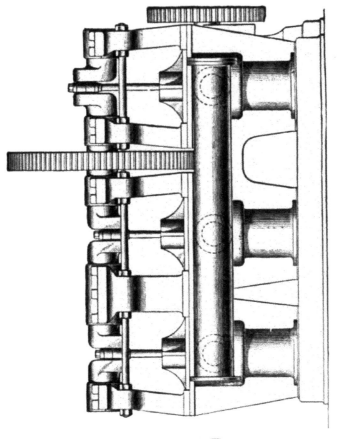

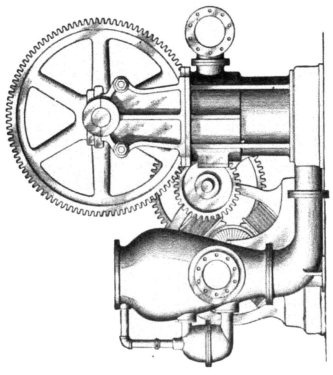

FIG. 80

an apparatus formed of bamboo tubes and mats of palm leaves, by which the Indian grandees used to cool the water for their baths. The water was slowly poured over the cooling mats, and drained below into a stone basin. The evaporation was effected by means of fans operated by hand.

75. Vaporization of the Cooling Water.—When water is freely exposed to the atmosphere, as in passing through a cooling tower, the stratum of air in contact with its surface becomes more or less charged with vapor. As evaporation takes place only from the exposed surface, the amount of vapor must therefore depend on the extent of the surface of water exposed to the air. Moderately dry air absorbs moisture, at first rapidly, but as it becomes saturated, the process proceeds more and more slowly, and finally ceases. Evidently, therefore, the more rapidly the air is circulated, the more rapidly evaporation proceeds. The capacity of air for carrying moisture increases in proportion to the temperature. At 202°, saturated air holds about 100 times the weight of water that would saturate air at 32°. For each temperature there is a maximum of density, and hence of pressure, that the vapor exerts, and it is found that the rate of evaporation at different temperatures is proportional to the differences of the elastic forces of the vapor at the surface of the liquid, and that of the vapor actually present in the surrounding air.

76. It is therefore evident that the efficiency of a cooling tower depends: (*a*) on the extent of the exposed water surfaces; (*b*) on the volume of air brought into contact with the exposed water surfaces and the rapidity of the air circulation; (*c*) on the difference of the pressure of vapors at the water surface, and that in the air—which is regulated by the accidental conditions of water and air as to temperature at the time of contact, and percentage of moisture—all three controlling the conditional rate of evaporation. The efficiency of any cooling tower must, therefore, be largely due to the uniform and perfect distribution of the water in fine spray or fine films as it passes through the tower. The more perfectly the water is subdivided, permitting the air to come in through contact therewith, the better efficiency will be attained in the tower.

77. Loss of Water by Evaporation.—The loss of water due to evaporation depends on the degree of heat extracted. As in every pound of water converted into vapor about 1,000 units of heat become latent, the loss evidently must be proportional to the total amount of heat thus absorbed and carried off, and consequently must depend on the difference of the temperature between the hot and cooled liquid, or in other words, on the degree to which the water has been cooled. If 100,000 pounds of water is cooled from, say, 130° to 70°, 60 units of heat have been absorbed from every pound, or a total of 6,000,000 units; and as in every pound of steam 1,000 units of heat become latent, 6,000 pounds, or 6 per cent., of the liquid will be evaporated. The loss of this small percentage of water by vaporization is a necessity, and must be estimated as one of the items of cost and will vary from 3 to 7 per cent.

It is clear that the quantity lost by evaporation and the reduction of temperature attained depends on the relative humidity in the atmosphere, because on this condition depends its readiness to absorb additional vapor; hence, with dry air the cooling will be more effective than with a high degree of humidity. This is shown in Table V, which gives the results taken from a test record during the operation of a pair of towers.

78. Reduction of Temperature.—The amount of heat extracted from the condenser discharge passing through the cooling tower will vary according to the atmospheric temperature, the surface over which the water is exposed, and the volume and velocity of the air passing through the tower; the reduction in temperature will vary from 30° to 45°, and with a sufficient volume of water in circulation, a vacuum of from 23 to 25 inches of mercury may be obtained. As the total water consists of the condensed steam from the boilers plus the circulating water, it will be necessary to add, from

time to time, a sufficient amount of fresh water to replace
the loss caused by evaporation. It will thus be seen that a
plant can operate almost entirely independent of a city water
supply. With any cooling tower, where the feedwater for
the boilers is taken from that circulating through the tower,
it will be necessary to utilize one or more of the best types of
grease extractors to remove the oil before the water is
pumped into the boilers.

TABLE V
RESULTS OF TESTS ON COOLING TOWERS

	First Test	Second Test	Third Test	Fourth Test
Temperature of air entering tower, degrees Fahrenheit	54.70	85.70	84.60	87.00
Atmospheric humidity, per cent. . .	37.50	51.00	63.00	42.60
Temperature of water delivered to tower, degrees Fahrenheit . . .	134.50	135.20	136.40	135.25
Temperature of water leaving tower, degrees Fahrenheit . . , ...	94.90	104.20	101.60	90.75
Number of degrees Fahrenheit water was cooled	39.60	31.00	34.80	44.50
Pounds of water supplied to tower .	12,531	12,508	10,713	6,304
Pounds of water lost by evaporation while passing through tower . . .	424	363	341	263

79. Air Circulation.—Three methods for obtaining
air circulation through cooling towers are commonly used:

1. Those in which the air is forced in rapid circulation
through the tower by means of a fan blower, which discharges fresh air into the lower part of the tower, whence
it is deflected upwards through the film or spray of condensing water.

2. Those in which a chimney or vertical flue, rising from
50 to 100 feet above the cooling compartment of the tower,
creates sufficient natural draft to draw a large volume of air
through the cooling compartment, and allows the vapor from
the cooling water to be carried off from the top of the

chimney. This type of tower entirely avoids cost for daily operation of fans.

3. Where the ground space is sufficient, the cooling surface of the tower is distributed over a larger area, allowing free circulation of the air naturally through the tower; this also avoids the use of any mechanical means to circulate the air, and saves the cost of driving fans.

EXAMPLES OF COOLING TOWERS

80. The Barnard-Wheeler Cooling Tower.—In this tower, shown in Fig. 31, the water is cooled by allowing it to flow over suspended galvanized-wire mats. The casing is usually made of steel plates. The pump discharge is led to the top of the tower and the warm water is there distributed by a suitable system of piping to the upper edges of the mats, over the surface of which it spreads in thin films, compelling a partial interruption of the flow and continuously bringing new portions of the water to the surface, thereby exposing it to the evaporating and refrigerating effects of the air-currents. To assist the cooling action, the air in immediate contact with the water is set in rapid circulation by means of the fan blowers a, a', Fig. 31, which force air into the lower part of the tower and upwards between the mats.

81. The Worthington Cooling Tower.—In this tower, shown in Fig. 32, the water runs over the inside and outside surfaces of a large number of cylindrical tubular tiles c, c', c'', which rest on a grating d supported by a brick wall e extending around the circumference of the tower. The warm discharge water from the condenser enters the tower through the pipe f, passes up the central pipe g, and is delivered on the upper layer of tiling and over the whole cross-section of the tower by the distributing device h, which consists of four pipes, radiating from the central pipe g, which are caused to revolve about the central pipe by the reaction of jets of water issuing from perforations on one side of each pipe. The water thus delivered spreads over the outside and inside

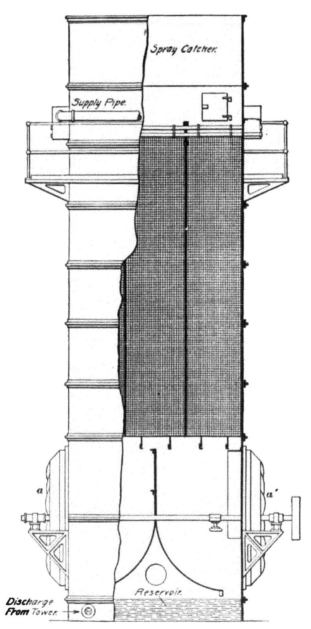

FIG. 31

surfaces of the walls of the tiling and forms a continuous
sheet, which is presented to the action of the air. The air
is circulated by the fan b, driven by a small engine or electric
motor. The air drawn in
by the fan is deflected upwards by the plate l. The
cooled water collects in
the reservoir i, from which
it is drawn off through the
pipe j; m is a manhole to
give access for inspection
or repairs. Fig. 33 shows
a fan-cooled tower as in-
stalled in relation to the
engine and condenser.

**82. Barnard's fan-
less self-cooling tower**
is shown in Fig. 34. In
this device the use of
mechanical means for cir-
culating air for cooling the
water is dispensed with,
thus avoiding the wear and
tear and the expenditure
of power that are always
associated with moving
parts. The hot circulating
water discharged from the
condenser is pumped up
through the central stand
pipe a, from which it is led
to the trough b and distributing pipes c, c, causing a constant flow of thin
films of water over the

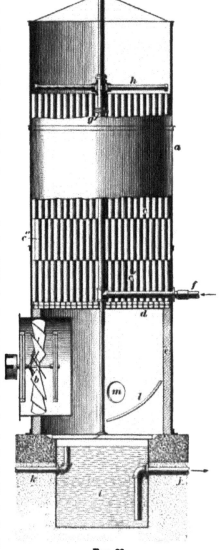

Fig. 32

meshes of the wire mats d, d', and finally draining into the
tank or reservoir f forming the foundation of the tower,

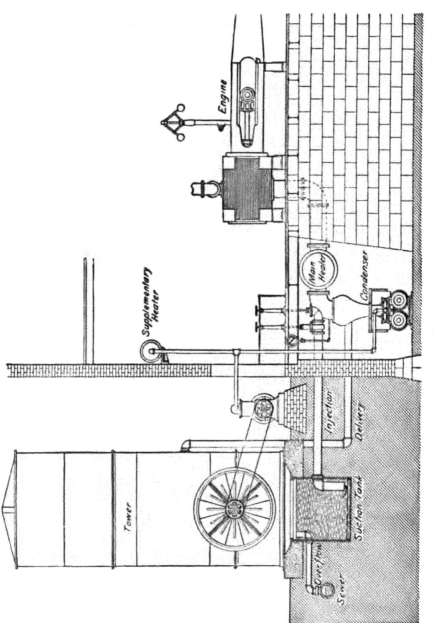

FIG. 88

from whence the cooled water is returned through the injection pipe *e* for use again in the condenser.

The mats are placed radially and are entirely exposed to the atmosphere; they are so arranged as to permit the air

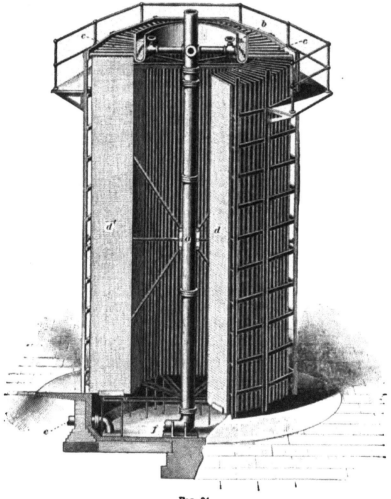

Fig. 34

to come into contact with the descending films of water by natural circulation, and the consequent evaporation is carried far enough to reduce the temperature of the injection water to a sufficiently low degree for condensing purposes.

83. The **Stocker cooling tower** is a structure built of wood, steel, or brick, according to circumstances. The cooling surfaces are built up of .checkerwork or crosspieces of boards in horizontal layers set at right angles to one another. At the intersections are placed upright partitions .diagonally across the square openings between the boards. The construction of this tower will be understood by referring to Fig. 35, where (*a*) is a plan view looking down on the latticework, (*b*) shows the crosspieces at right angles to each other, and (*c*) shows the upright diagonal partitions.

The water is pumped to the top of the tower and trickles down over these surfaces in thin films, which are broken up in falling at each intersection of the boards, and the water is thus brought into contact with the current of air that passes upwards through the tower. The air circulation may be set up by fans, or if the tower is made high enough natural draft may be used.

84. Location of Cooling Towers.—Cooling towers may be located in rear yards or on the roof of one of the power-station buildings, if ground space is not available. Where a large plant is to be provided for, a sufficient number of cooling towers may be erected in a battery with the piping so arranged that one or more towers may be operated as desired, according to the load.

85. Advantages of Cooling Towers.—The advantages that the addition of the cooling tower afford to an electric power station are that it may be possible to select a location for building the station more favorable to economy in the cost of copper for the system of distribution, and with highly desirable facilities for securing coal. Other advantages are, less coal consumed for the same power developed or more power for the same consumption of coal; saving in water by using the condensed steam for boiler-feed purposes; less boiler capacity necessary for a given amount of engine horsepower.

The knowledge that the results attained with condensers used in connection with cooling towers are practically the

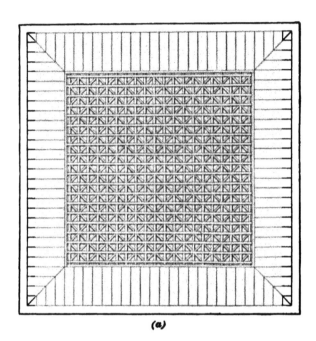

(a)

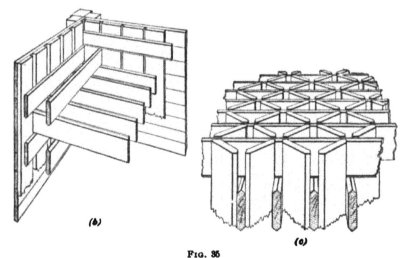

(b)

(c)

Fig. 35

same as those using water from the natural source of supply, makes it reasonable to recognize the operative advantages of the cooling tower. Stations that are located so
as to secure circulating water from the natural source of
supply are frequently interfered with by ice, low water, or
an excessive rise in the river. The cooling tower has
none of these disadvantages.

The conditions under which the use of a cooling tower
should be carefully considered are: (a) Where there is a
uniform load of, say, 300 horsepower, or upwards, for several
hours, such as a street-lighting load all night; (b) where the
cost of coal is, say, $2.50 per ton, or upwards; (c) where
boiler feedwater costs from $25 per month, up.

As an illustration of the saving effected by the use of a
compound engine in conjunction with a condenser and cooling tower in place of a simple non-condensing engine, the
following example may be given: An $18\frac{1}{2}'' \times 30''$ medium
high-speed engine was tested several times for its steam
consumption, and found under constant load to use an average
of 46.8 pounds of steam per horsepower-hour at 97.4 pounds
average boiler pressure per square inch. The high-pressure
cylinder was removed and a $14\frac{1}{2}'' \times 25'' \times 30''$ tandem compound cylinder put on. The steam consumption then proved
to be 16.5 pounds per indicated horsepower-hour with 110
pounds pressure on the same boilers as in the first test.

It is desirable (but not essential) to use a surface condenser in conjunction with a cooling tower, as with this type
of condenser there will be the minimum duty required of the
pump, the work being only due to the height of the tower, as
the ascending and descending columns of water will balance
each other below the tower reservoir. Where a cooling tower
is used in connection with a surface condenser with air and circulating pumps, the power required for operating the tower
and pumps will be from $2\frac{1}{2}$ per cent. to 4 per cent. of the total
indicated horsepower of the engines, much depending on
the size of the engines. The large engine will be the more
economical, and the fanless type of cooling tower will
require the least power for its operation.

ELECTRIC POWER STATIONS
(PART 4)

PIPE FITTING

LOSS OF PRESSURE IN PIPING

1. The proper designing of a system of **pipe fitting** for an electric power station requires a careful analysis of the conditions of service, a thorough knowledge of the methods for distributing and conveying steam and water, and of the quality and strength of materials employed. When steam leaves the boilers and starts to flow through a pipe of given diameter, several factors tend to change the form of the original energy possessed by the steam; among them are:

(*a*) **Condensation,** which may be divided into two parts—*static condensation*, which occurs when the steam fills the pipe, but is not flowing through it; and *dynamic condensation*, which takes place when a valve is opened permitting the steam to flow through the pipes. The latter should be less than the former, because of the fall in pressure and temperature that takes place at the delivery end of the pipe and the effort that is being constantly made to raise or maintain the original condition of the steam; but it is found that the amount of condensation is very nearly equal in both cases.

(*b*) **Friction** in the pipe causes a loss of pressure and requires work to be done to overcome the loss. The natural condition of condensation, combined with the loss in pressure, due to friction, cannot be wholly overcome; no .

matter to what extent the piping be covered with the best qualities of non-conducting covering there is still a loss of heat units.

(*c*) **Expansion** causes a change in the external latent energy, which is accounted for in the steam tables under different absolute pressures.

In horizontal pipes, there is less tendency to change the form of energy possessed by the steam than when flowing through inclined or vertical pipes. The force of gravity must be considered when long vertical pipes are required through which the steam will flow·upwards or downwards; this will be readily understood when consideration is given to the number of pounds of water in the form of steam to be transferred per hour.

2. Loss of Pressure by Friction.—Friction is greater through elbows of short radius than through elbows of long radius. Globe valves offer a serious ·impediment to the passage of steam, but the drop in pressure when passing through a gate valve is practically negligible. The loss of head due to getting up the velocity and caused by the friction of the steam entering the pipe and passing elbows and valves will reduce the flow below the estimated capacity of straight pipes. The resistance at a globe valve is about the same as that for a length of pipe equal to 114 diameters divided by a number represented by $1 + (3.6 \div \text{diameter})$. For example, a 3-inch globe valve would introduce as much friction as $\dfrac{114 \times 3}{1 + \dfrac{3.6}{3}} = 155$ inches of 3-inch pipe. The resistance at an elbow ·is approximately equal to two-thirds that of a globe valve. These equivalents—for openings, for elbows, and for valves—must be added in each instance to the actual length of the pipe, if the loss due to friction is to be calculated closely.

The rigid requirements of reliable and continuous operation of power stations have, within the past few years, revolutionized the simple methods of piping of earlier construction to a large degree, and have resulted in a great improvement in the design and quality of all the materials and fittings requisite for a complete system.

MOVEMENT OF STEAM

3. In preparing the plans for a system of piping it is essential to know not only the quantity of steam that must be transferred between the source of supply at the boiler and the points of demand at the engines, but also the quantity of steam that will flow through pipes of given diameters. The velocity of steam passing freely from one vessel into another is equal to the velocity of a body that has fallen through a distance equal to the difference in height of two columns of steam of 1 square inch cross-section, the entire weight of one column equaling the pressure per square inch in one vessel and the entire weight of the other column equaling the pressure per square inch in the other vessel, the density of steam being uniform and equal to that due to the heavier pressure.

EXAMPLE.—What will be the velocity of steam at 16 pounds pressure, absolute, passing into a steam of 15 pounds absolute?

SOLUTION.—The weight of a cubic foot of steam at 16 pounds absolute is .04111 lb. (see steam table); this divided by 144 = .0002855, which is the weight of a column of steam of 1 sq. in. cross-section and 1 ff. high. The total height due to 16 lb. pressure, therefore, is 16 ÷ .0002855 = 56,040 ft., and at 15 lb. $\frac{15}{16}$ of 56,040 = 52,540 ft. The difference in the height of columns, therefore, is 56,040 − 52,540 = 3,500 ft. In falling 3,500 ft., a body will acquire a velocity of 474 ft. per sec., which is the velocity of steam at 16 lb. pressure flowing into steam of 15 lb. pressure per square inch. The velocity, 474 ft. per sec., is obtained from the formula for falling bodies, $v = \sqrt{2\,g\,h}$. $\sqrt{2\,g} = 8.02$ and $h = 3,500$; hence, $v = 8.02 \times \sqrt{3,500} = 474$ ft. per sec. Ans.

4. Flow of Steam Under Pressure.—The amount of steam that will flow through a pipe of given diameter in 1 minute at specified pressures may be calculated by the following formula:

$$W = 87 \sqrt{\frac{D\,(p_1 - p_2)\,d^5}{L\left(1 + \dfrac{3.6}{d}\right)}} \qquad (1)$$

where W = weight of steam discharged, in pounds, per minute;

D = density of steam, weight per cubic foot, at initial pressure p_1;

p_1 = initial pressure;

p_2 = final pressure;

L = length of pipe, in feet;

d = diameter of pipe, in inches.

The difference $p_1 - p_2$ is equal to the drop in pressure in the pipe. If we consider the drop in pressure through the pipe to be 1 pound and the length L to be 100 feet, we have

$$W' = 87 \sqrt{\frac{D \times d^5}{100\left(1 + \frac{3.6}{d}\right)}} = 8.7\sqrt{\frac{D \times d^5}{d + 3.6}} \qquad (2)$$

where W' = weight, in pounds, of steam delivered per minute through 100-foot lengths of pipe.

EXAMPLE.—How many pounds of steam per minute can be delivered through a 3-inch pipe with an initial pressure, as shown by the steam gauge, of 100 pounds, if the loss of pressure in the pipe is 1 pound?

SOLUTION.—If the gauge pressure is 100 lb. per sq. in., the absolute pressure must be $100 + 14.7 = 114.7$. A cubic foot of steam at this pressure will weigh, approximately, .263 lb., $d = 3$ in., $d^5 = 729$ and

$$W' = 8.7\sqrt{\frac{.263 \times 729}{3 + 3.6}} = 46.7 \text{ lb. per min. Ans.}$$

These formulas for the flow of steam through pipes, like all similar formulas, give approximate results only, but the results are near enough for practical purposes. The actual inside diameter of pipe is very rarely the same as the nominal diameter, and it is seen from formula 1 that a slight change in the diameter has a great influence on the flow of steam. Table I gives the approximate weights of steam delivered per minute through 100 feet of pipe of various diameters with a drop in pressure of 1 pound. On the whole, these values are slightly higher that those given by formula 1.

If the allowable drop in pressure is to be other than 1 pound, multiply the values given in the table by the square root of the drop. If the length is other than 100 feet, divide

TABLE I

WEIGHT OF STEAM DELIVERED PER MINUTE THROUGH 100 FEET OF PIPE FOR STANDARD PIPE SIZES AND VARIOUS INITIAL PRESSURES

Initial Pressure by Gauge	Nominal Inside Diameter of Pipe, in Inches										Initial Pressure by Gauge
	3	3½	4	4½	5	6	7	8	9	10	
	Weight, in Pounds, of Steam Delivered per Minute Through 100 Feet of Pipe With 1 Pound Loss of Pressure										
70	43.2	64.5	91.7	124.3	168.7	277.2	410.5	577.2	793.3	1,051.7	70
80	45.5	68.0	96.6	130.9	177.7	292.1	432.5	608.2	835.5	1,108.4	80
90	47.6	71.2	101.2	137.2	186.3	306.0	453.3	637.3	875.9	1,161.3	90
100	49.7	74.3	105.7	143.2	194.4	319.8	473.0	665.1	914.1	1,211.8	100
110	51.7	77.3	109.9	148.9	202.1	332.2	491.8	691.5	950.4	1,259.9	110
120	53.6	80.2	113.9	154.4	209.5	344.3	509.9	717.0	985.4	1,306.3	120
130	55.4	82.9	117.8	159.7	216.7	356.1	527.4	741.6	1,019.6	1,351.1	130
140	57.2	85.5	121.5	164.7	223.6	367.4	544.6	765.7	1,052.9	1,393.9	140
150	58.9	88.1	125.2	169.6	230.2	378.3	560.2	787.7	1,082.6	1,428.1	150

100 by the length in feet and multiply the figures in the table by the square root of the quotient.

EXAMPLE.—How many pounds of steam will be discharged per minute, with 120 pounds initial gauge pressure, through a pipe 3 inches in diameter and 400 feet long, the allowable loss of pressure being 2 pounds?

SOLUTION.—From Table I, the amount discharged through 100 ft. of 3-in. pipe for a loss of 1 lb. is 53.6 lb. per min. for an initial pressure of 120 lb. With a loss of 2 lb. pressure the amount discharged will be 53.6 × $\sqrt{2}$. For a length of 400 ft. the discharge will be

$$53.6 \times \sqrt{2} \times \sqrt{\frac{100}{400}} = 37.9 \text{ lb. per min. Ans.}$$

5. Comparative Economy of High-Pressure Steam. When contemplating the use of steam at high pressures the actual gain in engine economy is worthy of close consideration. Table II shows the steam consumption of a perfect engine; i. e., an engine devoid of friction and which utilizes all the energy in the steam liberated by expanding from boiler pressure to the pressure of the exhaust. The table shows the steam consumption per indicated horsepower per hour for boiler pressures ranging from 100 to 250 pounds.

TABLE II

STEAM CONSUMPTION OF PERFECT ENGINE AT VARIOUS
STEAM PRESSURES

Boiler Pressure by Gauge Pounds	Temperature Degrees Fahrenheit	Pounds Steam per Indicated Horsepower per Hour	
		Non-Condensing	Condensing
100	337.7	16.24	7.71
125	352.6	14.71	7.37
150	365.6	13.63	7.09
175	377.1	12.81	6.87
200	387.6	12.16	6.68
250	405.9	11.19	6.39

It is at once seen that there is no gain beyond 200 pounds pressure sufficient to compensate for the extra costs incurred throughout the entire construction.

PRINCIPAL REQUIREMENTS OF THE STEAM-PIPING SYSTEM

6. A system of steam piping for an electric power station must be so designed as to insure reliability of service, economy of construction, and minimum of loss during transmission. The main lines of piping must be so interconnected that blow-outs will not cripple or derange the working of the plant. This continuity of operation is absolutely indispensable to a successful station.

The pipes and fittings must be so proportioned as to permit a free flow of the steam, water, etc., and so that undue loss shall not be caused by condensation, radiation, or friction. The steam piping should be so arranged that water pockets will be avoided, and where unavoidable they must be dripped free from water; the entrained water can be automatically returned to the boiler. By-pass pipes, with suitable placing of valves, should be arranged around feedwater heaters, economizers, pumps, etc. The system must be so designed as to give perfect freedom for expansion and

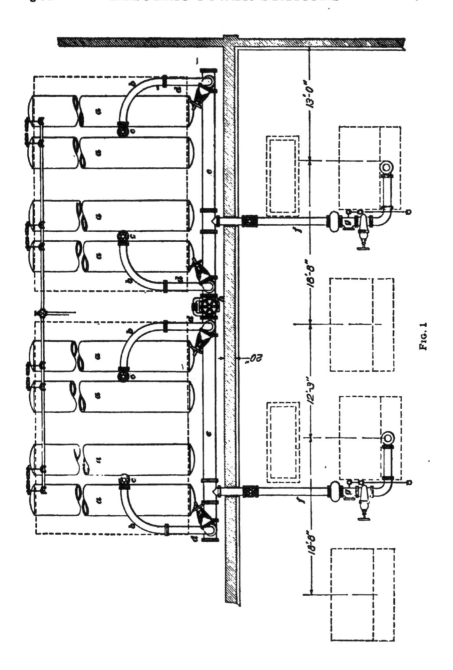

Fig. 1

contraction, without undue strain on any member of the system, and without opening joints that will cause leakage.

7. Perfect drainage must be provided to the end that all waters of condensation shall be fully separated from the steam, and by suitable traps or return systems delivered back to the boiler. The elaborate duplication of steam

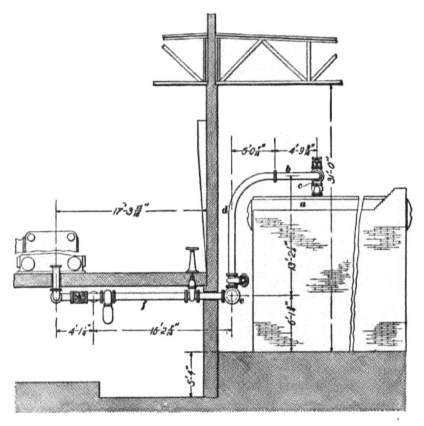

Fig. 2

mains and connections is not necessary. The double or duplicate system of piping was introduced a few years ago to overcome the deficiencies in valves, fittings, methods of workmanship, and to insure greater reliability, and thus became a fad for a short time. This method has no further reason for application because manufacturers now meet the demand with all materials for a first-class system.   Reliability is better insured by careful design and superior workmanship, combined with the use of high-class materials and fittings and the judicious placing of cut-out and by-pass valves, all of which will result in a system superior to elaborate duplication.

8.   Where several boilers of, say, 200 horsepower and upwards, are used, it will be found very convenient to place the *steam main*, or *header*, on or near the floor in the rear of the boiler; this brings all the large valves in accessible positions.   The steam lines leading to engines are placed below the engine-room floor.   This system is particularly applicable when horizontal engines are used.

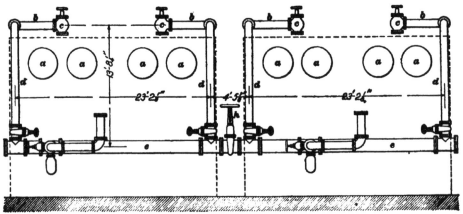

FIG. 3

The judicious use of long-radius bends, a most convenient arrangement of valves and location of live-steam header in accessible positions, and with steam connections to engines below engine-room floor, is shown in Figs. 1, 2, and 3.   From the cross-connection between the 42-inch steam drums *a, a* of the water-tube boilers leads an 8-inch connection *b*, starting from an automatic stop- and check-valve *c*; the long-radius bend *b* is placed horizontally and connects with a similar bend *d* leading vertically in a downward direction to the live-steam header *e*.   This arrangement gives great elasticity to a system of large piping, and the placing of the valves affords every convenience for ready manipulation.

47—15

In these figures, the main steam piping only is shown, the auxiliary piping for the boiler feedwater, heaters, etc. being omitted. Fig. 1 is a plan view, Fig. 2 an end view, and Fig. 3 a view showing the arrangement of the main steam pipes looking toward the rear of the boilers. The main steam pipes *f*, running from the header *c* to the high-pressure cylinders of the steam engines, are placed under the engine-room floor and a connection to the low-pressure cylinder is provided at *g* so that in case of emergency the low-pressure cylinder can be run with high-pressure steam. By examining the arrangement of valves between the boilers and engines, it is readily seen that it is possible to cut off any engine or boiler in case of accident and still run the plant with the remaining engines or boilers. The main steam header is divided into two sections by the large gate valve *h*, Figs. 1 and 3, so that one-half of the header can be entirely cut off from the other half.

PIPE BENDS

9. **Pipe bends** should be freely used to take up expansion strains and to reduce the number of joints where elbows would otherwise be required. Bends of short radius are undesirable, as they are so stiff that the object sought to be attained is not accomplished. Pipe bends of short radius with flanges can be used for elbows, but where it is intended to counteract the expansion strains on the system the radius of the bends should be liberal to allow the necessary elasticity. The radius of any bend should not be less than five diameters of the pipe, and a larger radius is preferable. The pipe should be carried completely through the flanges. The flanges should be refaced to bring them true with each other, and any protruding end of the pipe should be turned off to admit of the gasket having a full bearing. The best steam-fitting practice now uses these long double-swing bends. There is no special rule for the details of this method, which requires a knowledge of expansion of pipes when heated, and good judgment in placing the bends; the skill of the steam fitter puts the strain on the pipes when cold, with the result that when the steam pressure is put on, the expansion removes the tension and no strain remains on the pipe except that due to the initial steam pressure. Slip expansion joints of any style have no place in a good job of steam fitting; they are exceedingly undesirable and will surely be a source of trouble and expense.

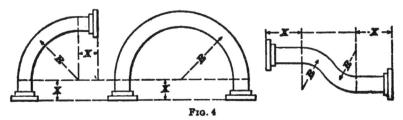

FIG. 4

10. Dimensions of Pipe Bends.—Table III, in connection with Fig. 4, shows the minimum radius of bends; all manufacturers prefer a longer radius than that indicated.

TABLE III
DIMENSIONS OF WROUGHT-IRON AND STEEL PIPE BENDS

Diameter of Pipe Inches	Dimension X Inches	Minimum Radius R Inches	Diameter of Pipe Inches	Dimension X Inches	Minimum Radius R Inches
$2\frac{1}{2}$	4	$12\frac{1}{2}$	7	8	35
3	4	15	8	9	40
$3\frac{1}{2}$	5	$17\frac{1}{2}$	10	12	50
4	5	20	12	14	60
$4\frac{1}{2}$	6	$22\frac{1}{2}$	15	16	75
5	6	25	16	20	80
6	7	30	18	22	90

FLANGED JOINTS

11. In all pipe fitting for sizes of 3 inches and upwards it is most desirable to use **flange joints**; while the first cost is slightly in excess of the usual threaded fitting, the facilities offered for repairs and alterations, the freedom from leaks, and the general security of the job are sufficient to warrant the additional cost.

TABLE IV

DIMENSIONS OF EXTRA HEAVY PIPE FLANGES FOR 250 POUNDS WORKING PRESSURE

Adopted by All the Leading Manufacturers

Size Inches	Diameter of Flanges Inches	Bolt Circle Inches	Number of Bolts	Size of Bolts	Length of Bolts Inches
1	4¼	3¼	4	½	2
1¼	5	3¾	4	½	2¼
1½	6	4¼	4	⅝	2¼
2	6½	5	4	⅝	2¼
2½	7¼	5⅞	4	¾	3
3	8¼	6⅝	8	⅝	3
3½	9	7¼	8	⅝	3¼
4	10	7⅞	8	¾	3¼
4½	10½	8½	8	¾	3½
5	11	9¼	8	¾	3¾
6	12½	10⅝	12	¾	4
7	14	11⅞	12	⅞	4
8	15	13	12	⅞	4¼
9	16	14	12	⅞	4½
10	17½	15¼	16	⅞	4¾
12	20	17¾	16	⅞	5
14	22½	20	20	⅞	5¼
15	23½	21	20	1	5½
16	25	22½	20	1	5¾
18	27	24¼	24	1	6
20	29½	26¾	24	1⅛	6¼
22	31¼	28¾	28	1⅛	6½
24	34	31¼	28	1⅛	6¾

NOTE.—Flanges, flanged fittings, valves, etc. are drilled in multiples of four, so that fittings may be made to face in any quarter and holes straddle center line.

12. Dimensions of Flange Fittings.—A majority of the manufacturers of flanged fittings and valves have agreed on standard dimensions for the thickness and diameter of the flanges, the diameter of bolts and holes, the number and size of bolts, and the diameter of the bolt circle; unless especially ordered otherwise, flanges will be made according to this standard. This standard has also been adopted by a joint committee of the American Society of Mechanical Engineers, and the Master Fitters Association. This is referred to as the Manufacturers' Standard. Fig. 5 shows the shape of the standard flanges: (*a*) is used where the pipe is simply screwed into the flange and not calked, and (*b*) where the pipe is calked, a calking seam being provided at *c*. Table IV shows the standard dimensions of extra heavy flanges and flange bolts used for central-station piping where the pressure runs up as high as 250 pounds per square inch.

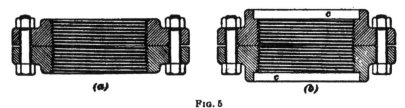

(a)   *(b)*

Fig. 5

13. Thickness of Flange Fittings.—Flange fittings are generally made in three weights: the least for low pressure, such as exhaust or condenser connections; the second for standard pressures, up to 125 pounds; and the heaviest for high pressures, up to 250 pounds per square inch. The thickness in the three styles is shown in Table V.

These flanges are designed to have an unusually large factor of safety in order to cover possible defects in the metal or imperfections in casting. In all cases where using iron or gun-metal fittings, it is important to specify that the castings shall be absolutely sound and free from flaws, blowholes, and shrinkage cracks.

14. Methods of Making Flanged Joints.—Several methods of making up flanged joints have been devised, tested, and abandoned; good practice has settled down to

TABLE V

THICKNESS OF PIPE FLANGES FOR VARIOUS PRESSURES

Diameter of Pipe Inches	Low Pressure, 50 Pounds Inch	Medium Pressure, 125 Pounds Inches	Pressures Up to 250 Pounds Inches
6	$\frac{5}{8}$	1	$1\frac{7}{16}$
8	$\frac{3}{4}$	$1\frac{1}{8}$	$1\frac{5}{8}$
10	$1\frac{1}{8}$	$1\frac{3}{16}$	$1\frac{7}{8}$
12	$1\frac{1}{4}$	$1\frac{1}{4}$	2
14	$\frac{7}{8}$	$1\frac{3}{4}$	$2\frac{1}{8}$
16	1	$1\frac{7}{8}$	$2\frac{1}{4}$

two methods of making a good job of securing flanges to pipes. The first and most economical method is the threaded joint having the pipe thoroughly peened or hammered in the flange, making a secure and lasting fit; Fig. 6 shows this style of joint and the method of peening over the end of

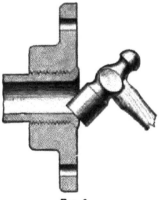

the pipe. The second and more expensive method is to have the flanges welded on the end of the pipe and then faced true in a lathe; this makes the most perfect and workmanlike joint in use at the present time.

Good practice for pressures up to 165 pounds is threaded flanges with the threads carefully cut, screwed on, and peened in the flange. A large number of tests have been made to determine the holding

FIG. 6

power of threads on piping, and results have shown that the strength of the threaded joint exceeds the strength of the flanged joint. Assuming that the shearing strength is one-half of the tensile strength of the metal, it is perfectly evident that the holding power of the threads is fully three times greater than the ultimate strength of the piping because of the number of threads that must be sheared before the joint gives way. It is desirable to make the threaded joints for flange fittings and high-pressure work longer than the standard, in order to reenforce the pipe by throwing a larger number of threads in contact, and thus guard against possible loosening by vibration.

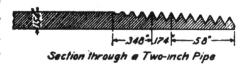

Section through a Two-inch Pipe

15. Pipe Threads. It is important that the threads be perfectly cut, with first-quality tools in good order, to standard sizes. Imperfect and careless work in the cutting and fitting of threads should not be accepted. The repairs required to threaded joints because of leakage or failure are frequently caused by imperfect and unworkmanlike construction. For example, standard work on threaded 4-inch pipes requires eight perfect threads occupying 1.05 inches along the pipe; two that are perfect at the bottom and slightly flat

Section through a Three-inch Pipe

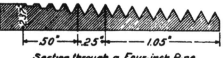

Section through a Four-inch Pipe

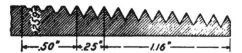

Section through a Five-inch Pipe

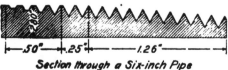

Section through a Six-inch Pipe

Fig. 7

on top, and four that are imperfect both at the top and bottom. The total length scored by the die on the pipe is approximately 1.8 inches. It is not uncommon to find only 1½ inches threaded on a 4-inch pipe and even on a 6-inch pipe, whereas the standard for the latter requires fully 2 inches. Fig. 7 shows the correct proportions of standard threads on the several sizes of pipe commonly used. All piping that is used for service around an electric power station should be of standard size, standard thickness, standard thread, and round, straight, and perfect in every respect.

For ordinary pressures in sizes below 3 inches, the fittings should be heavy, beaded, gray-iron casting, tapped accurately to standard gauge; and for pressures in excess of 100 pounds, it is desirable that the fitting be extra heavy and made of gun metal. It is important that the piping should be screwed completely into the fitting or through the flange, to guard against all vibration and leakage, and to make the thread

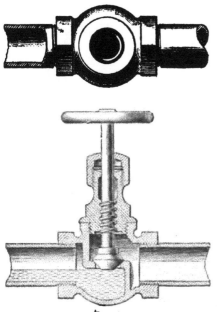

metal tight against oxidizing action by leaking steam or water. A properly erected system of pipe fitting should not show any indications of leakage under any conditions of expansion or contraction incident to daily service; where calking is required to stop leaks in threaded joints it is evident that the work is imperfect.

FIG. 8

VALVES

16. A valve is required not only to withstand its working pressure, but also the strains of expansion and contraction, the weight of piping, and the cutting effect of the steam on the seat disk.

17. Globe valves are objectionable because of their clumsiness, resistance to passage of steam, and the water pockets that they bring about in the system of steam piping; Fig. 8 shows how they may allow the accumulation of water when placed on a horizontal or vertical position. The use of globe valves should be avoided on all piping above 1¼ inches in diameter.

18. Gate valves are eminently desirable, and if frequently opened they will keep in good order and remain tight for many years; they are almost in universal use with the best standard of construction. They are manufactured for many lines of service, the principal being as follows: For low pressure not exceeding 50 pounds, and for exhaust

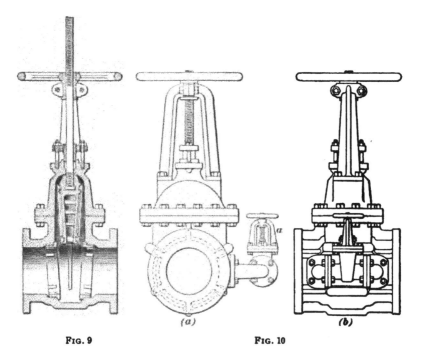

FIG. 9   FIG. 10

and condenser service; for pressure not exceeding 125 pounds; for medium pressure not exceeding 150 pounds; and for extra heavy pressures not exceeding 250 pounds. Standard valves from 4 inches to 8 inches should withstand a hydrostatic pressure of 600 pounds, and from 10 inches to 16 inches a hydrostatic pressure of 500 pounds without showing any defects. Valves for medium and extra heavy pressures should withstand from 1,200 to 1,500 pounds hydrostatic pressure.

For valves of 6 inches and upwards on steam lines, it is desirable to use the outside screw yoke, with stationary wheel and rising spindle, as shown in Fig. 9. The advantages of this type are that the extension of the stem shows the position of the gate; also, the screw can always be properly lubricated and does not come in contact with the steam.

By-passes are desirable on or around all valves, for live steam, of 6 inches and upwards. The by-pass permits of the easy equalizing of pressure on both sides, before opening the main valve. Fig. 10 shows a gate valve provided with a small by-pass valve *a*. By first opening the small valve the pressure on the two sides of the gate valve is equalized, thus making the valve easy to open.

RECEIVERS

19. Reduced Pipe Diameters With Receivers. The best-designed systems now use sizes of live-steam piping from 5 to 15 per cent. smaller than that called for by the engine builders for throttle-valve connections. These pipes deliver the steam into wrought-iron or steel receivers, which also act as separators, separating moisture from the steam. These receivers should have a capacity of from three to four times that of the high-pressure cylinder, and should be placed as close as possible to the cylinder.

In a system thus arranged for live steam, the action is as follows: At each stroke of the piston the engine obtains from the receiver the necessary volume of steam, the withdrawal of which for a fraction of a second reduces the pressure in the receiver. The boiler pressure behind the live steam in the pipe rushes new steam to the receiver at high velocity, thus restoring instantly the volume and pressure required for the next stroke of the engine. This process follows continuously, and a higher velocity in the flow of steam in the smaller pipes and larger volume for the engine to draw from, as well as separation of moisture, is thus obtained.

This arrangement also provides a cushion near the engine, which takes the reaction caused by the quick cut-off in the steam chest and prevents vibration from being transmitted through the system of piping. This plan of reduced diameters of live-steam pipe, combined with receivers, tends to produce a steady and rapid flow of steam in the direction of the engine. A system so arranged need not incur a loss of more than 1 or 2 pounds between boiler and engines, unless the engines are a long distance from the boilers, and even then the loss need not exceed 3 or 4 pounds for 400 to 500 feet transmission. The initial cost of the receiver system of piping will be considerably less than that using larger diameters of pipe without receivers.

EXHAUST PIPING

20. Most of the details previously described apply equally well to the design and fitting of **exhaust-pipe systems.** On account of the reduced pressure it is not necessary to use weights of pipe heavier than standard, due provision being allowed for expansion and contraction. Special care should be observed that the diameters of pipes are liberal, the bends of easy radius, and gate valves employed to the end that no back pressure be caused. When piping exhaust systems to condensers, the greatest care is necessary to secure air-tight work, as emphasized in connection with condensers. The placing of non-conducting covering on exhaust systems is usually done where the heat is desired for the feedwater, also to insure a lower temperature in the station. It is not necessary for this covering to be as thick as for live-steam systems.

STEAM JOINTS

21. The faces of flanges have been formed in many ways with the object of preventing the gasket from blowing out. The tongue-and-groove, raised faces, male-and-female ends have all been thoroughly tried, and cost somewhat in excess of a straight face and corrugated joint.

The straight-face joint slightly corrugated, if properly put together with a good gasket, will stand a test of 1,000 pounds without blowing out.

22. Gaskets.—Many styles of gaskets have been experimented with, but the most satisfactory thus far found is the corrugated copper gasket placed within the bolt circle and well pulled up; this makes a joint that is very durable. Gaskets containing rubber, asbestos, or alloys of soft metal require constant attention, and with superheated steam are quickly destroyed.

It is very essential that all the miscellaneous fittings, such as gauge-cocks, gauge-glasses, blow-off cocks, water-supply fittings, trap connections, lubricators, etc., should be extra heavy. While these represent very small items of cost in comparison to the total cost of the job of piping, a blow-out in any one will be the cause of a great deal of inconvenience and will sometimes result in far more expense than would have been the first cost of the kind of fittings suggested.

PIPE SUPPORTS

23. It is essential that the whole system of piping be supported in such a manner that there shall be perfect freedom for expansion and contraction, without undue strain on any section. The correct location of supports cannot be specified in advance and becomes a matter of good judgment as the system is erected. Several suitable appliances are illustrated in Figs. 11 and 12. Where large pipe systems are laid near the floor, it is desirable to build brick piers capped with iron plates, surmounted by iron rollers, say, 1 or 2 inches in diameter, on which the pipe rests.

24. Expansion of Wrought-Iron Pipes.—The amount by which a wrought-iron pipe will expand when heated can be determined by the following rule:

Rule.—*Multiply the length of pipe, in inches, by the number of degrees to which it is heated, and divide by 150,000; this gives the expansion, in inches.*

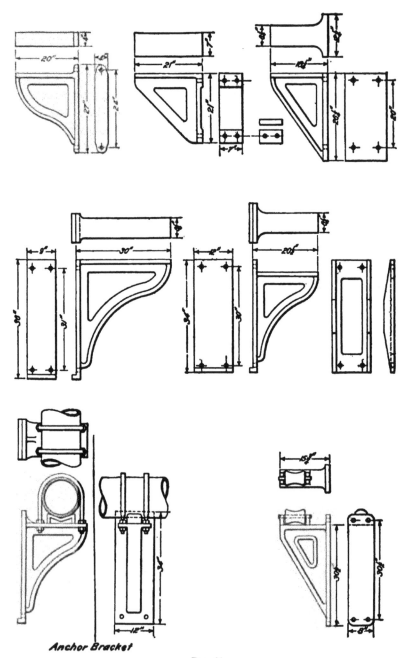

Anchor Bracket

FIG. 11

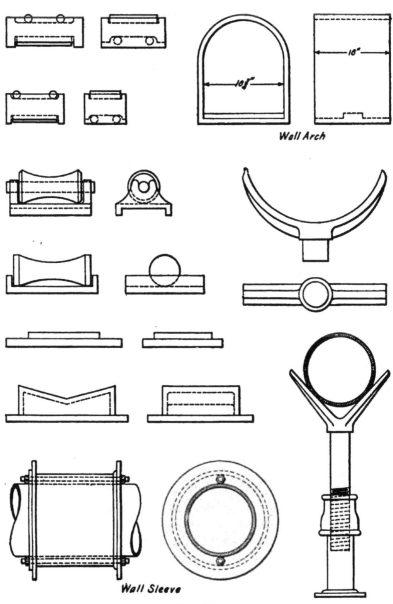

Wall Arch

Wall Sleeve

Fig. 12

Cast-iron pipe expands $\frac{1}{162000}$ of its length for each degree Fahrenheit it is subjected to under ordinary conditions; wrought-iron pipe $\frac{1}{156000}$. A 2-inch pipe when heated to a temperature of 338° F. (temperature of steam at 100 pounds pressure) exerts an expansive force of 25 tons.

QUALITY OF PIPING AND FITTINGS

25. The standard grades of wrought-iron pipe are known as *butt-welded, lap-welded, extra strong,* and *double extra strong.* Table VI shows the standard dimensions of wrought-iron piping from $\frac{1}{8}$ inch to 12 inches diameter, inclusive. Standard wrought-iron pipe lap-welded is always tested at the mills to a pressure of 500 pounds per square inch, by hydrostatic pressure. Commercial piping in sizes of 12 inches and under with perfectly welded joints has been tested up to 1,500 pounds per square inch without bursting; 8-inch and 10-inch piping taken from stock at random has been tested to 2,000 pounds per square inch; 16-inch pipe to 800, and 24-inch pipe to 600 pounds without rupture or fracture. It would appear that there is, therefore, no reason why, for diameters less than 15 inches and pressures up to 200 pounds per square inch, piping heavier than standard should be used for the live-steam system in an electric power station.

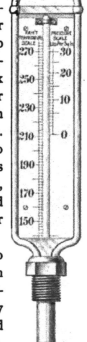

Fig. 18

Wrought-steel piping has now become so reliable and is so generally used, especially in the larger sizes, that it is preferable to wrought-iron pipe. For expansion bends, it is equally well made and reliable, and copper bends need only be used for very heavy work. All pipes, valves, fittings, and materials used on a first-class job are now readily obtained to withstand the required working pressure of 200 pounds per square inch. Any modern job of steam fitting necessarily requires good design, good material, and skilful workmanship.

TABLE VI
STANDARD DIMENSIONS FOR WROUGHT-IRON AND STEEL STEAM, GAS, AND WATER PIPE

Diameter			Nominal Thickness Inches	Circumference		Transverse Areas			Length of Pipe per Square Foot of		Length of Pipe Containing 1 Cubic Foot Feet	Nominal Weight per Foot Pounds	Number of Threads per Inch of Screw
Nominal Internal Inches	Actual External Inches	Approximate Internal Diameter Inches		External Inches	Internal Inches	External Square Inches	Internal Square Inches	Metal Square Inches	External Surface Feet	Internal Surface Feet			
1/8	.405	.270	.068	1.272	.848	.129	.0573	.0717	9.440	14.15	2,513.0	.241	27
1/4	.54	.364	.088	1.696	1.144	.229	.1041	.1249	7.075	10.49	1,383.3	.420	18
3/8	.675	.494	.091	2.121	1.552	.358	.1917	.1663	5.657	7.73	751.2	.559	18
1/2	.84	.623	.109	2.639	1.957	.554	.3048	.2492	4.547	6.13	472.4	.837	14
3/4	1.05	.824	.113	3.299	2.589	.866	.5333	.3327	3.637	4.635	270.0	1.115	14
1	1.315	1.048	.134	4.131	3.292	1.358	.8626	.4954	2.905	3.645	166.9	1.668	11 1/2
1 1/4	1.66	1.380	.140	5.215	4.335	2.164	1.496	.6680	2.301	2.768	96.25	2.244	11 1/2
1 1/2	1.9	1.611	.145	5.969	5.061	2.835	2.038	.7970	2.010	2.371	70.66	2.678	11 1/2
2	2.375	2.067	.154	7.461	6.494	4.430	3.356	1.074	1.608	1.848	42.91	3.609	8
2 1/2	2.875	2.468	.204	9.032	7.753	6.492	4.784	1.708	1.328	1.547	30.10	5.739	8
3	3.50	3.067	.217	10.996	9.636	9.621	7.388	2.243	1.091	1.245	19.50	7.536	8
3 1/2	4.0	3.548	.226	12.566	11.146	12.566	9.887	2.679	.955	1.077	14.57	9.001	8
4	4.5	4.026	.237	14.137	12.648	15.904	12.730	3.174	.849	.949	11.31	10.665	8
4 1/2	5.0	4.508	.246	15.708	14.162	19.635	15.961	3.674	.764	.848	9.02	12.490	8
5	5.63	5.045	.259	17.477	15.849	24.306	19.990	4.316	.687	.757	7.20	14.502	8
6	6.625	6.065	.280	20.813	19.054	34.472	28.888	5.584	.577	.630	4.98	18.762	8
7	7.625	7.023	.301	23.95	22.063	45.664	38.738	6.926	.501	.544	3.72	23.271	8
8	8.625	7.982	.322	27.096	25.076	58.426	50.040	8.386	.443	.478	2.88	28.177	8
9	9.625	8.937	.344	30.238	28.076	72.760	62.730	10.030	.397	.427	2.29	33.701	8
10	10.75	10.09	.366	33.772	31.477	90.763	78.839	11.924	.355	.382	1.82	40.065	8
11	11.75	11.		36.914	34.558	108.434	95.033	13.401	.325	.347	1.51	45.028	8
12	12.75	12.		40.055	37.700	127.677	113.098	14.579	.299	.319	1.27	48.985	8

26. Piping for Feedwater System.—Experience has shown that hot feedwater corrodes wrought-iron pipe quite rapidly, due to the amount of impurities in the water. It is therefore most desirable to use extra thick pipe for this purpose, but it is far better to use brass pipe than wrought-iron pipe if the expense is not too great. The brass pipe is not subject to corrosion, and even the extra thick wrought-iron piping will require replacing within a few years.

27. In order to test the temperature of feedwater, it is desirable to locate pockets for testing thermometers at

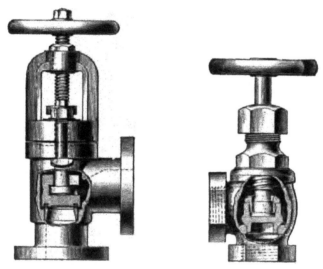

FIG. 14

suitable points; Fig. 13 shows a thermometer provided with a pocket for screwing into the pipe.

The feedwater-pipe connections to every boiler should be duplicated or so interconnected that the boiler can be supplied at either of two feeding-in points, or from either of two directions, to the end that the supply of feedwater shall be sure beyond all question of failure. This is particularly important if brass feedwater piping is not used. The valves set in the feedwater pipe leading to each boiler should consist of a flanged gate valve with companion flanges, a flanged check-valve, and a second flanged gate valve. For double-deck water-tube boilers, it is also desirable to have a duplicate steam pressure gauge at about 7 feet above floor level.

28. Blow-Off Valves.—The ordinary blow-off valve usually supplied with boilers is of little or no practical value in power-station service. A valve should be used that will withstand the cutting effects of scale and sediment. This is a point that should be unusually well guarded, otherwise continual repairs will be necessary because of leaky blow-off valves. Fig. 14 shows two types of blow-off valves of satisfactory design.

_____

PIPE COVERING

29. The importance of protecting steam pipes from loss by radiation is so generally acknowledged that we shall only touch on the salient points to emphasize the reasons for loss if not covered, and the importance of the saving effected when the work is properly done.

Between two bodies near each other and at different temperatures, there exists a tendency to temperature equalization by radiation, conduction, and convection. The very great difference in temperature between the surrounding atmosphere and a pipe containing steam or water at a high temperature is a cause for rapid radiation from the surface of the pipe to the atmosphere. This rapid radiation is a direct loss of the heat units derived from the fuel and stored up in the steam. To prevent this loss, it is essential that all live-steam pipes, and also those containing hot water for boiler feed, should be protected by a covering that is a non-conductor of heat. The value of a pipe covering will be represented by the saving in the money expended for fuel that would otherwise be wasted.

The following items must be considered in order to properly determine the economic value of a pipe covering: (a) The cost of the covering; (b) the cost of coal required to supply the loss of heat if the pipes are not covered; (c) the extra capital required to be invested in boilers to make up for the loss and the interest on the investment; (d) the guaranteed life of the covering.

ground and compressed cork, and other coverings of similar composite character. Such coverings as wool, hair felt, wool felt, sawdust, or paper pulp are liable to become charred and carbonized by heat from the pipe, and finally ignited; such coverings are not worth the money and labor expended on them. Sectional covering is most desirable, as it can be removed when repairs are needed, and replaced with but slight waste of material, and only the cost of labor.

For a specified result of minimum radiation, different grades and qualities of covering will require to be of different thicknesses, and if first cost and real utility are to be compared under fair conditions, the specification should require a guarantee that the covering around a pipe carrying steam at a given pressure and temperature shall not show a loss of British thermal units per square foot of pipe surface per minute exceeding a percentage to be determined according to the quality and thickness of the covering selected.

EXAMPLES OF STEAM PIPING AND GENERAL ARRANGEMENT OF PLANT

MEDIUM-SIZE PLANTS

32. In order to illustrate the general arrangement of steam plants, particularly with reference to the arrangement of steam boilers, steam piping, and engines, a few typical cases are given.

Fig. 15 shows the arrangement of a plant of comparatively small output in which the steam piping is of about the simplest possible character. In this plant three Babcock and Wilcox water-tube boilers supply steam to two tandem compound engines direct-coupled to polyphase alternators. The total rated boiler capacity is 500 horsepower. Condensers are not used and the steam exhausts into the atmosphere after passing through the feedwater heater.

33. Fig. 16 shows a larger plant with cross-compound condensing engines. This plant has 900 rated horsepower boiler capacity and 750 kilowatts in generator capacity. The

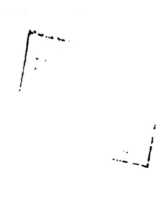

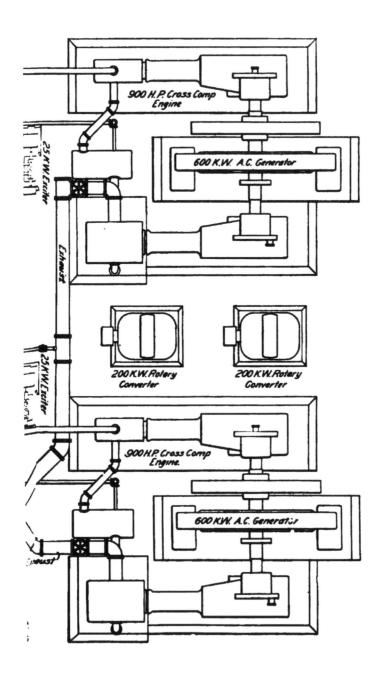

900 H.P. Cross Comp Engine

600 K.W. A.C. Generator

25 K.W. Exciter

Exhaust

200 K.W. Rotary Converter

200 K.W. Rotary Converter

25 K.W. Exciter

900 H.P. Cross Comp Engine.

600 K.W. A.C. Generator

Exhaust

Y

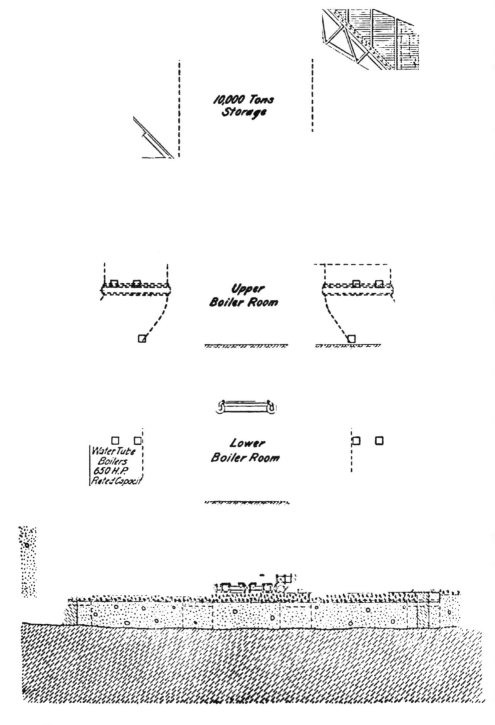

10,000 Tons
Storage

Upper
Boiler Room

Lower
Boiler Room

Water Tube
Boilers
650 H.P.
Rated Capacity

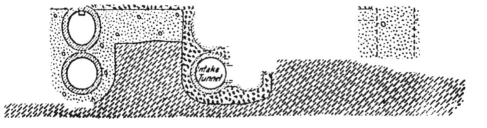

5200 H.P. Three Cylinder
Compound Engine

4500 K.W.
3-Phase Alternators
Volts

Intake
Tunnel

31

exhaust steam is condensed by means of jet condensers and the feedwater heated by means of a Cochrane feedwater heater and purifier. The condensers and pumps are placed in a pit somewhat below the engine-room floor line. Fuel is supplied to the boiler by means of Roney stokers, which are fed from a coal bin, as indicated in the figure.

34. Fig. 17 shows a plan of a traction plant in which alternating current is generated for transmission to distant parts of the line. The generating equipment consists of two 900-horsepower cross-compound engines direct-connected to 600-kilowatt alternators. The direct current required for operating the cars is obtained from substations by stepping-down the high-tension alternating current and passing it through rotary converters. The two rotary converters shown in the station are for supplying the near-by portions of the road with direct current. If a large proportion of the output were to be used as direct current, it might be advisable to install double-current generators instead of plain alternators and thus dispense with the rotary converters in the power station. The exhaust steam from the engines is condensed by means of a siphon or barometric condenser, and in order to maintain a good vacuum, a dry-air pump (i. e., an air pump that pumps out vapor and air only and not a mixture of these with water) is attached to the condenser chamber. The pumps used for handling the feedwater and injection water are located as shown in the figure, together with the usual fire-pump for furnishing a water pressure for fire-protection. The fields of the alternators are excited by means of two independent exciters driven by small direct-connected steam engines.

LARGE PLANTS

35. As examples of the very large plants that are now becoming common in large cities, we will take the Waterside station of the New York Edison Company, and the power station of the Manhattan Elevated Railway, New York. A cross-section of the Waterside station is shown in Fig. 18, and a skeleton plan, showing the arrangement of boilers,

engines, and steam piping, in Fig. 19.　There are two tiers of
boilers with twenty-eight boilers in each, fifty-six altogether.

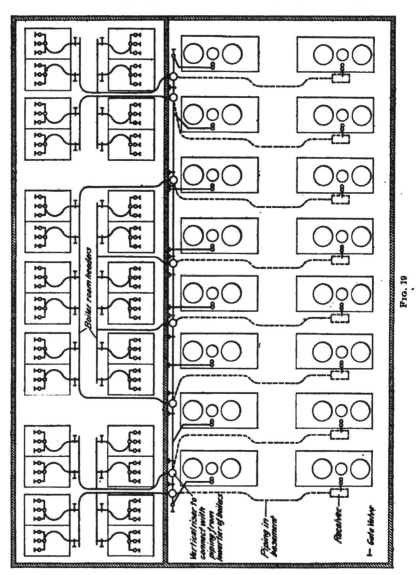

FIG. 19

These boilers supply sixteen vertical engines, which work
most economically at about 5,200 horsepower each or a total
output of 83,200 horsepower; the station is, however, capable

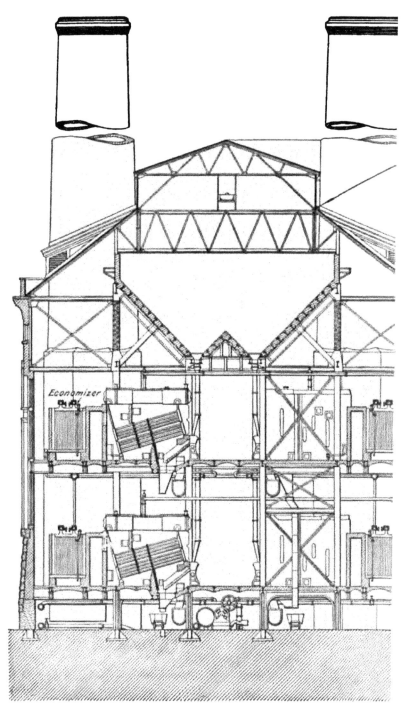

Switchboard

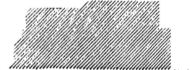

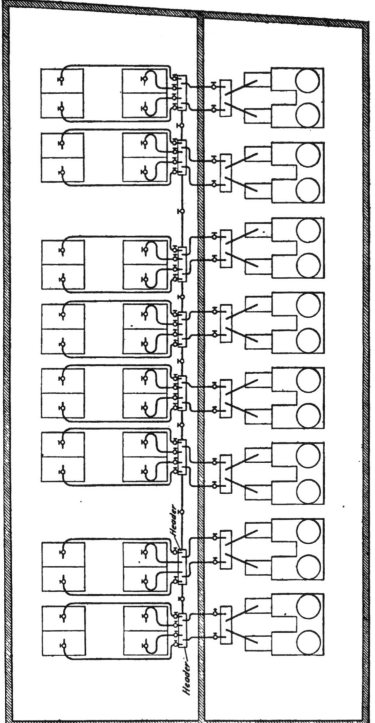

FIG. 21

of a maximum output of 128,000 horsepower. The condensers are arranged in the basement below the engines, as
shown, and are of the surface type, salt water being used
for cooling purposes. The various switchboard appliances
are arranged in the galleries at the right-hand side, Fig. 18.
The engines are of the cross-compound vertical type and the
piping for the right-hand row of engines is carried under
the floor, as indicated. The general arrangement of the

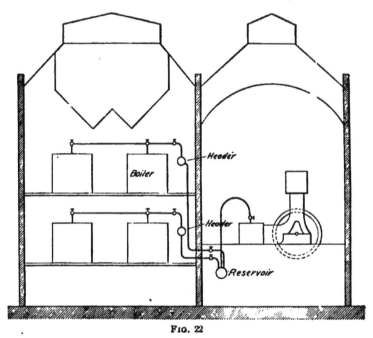

Fig. 22

piping, coal storage, etc. is evident from the figures, so that
an extended description will be unnecessary.

36. Fig. 20 shows a cross-section of the power plant of
the Manhattan Elevated Railway, which has a rated capacity
of 64,000 horsepower and a maximum capacity of 100,000
horsepower. There are eight generating units each consisting of an 8,000-horsepower, vertical-and-horizontal, duplex,
compound engine coupled to a 5,000-kilowatt, revolving-field,
three-phase alternator. Figs. 21 and 22 show the arrangement
of the steam piping. Each engine is provided with a reservoir

from which it draws its supply of steam, thus allowing the use of smaller steam piping than would otherwise be needed. Jet condensers are used, as shown in Fig. 20, and all auxiliary pumps are driven by electric motors instead of steam engines. The exciters for supplying current to the fields of the alternators are located as shown in Fig. 20, and are driven by direct-connected high-speed engines. The arrangement of the coal-storage plant, boilers, etc. will be apparent from Fig. 20, so that a detailed description is not necessary.

37. Steam-Turbine Plant.—The plant shown, Figs. 23 and 24, is of interest, as it represents a steam-turbine installation. The turbines are of the Curtis vertical type, 1,000 horsepower each. On account of the comparatively high speed and vertical arrangement of the turbines, the dimensions of the engine room of the plant are much smaller than would be required for reciprocating engines. The generators are mounted on top of the turbines, the latter being under the engine-room floor. The generators are the only part of the equipment above the floor. In this respect the steam plant bears considerable resemblance to a water-power plant using vertical turbines. Each turbine is about 12 feet in diameter at the base and is operated at 175 pounds steam pressure.

ELECTRIC GENERATORS AND SPECIAL APPARATUS

38. In connection with electric power stations, it will be necessary to say very little about the **generators,** as these are treated more in detail elsewhere. The only points, therefore, that will be considered are a few relating to the selection of machines and the general requirements to which they should conform.

In all remarks about electric generators it is well to emphasize the point that the design of the generator should be left entirely with the manufacturer thereof; there should be no divided responsibility. The conditions demanded by the service and the contract requirements should be clearly specified and incorporated in the contract, and rigidly

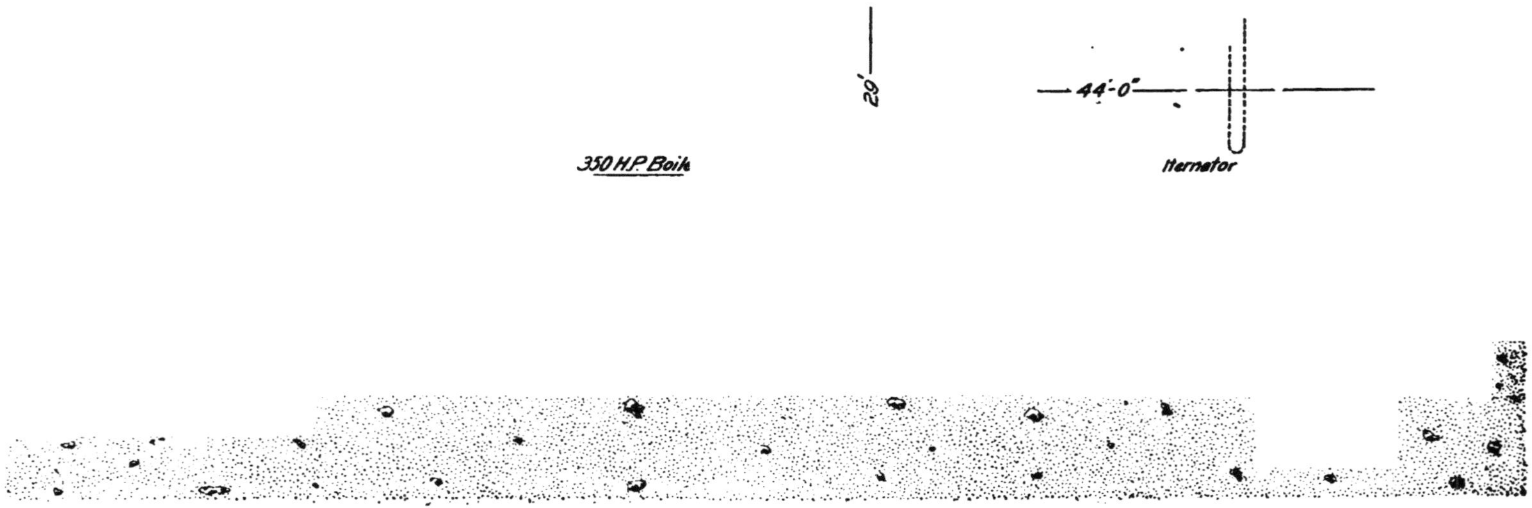

350 H.P. Boik

Hernator

44'-0"

20'

Fig. 23

engines, and steam piping, in Fig. 19.   There are two tiers of boilers with twenty-eight boilers in each, fifty-six altogether.

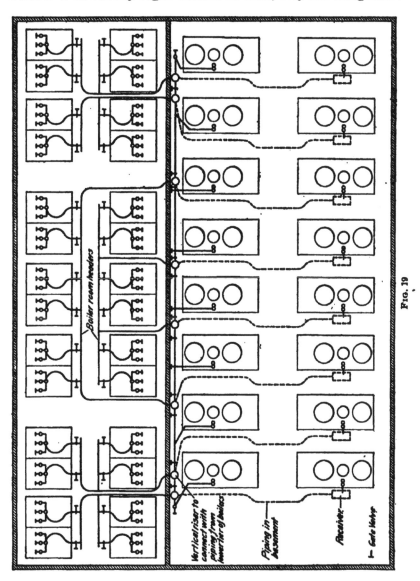

FIG. 19

These boilers supply sixteen vertical engines, which work most economically at about 5,200 horsepower each or a total output of 83,200 horsepower; the station is, however, capable

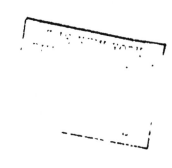

adhered to. Any local engineer who undertakes to specify the design and proportions and windings of an electric generator, and insists on contracting for such, must thereby assume the responsibility for its success or failure, and the manufacturer cannot legally be held liable for satisfactory performance under a divided responsibility. The manufacturing companies, with their broad experience, combination of engineering talent, and splendid facilities are far better equipped to design and furnish a generator that will successfully fill all reasonable conditions of service than any individual engineer can possibly be.

REQUISITE CHARACTERISTICS OF GENERATORS

39. In selecting a dynamo, it must be remembered that although it is a simple form of machine, it is subject to more sudden strains and overloads than ordinary machines, and is expected to run for long periods with very little attention of any kind. While repairs should very seldom be needed in any generator, necessity is extreme when it does come, and the parts subject to wear and renewal must be so arranged that the work may be done by any good mechanic or armature winder, to the end that the delay, trouble, and expense of sending large parts of the machine to the manufacturer may be avoided.

Certain salient features of construction that apply to any type of generator, motor, or rotary converter should be fulfilled by the manufacturer. All machines must be so designed that an armature coil, a commutator bar, or a field coil may be easily removed and replaced, and, for large machines, without having to lift massive parts or excessive weights. The bearings on all modern generators should be of the continuous, self-oiling type, and must be of the best design. The best practice is in the direction of shafts of large diameters. The shaft should be of the best hammered steel and the journals polished. The length of the journal should be ample to carry the total weight and strain without undue pressure per square inch on the bearing, and the shells of the bearing should be lined with the best Babbitt metal. The most satisfactory bearings will be so proportioned that the length is about four times the diameter of bore.

40. The armature and spider should be stiff and strong, and as light as consistent with the work required, with ample bearing on the shaft and rigidly keyed to it. Modern armatures are of the iron-clad type; the windings are protected from mechanical injury, and the insulating material is placed on the coil in a superior manner.

41. Insulation of Windings.—The insulation should withstand an alternating E. M. F. of two to five times the rated E. M. F., tested when heated after 2 hours' constant operation under full load. For example, machines generating up to 400 volts should stand a high-potential test of 1,500 volts; those generating from 400 to 800 volts, a test of 2,000 volts; from 800 to 1,200 volts, a test of 3,500 volts; from 1,200 to 2,500 volts, 5,000 volts; and above 2,500 volts, a test of twice the rated voltage. It must be remembered that high initial insulation, as measured by a galvanometer, does not necessarily imply durability. Such insulating material as carbonizes or cracks under high temperature long continued, and such paints or varnishes as injure the fabric on which they are put, may show very high initial insulation, but a breakdown always comes soon in actual service.

Field coils should be arranged to slip over the magnet cores, or, in large machines, the poles should be so designed that any one can be removed, field coil and all, without disturbing other parts of the machine. The field winding should be so proportioned that the rise of resistance occasioned by rise of temperature will be so slight as to necessitate very little attention to the regulation.

42. Commutators should be of sufficient length to afford ample brush surface and of sufficient depth to insure long life without renewal. Half the trouble experienced with carbon brushes is due to insufficient contact surface at the commutator. Only the very best clear mica should be accepted for commutator insulation. Poor mica will gradually

chip away between the bars, no matter how much care
and attention is given to the commutator. Very hard mica
sometimes wears slower than the bars, causing ridges, which
give rise to sparking at the brushes. Commutators must be
designed so that oil cannot possibly give trouble.

43. Brush Holders and Connections.—The connections to brush studs and cables should be such that none of
the current need pass through the spring, otherwise this will
soon be annealed by heat and rendered useless. All brush
holders, carrying bars, cables, etc. conducting current should
be as massive as is consistent with their due proportion of
current conducted, and should not show any perceptible rise
of temperature under 25 per cent. overload.

44. The **field terminals** should be arranged so as to
be readily accessible, when the machine is assembled complete, so that they can be easily tightened or disconnected if
desired. The frame of the whole machine should be well
braced by ribs, and sufficiently strong and rigid to withstand
any strain without the slightest spring or vibration.

45. Capacity and Temperature Rise.—In specifying
the capacity of a generator, or other type of machine, the
rise of temperature permitted above the surrounding atmosphere must be clearly stated. Each machine should in every
respect be constructed for and guaranteed to carry its full-
rated load for a continuous run of 24 hours. The rise in
temperature on armature or field windings should not exceed
40° C., above the surrounding atmosphere after a continuous
run of 12 hours at full-rated current, voltage, and speed.
Immediately succeeding the 12-hour test the machine should
be capable of carrying 25 per cent. overload for 2 hours, and
during this period should not show a rise in temperature in
armature of field windings exceeding 45° C., above the surrounding atmosphere. An allowance of 10° extra should be
made for commutator temperature. Each machine should be
capable of carrying an overload of 50 per cent. for 30 minutes,
and a momentary overload of 75 per cent. at any time during
its 24 hours' use under conditions of regular service.

46. Efficiency.—The efficiency of a well-designed generator will depend to some extent on the size of the generator, but for a machine of fairly good size will run close to 95 per cent. The total loss of 5 per cent. will be made up about as follows: core and eddy-current loss about 2.27 per cent., I^2R losses about 2.57 per cent., and frictional loss about .16 per cent. The efficiency of generators will vary according to type, capacity, and load and should be clearly set forth in the manufacturer's contract, showing what will be the efficiency at one-fourth load, one-half load, three-fourths load, full load, and 25 per cent. overload.

47. When machines are tested to prove the fulfilment of guarantees, such test should be made in the presence of the engineers, the purchaser, and the manufacturer, and should be made in the works of the manufacturer as far as practicable. All instruments required for testing should be furnished by the manufacturer, and correctly calibrated to agree with accepted standards. When machines are tested on the premises of the purchaser, he should provide the load and motive power free of expense to the manufacturer.

WATER-POWER ELECTRIC STATIONS

48. The steam-power station has a decided advantage over the water-power station in the fact that it may be located at any point where fuel is available for raising steam, while water-powers have a restricted value because they are a natural power confined to locations fixed by nature, and capable of development only in the immediate or near vicinity of a fall of water. The energy of water-powers rightly developed and electrically transmitted to centers of population may largely reduce the consumption of coal for electric-supply systems, or under favorable conditions render the use of steam unnecessary.

When contemplating a water-power development for electric transmission certain salient features that are clearly defined, must be fully investigated if one would avoid mistakes. They are as follows:

(*a*)   The quantity of water flowing, in cubic feet per second, both in the dryest season and at the maximum flow.

(*b*)   The most advantageous location for a dam, assuming that one must be constructed.

(*c*)   The most advantageous location for the power station.

(*d*)   The total height of the fall of water from the normal level in the forebay at the dam, to the normal level in the tailrace at the discharge.

(*e*)   What effects (if any) floods or extra high water will have on the water level, either at the dam or at the tailrace.

(*f*)   The market for power if electrically transmitted.

(*g*)   The estimated cost of the power development and the cost of transmission.

Too much stress cannot be laid on all these points at the beginning of the enterprise, as the data necessary to insure the capacity and reliability of a water-power requires that measurements should have been made, and that statistics should have been compiled for a period of many years prior to the time when the enterprise is first initiated.   The financial success of harnessing a water-power, to the average business man, seems to be a self-evident proposition.   He sees millions of foot-pounds of energy which, when transformed by electric generators, is to him the same as is developed by coal when burned to generate steam for driving engines. Suitable mechanism must be interposed to deliver the force of the falling water to a revolving shaft, and apparently there is a perpetual supply of power, but between this proposition and the actual attainment of the results there are many details that the business man does not readily grasp, and for the success of the enterprise he must rely on the experienced engineer.   The price paid for electric power depends on the demand, the same as for any other commodity, and there are many locations of excellent water-powers where there is not a market for power even within the commercial limits of successful electric transmission.

The margin of profits in water-powers within the territory where coal can be distributed economically is often doubtful and must be investigated with great care.   It is of course recognized that in exceptional instances, such as the flow of water through Niagara Falls, there can be no doubt regarding reliability. Almost all rivers and streams have variations in flow during certain seasons of the year, that are occasioned by local rains, melting of snows, or other meteorological conditions, which must be thoroughly and exhaustively investigated in every individual instance.

49. The area and characteristics of the watershed supplying a river or other stream should be a subject of most careful investigation and study. For instance, a watershed of 5,000 square miles area, having gentle slopes well covered with forest growths and subject to copious rainfall throughout almost every month of the year, will be more certain to give a fairly constant flow of water than would an equal area of steep and rocky mountain slopes with little verdure, where the rainfall would quickly run off in sudden floods and periods of low water would be experienced between times. Again, where high mountain ranges perpetually covered with snow exist, the melting snows afford quite a reliable flow of water, except during the winter months if the locality be in the northern latitudes.

It is quite customary to assume that the ordinary water-power has the possibility of a much larger power development than that for which it is really reliable, whereas its absolute reliability and capacity must be based on the minimum flow of water during the dryest season of the year, and such minimum flow should be verified by the most exhaustive investigation, which must be entirely devoid of all personal views and based solely on practical and unprejudiced statistics.

50. Storage of Water.—It is frequently possible in favorable localities to build a system of storage reservoirs that may hold sufficient water to carry over a short period of dry weather, or that will afford the extra water supply requisite to carry the station over the peak loads of daily service. If the water is held for 12 hours, and the whole quantity used the next 12 hours, the power of the stream will be doubled for that time.

51. Definitions.—Following are definitions of some terms used in connection with certain features of water-power development between dam and power station.

Forebay, that part of a mill race where the water starts to flow in the direction of the wheel.

Flume, an open or covered artificial channel or conduit for conducting the water from the forebay to the wheel.

Penstock, that part of the channel, conduit, or trough supplying water to the wheel that extends between the raceway and the gate through which the water flows to the wheel.

52. Combined Water- and Steam-Power Plants. Where it is undertaken to develop a greater power than can be reliably attained during the period of minimum flow it will be found necessary in many instances to reenforce the water-power station by steam power. Fig. 25 illustrates a station of this type, the generator being so arranged that it can be driven either by the engine or waterwheels. The main object to be attained in the development of a water-power is greater economy in the production of power than can be secured by steam.

This object is frequently attained with satisfactory results, and where water-power and steam power are combined the use will be varied according to the seasons, as shown in the following schedule for 1 year's service in a station in Massachusetts:

1899	PER CENT. ENERGY FROM WATER	PER CENT. ENERGY FROM STEAM
July	92.5	7.5
August	92.0	8.0
September	90.5	9.5
October	54.0	46.0
November	94.0	6.0
December	75.0	25.0
1900		
January	76.0	24.0
February	94.0	6.0
March	100.0	0.0
April	100.0	0.0
May	100.0	0.0
June	100.0	0.0

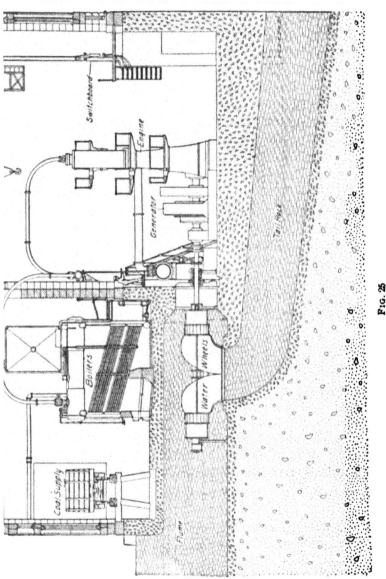

FIG. 25

53. As a general rule, those who attempt water-power development should expect to make a larger initial investment to place the enterprise on a working basis than will be necessary with steam power of equal capacity. A good steam plant, according to its type, degree of economy, and market prices, can be installed at a cost of from $30 to $75 per horsepower, whereas the expenditures involved to develop a water-power based on utilizing 70 per cent. of the power of the water, including the wheels and necessary appurtenances, will cost from $75 to $150 per horsepower, and if the water-power is deficient in capacity the cost of steam power to reenforce and insure its reliability should be added thereto. The total investment in each case should be carefully compared, and the interest on extra cost added where properly chargeable. After the initial cost of a water-power plant has been met, even though it be larger than for a steam plant, the recompense comes in reduced annual operating expenses. Where a steam plant must be used as an auxiliary, the operating expenses are increased by the additional cost of fuel, and for labor of firemen and engineers.

54. Between the water flowing in the stream and the point where the power may be utilized, the following elements must be considered:

(*a*) The best method of developing the power in the building of a dam and canal sluiceway or pipe, the placing of the power station, and the type of wheels to be employed.

(*b*) The type of generator and dynamo to be used and the method of imparting the power of the wheel to the shaft of the dynamo, whether by direct connection or by belting or gearing.

(*c*) The potential at which the current shall be generated; the distance over which the power must be transmitted will determine whether the original potential from the generator can be delivered at the point where the power is used, or whether step-up transformers are necessary to raise the potential of the current at the power station and step-down transformers to reduce it at the point where it is utilized.

(*d*) The electrical conductors necessary for conducting the energy developed at the station to the point where it is used.

(*e*) The step-down transformers that for long-distance transmission receive the current from the transmission lines and reduce the high potential to that available for commercial distribution.

(*f*) The method of reconverting the energy transmitted into power that may be applied to the driving of machinery or other appliances for industrial purposes.

Each one of these successive stages of conversion and transformation of energy involves a certain percentage of loss.

55. The most economical point for the development of a water-power may be so distant from any point of possible utilization or so undesirable for the location of any industry of considerable size that it cannot be made remunerative; therefore, the distance, which involves an expense in electrical conductors and pole-line construction, and the cost thereof require to be very accurately estimated, to the end that the true comparison of cost between a water-power station at its distant location and the cost of a steam-power station at the point where the power is desired may be arrived at. These points are not mentioned to discourage the development of water-powers, but to put the engineer on his guard that he may analyze with care the possible objections, and if all features are favorable he is then warranted in approving the proposed development.

MEASUREMENT OF WATER-POWER

56. Measuring the Volume of Water.—In the improvement or development of a water-power it is necessary to know the amount of power that can be depended on as reliable; it is not advisable to rely on any superficial examination, and it is very necessary that accurate measurements of capacity and fall should be made. The quantity of water flowing in a small stream can be estimated by the use of the weir, which is a partially submerged orifice through which the stream is forced to overflow. Use a board long enough to reach across the stream, with each end set in the bank, as shown in Fig. 26. Cut a notch in the board deep enough to pass all the water, and long enough to reach about two-thirds across the stream. This is called *weir dam*. The bottom and ends of the notch *a* in the board should be beveled on the down-stream side, leaving the upper edge almost sharp. The stake *b* should be driven in the bottom

FIG. 26

of the stream, several feet from the board, on a level with the sharp edge of notch *a*, this level being easily found when the water is beginning to spill over the notch.

When the water has reached its greatest depth, a careful measurement can be made of the depth over the stake. The dotted lines represent the level *d* of the running water, and the level *e* of the top of the stake. The distance between these lines gives the true depth, or spill over the weir board; if measured directly on the notch, the currents of water would reduce the depth. The surface water after passing from the board should not be nearer the notch *a* than 10 inches.

The nature of the channel above the board should not be such as to force or hurry the water; but must be amply wide and deep to allow the water to approach the notch quietly. For convenience of calculation Table VII is given.

TABLE VII

WEIR TABLE—FLOW FOR EACH INCH IN WIDTH

Inches		$\tfrac{1}{8}$	$\tfrac{1}{4}$	$\tfrac{3}{8}$	$\tfrac{1}{2}$	$\tfrac{5}{8}$	$\tfrac{3}{4}$	$\tfrac{7}{8}$	Inches
1	.40	.47	.55	.65	.74	.83	.93	1.03	1
2	1.14	1.24	1.36	1.47	1.59	1.71	1.83	1.96	2
3	2.09	2.23	2.36	2.50	2.63	2.78	2.92	3.07	3
4	3.22	3.37	3.52	3.68	3.83	3.99	4.16	4.32	4
5	4.50	4.67	4.84	5.01	5.18	5.36	5.54	5.72	5
6	5.90	6.09	6.28	6.47	6.65	6.85	7.05	7.25	6
7	7.44	7.64	7.84	8.05	8.25	8.45	8.66	8.86	7
8	9.10	9.31	9.52	9.74	9.96	10.18	10.40	10.62	8
9	10.86	11.08	11.31	11.54	11.77	12.00	12.23	12.47	9
10	12.71	13.95	13.19	13.43	13.67	13.93	14.16	14.42	10
11	14.67	14.92	15.18	15.43	15.67	15.96	16.20	16.46	11
12	16.73	16.99	17.26	17.52	17.78	18.05	18.32	18.58	12
13	18.87	19.14	19.42	19.69	19.97	20.24	20.52	20.80	13
14	21.09	21.37	21.65	21.94	22.22	22.51	22.79	23.08	14
15	23.38	23.67	23.97	24.26	24.56	24.86	25.16	25.46	15
16	25.76	26.06	26.36	26.66	26.97	27.27	27.58	27.89	16
17	28.20	28.51	28.82	29.14	29.45	29.76	30.08	30.39	17
18	30.70	31.02	31.34	31.66	31.98	32.31	32.63	32.96	18
19	33.29	33.61	33.94	34.27	34.60	34.94	35.27	35.60	19
20	35.94	36.27	36.60	36.94	37.28	37.62	37.96	38.31	20
21	38.65	39.00	39.34	39.69	40.04	40.39	40.73	41.09	21
22	41.43	41.78	42.13	42.49	42.84	43.20	43.56	43.92	22
23	44.28	44.64	45.00	45.38	45.71	46.08	46.43	46.81	23
24	47.18	47.55	47.91	48.28	48.65	49.02	49.39	49.76	24

The figures 1, 2, 3, etc. in the first vertical column of the table are the inches depth of water running over the notch of the weir; the top horizontal row of figures shows fractional parts of an inch, and the body. of the table shows the cubic feet and the fractional parts of a cubic foot that will pass each minute, for each inch and fractional inch depth of water in notch, from 1 to 25 inches. For example, if the depth of water were 6⅝ inches there would be 6.85 cubic feet per minute flowing over the weir, the number 6.85 being found in the ⅝ column opposite 6 in the left-hand column. Each of these results is for 1 inch only in width of weir; for any particular number of inches width of weir notch, the result obtained in the table must be multiplied by the number of inches of breadth or length the weir notch may be.

EXAMPLE.—The notch in the board is 20 inches wide, and the water at the stake 5¼ inches deep. How many cubic feet of water will flow over the weir per minute?

SOLUTION.—Take the figure 5 in the first vertical column, Table VII, and follow the horizontal line of figures until a vertical column is reached, containing ¼ at the top. At the intersection is found 5.18 cu. ft. This is the quantity of water passing for each inch in width; but the supposed weir is 20 in. in width; therefore, this result must be multiplied by 20, which gives 103.6 cu. ft. per min. Ans. The same method may be applied to any depth from 1 to 25 inches.

57. Measurement of Flow in Large Streams. Where it is impossible to construct a weir, the simplest method is to ascertain the mean velocity of the stream, in feet per minute, and its cross-section in square feet, then multiply these amounts together and thus obtain the required flow in cubic feet per minute. The velocity can be estimated by throwing floating bodies into the stream and noting the time required for them to pass over a measured distance between two lines a', b', Fig. 27. But it must be remembered that the velocity is greatest in the center of the stream, and near the surface, and that it is least near the bottom and sides. However, the velocity at the center can be measured and from this the mean velocity can be estimated, as it is known from reliable experiments that the mean velocity will be 83 per cent. or approximately four-fifths of the velocity of the surface.

The cross-section may be estimated by measuring the depth of the stream, at a number of points an equal distance apart, as illustrated in Fig. 27, at *a*, *b*, *c*, *d*, etc. The sum of these depths multiplied by the distance, in feet, between any two points gives the cross-sectional area of the stream. The product of the cross-section of the stream, in square feet, and the average or mean velocity of the water, in feet per minute, gives the quantity of water that the stream affords, in cubic feet per minute. The measurements at *a*, *b*, *c* must be in feet or fractions of a foot.

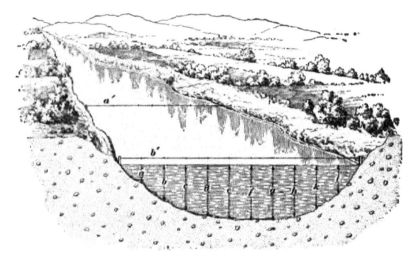

Fig. 27

58. Measurement of Head or Fall.—The next point necessary in the improvement of a water-power is to ascertain the amount of *head* or *fall*. The head of water is the difference between the level of the water in the forebay, or headrace, and the level of the water in the tailrace, or aqueduct, that carries the water from the wheel. The head is usually determined with an engineer's level, by running a level from a point at the upper level of water to a point at the lower level of water. The vertical distance between the two points is the standing head or fall. If the stream is liable to floods backing up the water in the tailrace, the difference between the normal level and flood level in the tailrace must be allowed for, since under such abnormal conditions the net value of the power will be reduced in proportion.

A head of water can be utilized in any of the following ways: by its weight, as in the overshot wheel; by its pressure, as in the turbine wheel; by its impulse, as in the undershot, Pelton, or other type of impulse wheel. The *gross power* of a fall of water is the product of the weight of water discharged in a unit of time, and the total head.

59. Estimation of Horsepower.—The horsepower of a stream may be found by multiplying the volume of flow per minute of the stream by 62.5 pounds, which is the approximate weight of water per cubic foot, and this product again by the head or fall, in feet, which will give the foot-pounds for the stream in question. This product divided by 33,000 will give the theoretical horsepower of the stream. This may be expressed as follows:

$$\text{Theoretical horsepower} = \frac{62.5\,VH}{33,000} \qquad (3)$$

where V = volume of flow, in cubic feet per minute;

H = effective head, in feet.

To determine the net mechanical horsepower that may be obtained at the shaft of the wheel, there must be deducted from the theoretical horsepower the loss in head by the friction of the water in flowing through the channels between the forebay and the wheels and the loss in the wheel itself, which will vary according to the efficiency of the wheel.

EXAMPLE.— 10,500 cubic feet of water are delivered each minute by a stream under a head of 50 feet. How many horsepower will be available at the shaft of the wheel if 5 per cent. is lost through friction before the water reaches the wheel and if the wheel has an efficiency of 75 per cent.?

SOLUTION.—From formula **3**, the theoretical horsepower will be equal to $\dfrac{62.5 \times 10,500 \times 50}{33,000}$ = 994 H. P., approximately. Of this, 5 per cent. is lost in friction so that the horsepower delivered to the wheel will be 994 × .95. The wheel has an efficiency of 75 per cent.; hence, the power delivered at the shaft of the wheel will be 994 × .95 × .75 = 708 H. P. Ans.

TABLE VIII

**HORSEPOWER PER CUBIC FOOT OF WATER PER MINUTE
FOR VARIOUS HEADS**

Heads Feet	Horsepower	Heads Feet	Horsepower
1	.0016098	320	.515136
20	.032196	330	.531234
30	.048294	340	.547332
40	.064392	350	.563430
50	.080490	360	.579528
60	.096588	370	.595626
70	.112686	380	.611724
80	.128784	390	.627822
90	.144892	400	.643920
100	.160980	410	.660018
110	.177078	420	.676116
120	.193176	430	.692214
130	.209274	440	.708312
140	.225372	450	.724410
150	.241470	460	.740508
160	.257568	470	.756606
170	.273666	480	.772704
180	.289764	490	.788802
190	.305862	500	.804900
200	.321960	520	.837096
210	.338058	540	.869292
220	.354156	560	.901488
230	.370254	580	.933684
240	.386352	600	.965880
250	.402450	650	1.046370
260	.418548	700	1.126860
270	.434646	750	1.207350
280	.450744	800	1.287840
290	.466842	900	1.448820
300	.482940	1,000	1.609800
310	.499038	1,100	1.770780

Table VIII gives the horsepower of 1 cubic foot of water per minute under heads of from 1 foot up to 1,100 feet, based on a wheel efficiency of 85 per cent. and exclusive of any loss due to friction after leaving the forebay.

DAMS

60. The construction of dams may be considered under four classes of work: (*a*) The *log*, or *brush, dam*, (*b*) the *frame dam*, (*c*) the *crib dam*, (*d*) the *masonry dam*. Under each class of construction there are many variations in detail. The first two classes are not worth considering in connection with the subject in hand.

In the building of dams the following six essential features must be thoroughly provided for:

(*a*) The foundation must be absolutely substantial and secure.

(*b*) The thickness and weight of the structure and its attachment to the foundation must be such as to withstand the pressure or thrust of the water in the reservoir above the dam.

(*c*) The shape of the dam must be such that it will not in any manner be injured by ice expansion when the water is frozen.

(*d*) Discharge from the slope or face of the dam must be such that the sheet of water flowing over will maintain a continuous and unbroken column free from atmospheric interferences; there must not be a broken flow followed by a rebound that causes vibration and tremulous motion in the dam.

(*e*) The shape of the dam must be such that the discharge of the water from the slope will not wear away the bed of the stream, or the footing at the base of the dam.

(*f*) The entire structure must be tight against leakage; capillary attraction must be avoided, or leaking currents of water may in time wear away the banks or undermine the foundations.

Fig. 28 shows a sectional view of a masonry-and-concrete dam. Improper discharge of water over the crest of a dam may set up swirling, eddy, direct, and reversed currents, which are sure to produce centrifugal and centripetal currents that will scour the banks and abutments and deepen the channel.

Arched dams have, by many, been preferred to straight dams, because of the mistaken notion that they obtain an additional support from their abutments, which is not a fact. In architecture, the weight of the arch and its load is thrown against the abutments, and they to a large extent support the masonry above and extending over the arch. With the arched dam the question is not one of gravity, but of force exerted horizontally. The horizontal pressure of the water

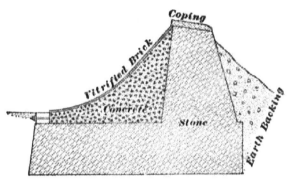

Fig. 28

is about the same at all points throughout the length of the arch and there can be no end thrusts. The water flowing from an arched dam is liable to create more friction than would occur with a longitudinal flow from a straight dam.

61. The first cost of a dam whether of stone or timber, its rapid impairment, the important part that it bears in relation to the whole enterprise, the cost of maintenance, and the uncertainty of its stability are matters of such vital importance that the work should be left to an engineer especially skilled in this line. The dam may be regarded largely as the very foundation of a water-power enterprise, and it is poor economy that does not employ the best skill in its design and construction.

VELOCITY OF FLOWING AND FALLING WATER

62. The theoretical velocity of water flowing from an orifice is the same as the velocity of a falling body that has fallen from a height equal to the head of water. The quantity of water discharged through the pipe or open channel depends on the head (that is, the vertical distance between the level surface of the still water at the entrance end of the channel and the level of the center at the discharge end), the length of the channel, the character of its interior surface as to smoothness, and the number and radius of the bends.

The total head operating to cause a flow is divided into three parts: the *velocity head*, which is the height through which a body must fall to acquire velocity; the *entry head*, or that required to overcome the resistance at the entrance to the channel; and the *friction head*, or the head required to overcome the frictional resistance within the channel or pipe. In ordinary cases where the work is properly designed, the sum of the velocity and entrance heads can be disregarded, as they rarely exceed 1 foot.

As already indicated, there are certain losses that must be deducted from the gross theoretical horsepower of a falling stream to determine the net mechanical horsepower available for useful work. Much will depend on the local conditions and the nearness of the power station to the forebay. If immediately adjacent, as shown in Fig. 37, the loss by friction through connecting flumes or penstocks will be almost negligible. If the water is conducted from the forebay to the wheels through a closed pipe or flume, the loss of head in a pipe increases directly with the length, with the square of the velocity, and with the roughness of the pipe; it decreases as the diameter of the pipe increases, and is independent of the pressure or head of water.

LOSS OF HEAD DUE TO VELOCITY IN OPEN CHANNELS

63. When a stream leaves the still water of a lake or reservoir at a given velocity, there will be a certain loss of head to generate that velocity; that is to say, the stream in the conduit must be lower than the still water in the reservoir in order to create the velocity required. The velocity will vary at the entrance according to the shape of the entrance and the position of the gate, and there is a loss at the entrance due to contraction at the sides of the aperture. When the channel is long, there is not only a loss of head due to the velocity, but also a loss by friction against the sides and the bottom. If the channel is maintained of equal cross-sectional area from end to end the loss of head increases uniformly throughout the length, but if the cross-sectional area of the channel is increased from the entrance to the discharge end, the friction will be decreased.

The discharge of open-water courses may be determined experimentally by observing the velocity of the current, and measuring the cross-sectional area of the stream. To do this correctly requires the mean velocity throughout the section, which is not obtainable by observation and must be determined as described in Art. **57.**

64. Influence of Inside Surface of Flumes or Pipes.—The shape and inner surface of the channel, conduit, or pipe through which the water flows, and which, by friction, reduces its velocity, is a highly important factor in the development of a water-power. Extreme care should be observed to have these surfaces as smooth as practicable, and with bends the fewest possible and of long radius. Channels having their inner surfaces made of well-planed timber in perfect order and alinement, of glazed or enameled stoneware, iron pipes, cement finish with stone surface in good condition, unplaned timber, well-laid brickwork, rough-faced brickwork, well-dressed stonework, or rubble masonry in cement, will afford friction increasing in the order in which the surfaces have been given. Furthermore, a channel or sluiceway composed of earth with the banks trimmed uniformly and laid smooth and in good order will offer more friction than any of the surfaces above mentioned, and as these earthen banks are obstructed by stones, rocks, weeds, and other surfaces of a rough character, and as they become somewhat disintegrated with alternate freezing and thawing in northern latitudes, the friction will be increased and the velocity of the flow of water reduced.

65.   Loss of Head in Smooth Straight Pipes.—The loss of head, expressed in feet, due to the flow of water through smooth straight pipes may be found approximately by means of the formula

$$h = \frac{k L v^2}{d} \qquad (4)$$

where h = loss of head, expressed in feet;

  L = length of pipe, in feet;

  v = velocity, in feet per second;

  d = diameter of pipe, in inches;

  k = .0056 for pipes up to 6 inches diameter; .0047 for pipes between 6 inches and 21 inches; .0037 for pipes between 21 inches and 48 inches; .0028 for pipes between 48 inches and 72 inches; and .0019 for pipes larger than 72 inches.

This formula will give approximate results within at least 10 per cent. of correctness for smooth straight pipes.

Example.—What will be the loss of head in 500 feet of 36-inch pipe through which water is flowing at the rate of 10 feet per second?

Solution.—We have $L = 500$, $v = 10$, $d = 36$, and $k = .0037$;

$$h = \frac{.0037 \times 500 \times 10^2}{36} = 5.14 \text{ ft.   Ans.}$$

WATERWHEELS

66. Waterwheels may be divided into the following classes: (*a*) *Overshot wheels*, (*b*) *undershot and breast wheels*, (*c*) *turbines*, (*d*) *impulse wheels*. The first two classes may be passed over without comment, as they are seldom, if ever, used in modern electric water-power plants.

Overshot wheels have been known to give an efficiency of 75 per cent., but the average performance does not exceed 60 per cent. The efficiency of turbines will range, according to their design and the speeds at which they are operated, from 50 to 75 per cent., very rarely reaching as high as 83 per cent. The *draft tube*, or *suction tube*, taking the discharge between the turbine and the tailrace, helps to improve the efficiency, and its effect is included in the above figures. The falling column of water in the draft tube creates a suction on account of the vacuum formed, but perfect vacuum is not obtained. Impulse wheels of the Pelton or similar types are claimed to have an efficiency of 80 to 85 per cent.

TURBINES

67. In the **turbine waterwheel,** the water acts on a series of curved vanes and the rotating part is acted on by a continuous flow of water through the turbine. The passages between the wheel vanes are always completely filled with water and the forces that act on the wheel vanes are: first, a certain amount of static pressure; second, the pressure caused by the change in direction of the moving water; and third, a pressure due to the reaction of the water as it issues from the wheel vanes. In most cases, the greatest of these forces is the pressure caused by the change in the direction of the moving water in its passage through the wheel.

The turbine wheel is a thoroughly reliable hydraulic motor, and under practical use for many years has been improved to such an extent as to incorporate the largest durability with high efficiency and the greatest ease of operation. Turbines are made in many different forms; they may be arranged for vertical or horizontal shafts and connected single or double on the same shaft; they may or may not be enclosed in casings and may be used with or without draft tubes.

EXAMPLES OF TURBINES

68. Turbines are frequently classed according to the direction of the flow of water through them. For example, the water may enter parallel to the shaft and be discharged parallel to the shaft also; such wheels are known as *parallel-flow*, or *axial-flow*, *turbines*. Or, the water may flow in at the center and be discharged outwards at the circumference of the wheel, constituting what is known as an *outward-flow turbine*. In others, the flow may be radial but reversed in direction, flowing from the circumference to the center and making an *inward-flow turbine;* in many American turbines, the vanes are so shaped that the water is discharged in an axial direction, thus making a *mixed-flow turbine.*

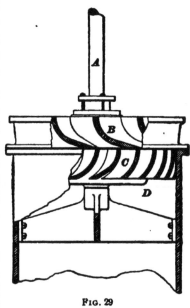

FIG. 29

69. No matter what the direction of flow may be, turbines consist of two essential parts: first, a series of curved guide vanes, which are fixed and serve to guide the water to the wheel in the most efficient manner so as to deliver it in a direction where it will act to the best advantage; and second, a wheel, or *runner*, which also carries curved vanes that change the direction of the water passing through the wheel and take up the pressure, which is effective in driving the wheel around.

Fig. 29 illustrates the essential parts of a vertical-shaft, axial-flow turbine. The water flows in, from above, through the fixed guide vanes B, acts on the curved vanes of the runner C, and is discharged into the draft tube D. The general direction of the water is always parallal to

the shaft A and the spaces between the vanes is at all times filled with water.

Fig. 30 shows the runner for a *Risdon turbine*, which is of the mixed-flow type. The vanes have a double curvature, and the water is delivered to the upper part in a radial direction by means of suitable guide vanes, but owing to the curvature of

FIG. 30

the runner vanes it is discharged in an axial direction. The band a serves the double purpose of strengthening the wheel and of making the proper form for the passage of the water through the lower part of the wheel, confining it on all sides.

70. Fig. 31 shows a single horizontal turbine intended for placing in an open penstock. The water is delivered in a radial direction between the guide plates a, a; these plates are hinged and the amount of flow can be regulated by varying the opening between them. The opening or closing of the guide plates a, a is effected by turning the shaft b. The water is discharged from the wheel in an axial direction and passes through the casing c to the tail-race or draft tube.

71. Fig. 32 shows two horizontal turbines with closed penstock. The water is delivered through the pipes a, a to the closed penstocks, or casings *b, b,* and having passed through the wheels is discharged into the draft tubes *c, c.* The governors *d, d* regulate the gate opening so as to keep the speed approximately constant.

72. Draft Tubes.—The draft tube is the discharge pipe leading from the turbine to the water level in the tailrace. The descending column of water in this tube creates a vacuum ʼ or suction (if all is tight against air leaks) and thereby makes the effective head of water acting on the wheel the same as if the wheel were placed at the level of the water in the tailrace, or it augments the power obtained over and above that

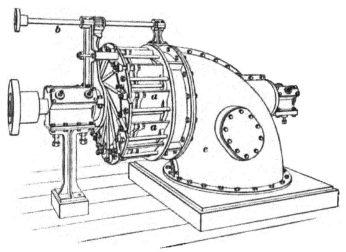

Fig. 31

which would be obtained if the wheel were placed above the lower water level and provided with an open discharge. In fact, the draft tube has a relation to the turbine wheel somewhat similar to that of a condenser to an engine. The end of the draft tube must, of course, always be below the surface of the water in the tailrace.

Theoretically, the turbine wheel with draft tube could be set at an elevation of 33 feet above the level of the water in the tailrace, the end of the draft tube being submerged sufficiently to prevent the air entering the tube, and thereby displacing the head-pressure, but practically, it is found

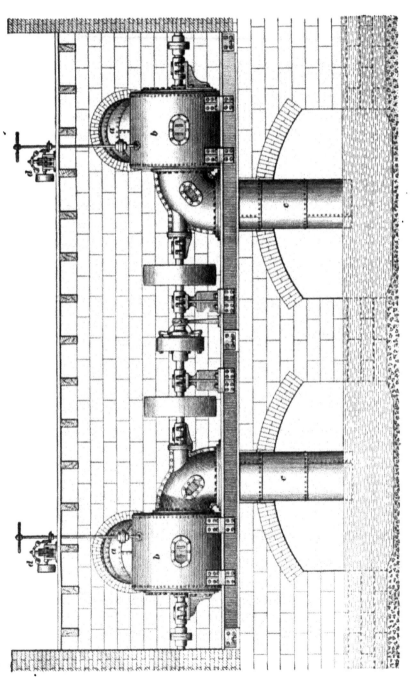

FIG. 32

desirable to set the wheels at an elevation not exceeding 18 feet above the level of the water in the tailrace.

73. Relation Between Head and Discharge.—With turbines of the same diameter and vent under different heads, the speed and discharge increases directly as the square root of the head. For example, if the head is made four times as great, the speed will be $\sqrt{4}$, or twice as great, and the wheel will discharge twice as much water per hour. The horsepower increases with the increase in head and also in proportion to the discharge. In the above case the head is increased fourfold and the amount of water discharged is doubled; hence, the horsepower will be eight times as great. A turbine that under 15 feet head gives 40 horsepower and vents 1,760 cubic feet at 160 revolutions per minute, should under 60 feet head, or an increase of four times, discharge $1,760 \times \sqrt{4} = 3,520$ cubic feet of water at 320 revolutions and develop eight times as much power, or 320 horsepower.

74. Turbine wheels are ordinarily used for heads of water ranging from 6 to 175 feet. For low and medium heads, wheels of ordinary weight and strength are sufficient, but for heads above 50 feet it is frequently necessary that the wheels shall be designed and built especially for each individual installation. It is desirable in applying turbines for electric power station work that they should be of the horizontal type and arranged as to speed and position in the station for direct connection to the shaft of the generator. It is extremely undesirable to introduce any intermediate pulleys, gearing, or belting, as the loss of power through such will vary from 10 to 30 per cent.

EXAMPLES OF TURBINE INSTALLATIONS

75. Fig. 33 shows a typical arrangement of horizontal turbines direct-connected to alternators. This arrangement is for a low head and in order to avoid using a large number of small generating units, each generator is driven by four turbines a, b, c, d, mounted on a common shaft extending

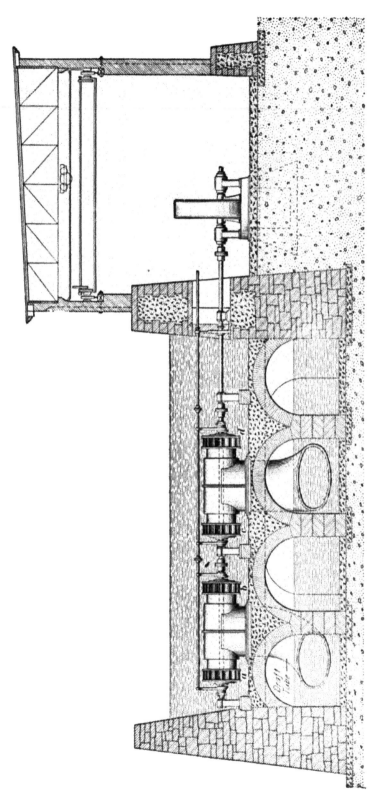

F_{IG.} 38

through the wall or bulkhead into the dynamo room. Fig. 34 shows a somewhat similar arrangement. The total head is 27 feet and there are three wheels coupled to each generator. The bottom of the wheel casing is 10 feet 3 inches above the level of the tail-water, and draft tubes are provided as shown.

76. Figs. 35, 36, and 37 show three views of a station equipped with horizontal turbines direct-connected to 650-kilowatt alternators and operating under 28 feet head. In

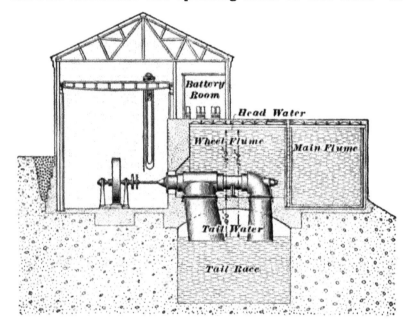

FIG. 34

this installation the wheels are encased in water-tight casings made of boiler plate and are placed in a wheel house. Water flows from the canal, through the end of the casing, and, passing through the turbines, flows through the draft tubes. The exciters *a, b* are driven by a small independent turbine *c*, so that their speed is constant irrespective of fluctuations in the speed of the main wheels, thus securing better voltage regulation than where the exciters are driven from the main wheels. .

Coal Bin.

 Iron
Platform.

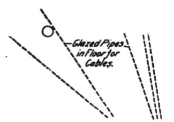

Glazed Pipes
in Floor for
Cables.

FIG. 85

FIG. 86

77. As a prominent example of vertical turbine installation, the plant at Niagara Falls may be chosen. Fig. 38 shows a cross-section of the later power house and wheel pit. The

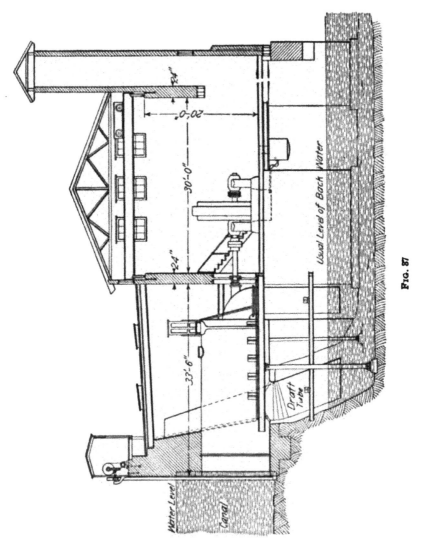

Fig. 87

water is taken from the canal *a* and flows down the vertical penstock *b* to the wheel *c*; after passing through the wheels, it flows down through the draft tubes *d* and off through the tunnel at *e*. In the first installation at Niagara, draft tubes

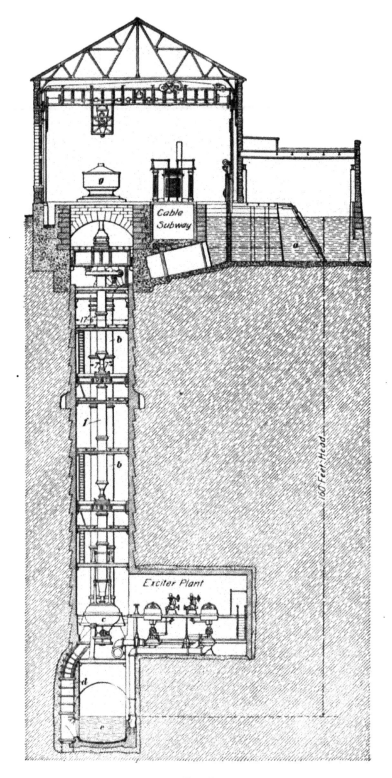

Fig. 38

were not used. The vertical shaft *f* is in the form of a large, hollow, steel tube, except at the bearings, and on its upper end carries the revolving field of the alternator *g*. The alternators are of 5,000 horsepower capacity at 250 revolutions per minute, two-phase, 25 cycles, and wound for 2,300 volts. In order to avoid long shafts, the independently driven exciters are located in an underground chamber and are driven by small wheels supplied from the main penstock. The head of water from the surface of the canal to the surface of the water in the tunnel is in the neighborhood of 160 feet.

IMPULSE WHEELS

78. Where water-power of any considerable head is obtained, say from 100 feet upwards, it frequently becomes desirable to use a type of wheel different from the turbine. A successful development in waterwheels for extra high heads is the *Pelton wheel*, which will serve as an example of **impulse wheels** in general.

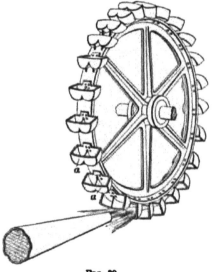

Fig. 39

The **Pelton wheel**, Fig. 39, consists of a series of peculiarly shaped buckets *a, a* mounted on the rim of a wheel. The buckets are wedge-shaped in the center for the purpose of dividing the stream and deflecting it backwards so as to develop its full force. While passing out from the bucket, the water sweeps against the curved sides and gives the effect of a prolonged impact. It is thus deflected to each side from the course of the wheel and offers no resistance to the rotary motion. In Fig. 39, the wheel is shown with the surrounding casing removed. In most electric power stations where these wheels are used, they are mounted on an iron framework and surrounded by an iron casing. Wheels of the Pelton type may be fitted with one or more sets of buckets on the same shaft, each set of buckets being supplied from a separate nozzle, the flow through which can be regulated by a valve. Where automatic regulation is desired the valve, through a series of suitable connections, can be controlled by a waterwheel governor. The wheel

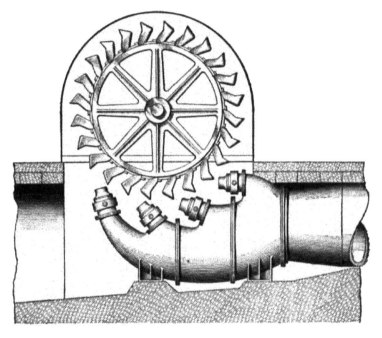

FIG. 40

is naturally of a high-speed type and is particularly well adapted to direct-connected generators. It is claimed that impulse wheels of this type, under favorable circumstances, can develop an efficiency of 85 per cent.

79. Impulse Wheel for Low Heads.—Fig. 40 illustrates an impulse wheel driven by water delivered through multiple nozzles. As all the streams have a distinct line of impingement, they do not conflict and there is therefore no appreciable loss of efficiency. By this means adaptation can be made to almost any requirement as to power for heads ranging from 25 feet upwards. Each nozzle has an independent gate valve to facilitate regulation and adapt the wheel to varying supplies of water. Where automatic regulation is desired, the valves, by suitable connections, can be controlled by one governor.

Fig. 41 shows a Pelton wheel, with the top half of its casing removed, designed for dynamo driving and provided

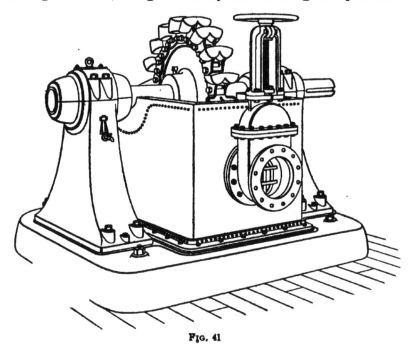

Fig. 41

with a base and self-oiling bearings of the same general style as used for dynamos.

The impulse wheel can be successfully used under vertical heads of water ranging from 25 feet to 2,500 feet, though its chief use lies in connection with heads over 100 feet. A remarkable installation has been made near Virginia City, Nev., where a wheel of 36 inches diameter and made of a solid steel disk with buckets riveted to the periphery, is used. This wheel operates under a vertical head of 2,100 feet of water equal to 911 pounds pressure per square inch. The wheel runs at 1,150 revolutions per minute, giving a speed at the circumference of 10,804 feet per minute, or over 120 miles per hour. The water issues through a ⅜-inch nozzle and the wheel develops 100 horsepower.

OPERATION OF IMPULSE WHEELS

80. The function of a waterwheel operated by a jet of water escaping from a nozzle is to convert the energy of the jet, due to its velocity, into useful work. In order to utilize this energy to the fullest extent, the shape of the wheel bucket must be such that, after receiving the jet, the water will be brought to rest before discharge. This, of course, cannot be fully effected, and unavoidably necessitates the loss of a portion of the energy. The principal losses occur as follows: First, in sharp or angular diversions of the jet in entering, or in its course through the bucket, causing impact, or in the conversion of a portion of the energy into heat instead of useful work; second, in the so-called frictional resistance offered to the motion of the water by the wetted surfaces of the buckets, causing also the conversion of a portion of the energy into heat instead of useful work; third, in the velocity of the water, as it leaves the bucket, representing energy that has not been converted into work. Hence, in seeking a high efficiency the following features must be provided for: First, the bucket surface at the entrance should be approximately parallel to the relative course of the jet and the bucket should be curved in such a manner as to avoid sharp, angular deflection of the stream. If, for example, a jet strikes the surface at an angle and is sharply deflected, a portion of the water is backed, the smoothness of the stream is disturbed, and there results considerable loss by impact and otherwise. The entrance and deflection of the water in the Pelton bucket are such as to avoid these losses in the main. Second, the number, of buckets should be small, and the path of the jet in the bucket short; in other words, the total wetted surface should be small, as the loss by friction will be proportional to this.

Third, the discharge end of the bucket should be as nearly tangential to the wheel periphery as is compatible with the clearance of the bucket that follows, and great differences of velocity in the parts of the escaping water should be avoided.

WATERWHEEL GOVERNORS

81. For satisfactory service in water-driven electric power stations, uniformity of speed has required the development and application of types of **governors** designed on theoretically correct lines, and fully adapted to all the requirements of perfect regulation under sudden load variations. The governing of waterwheels for such service has been a difficult problem, and has required a complete revolution in the design of governors, which were previously suitable for manufacturing purposes only. The result has been accomplished to such a point of success that the regulation is equal to the best steam practice, and in some instances, particularly for generators of the alternating-current type, is eminently satisfactory. Less difficulty is experienced in running waterwheel-driven alternators in parallel than those driven with reciprocating engines.

In connection with waterwheel regulation, the difficulty has been to produce a governor that would accomplish the required results without allowing the wheel to race or hunt. Most of the earlier types consisted of a flyball governor, similar to that used on a steam engine, arranged so as to control the opening and closing of the gates. When the speed rises above normal, the balls fly out, thus throwing into action the appliances necessary for closing the gate. If the speed falls below normal, a reverse action opens the gates. With this type of governor, if the load is, say, suddenly reduced, the balls fly out beyond their normal position and the gate opening is reduced. Owing, however, to the inertia of the various parts, the speed does not drop at once and before the speed has come back to its normal amount, and the governor balls assumed their normal position, the gates have been closed too much and the speed drops below the normal. In other words, the regulation overshoots the mark and there is a seesawing or hunting action until the speed finally settles down to the normal. This hunting or racing action is not so objectionable where the wheels are used to run ordinary machinery, but it cannot be tolerated where the wheels are used to drive alternators. In a number of the more recent types of governor the effects of inertia and consequent hunting are overcome. There are a number of successful types of governor on the market, but for purposes of illustration we will select the *Lombard governor*, which has been widely and successfully used in electric water-power plants where the waterwheels are direct-connected to alternators and where the requirements as to speed regulation are, therefore, very exacting.

LOMBARD GOVERNOR

82. In the **Lombard governor**, Fig. 42, the movement of the gate is effected by means of the oil or water pressure acting on a piston, and the admission of oil or water to the operating cylinder is controlled by a centrifugal governor, the action of which is modified by the gate movement in such a manner as to compensate for the effects of inertia. Water pressure is used for moving the gates only in those cases where the wheels are operated under high heads and where a high water pressure is thus easily obtained. Most of the Lombard governors are operated with oil pressure; that shown in Fig. 42 is known as the *Type B governor*, and is the kind used for most installations operating under low or moderate heads.

83. In Fig. 42, the oil used for operating the governor is stored in the tank located under the governor. This tank is divided by an air-tight partition into two compartments *1, 2*; compartment *1* is smaller than *2*, the partition being at the point indicated by the row of rivet heads. The larger compartment is partly filled with thin, petroleum. engine oil and over this oil is air under a pressure of about 200 pounds per square inch. A vacuum is maintained in compartment *1* and

air is compressed into *2* by means of a pump *3* driven by a
pulley *4* belted to the waterwheel shaft. Gauges *5* and *6*
indicate the pressure and vacuum, respectively.   A pipe *7*
leads from the oil in *2* to the oil-filled valve *8*, which is
opened or closed by the centrifugal governor *9*. This governor is belted to the waterwheel by means of pulley *10*. A

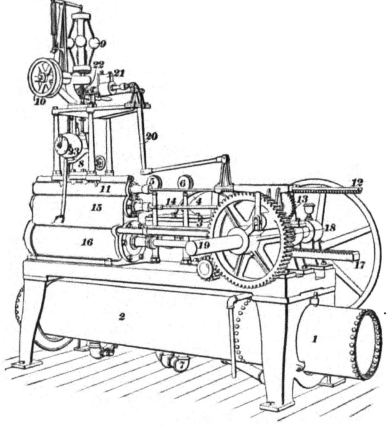

small, balanced, piston valve located at *8* allows oil under
tank pressure to flow into either end of the small horizontal
cylinder *11*.   The piston and piston rod in this cylinder
terminate in a rack *12* that runs on top of a floating gear *13*.
This gear has no fixed center, but to its axis is attached the
piston rod *14* that actuates a large piston valve in cylinder *15*.

This valve, according to its position, allows oil to flow into either end of the largest cylinder 16 in which is a piston attached, through its piston rod, to the large rack 17. This rack engages with a gear 18 on shaft 19, which is connected to the gate-operating mechanism of the wheels. One complete inward stroke of the main piston and rack closes the gates, while a complete outward stroke gives a full-gate opening. When the wheel is running at normal speed, valve 8 occupies a neutral position and there is no movement of the governor. Suppose that part of the load on the wheel goes off suddenly. The speed will immediately begin to accelerate and the balls of the governor 9 will fly out, thus depressing the valve at 8. This admits oil into the rack end of cylinder 11 and its piston travels inwards, carrying with it rack 12; the floating gear 13 that meshes with 12 also travels inwards, carrying with it the rod 14 and opening the valve in 15. This admits oil to the front end of the cylinder 16 and allows it to flow out of the back end, thereby moving rack 17 inwards. Now, as soon as rack 17 moves inwards, it turns gear 19 and this in turn rolls the floating gear outwards, thus bringing the valve in cylinder 15 to its central position. This stops all motions of the governor and the gates are at the correct position for the new value of the load.

The above actions have been described as occurring consecutively, but as a matter of fact they occur so rapidly, one after the other, that the action appears almost simultaneous. The oil discharged from cylinders 11, 15, and 16 is returned to the vacuum tank 1 from which it is returned to tank 2 by means of pump 3. It will be noted that no air circulates through the cylinder of the governor. The compressed air on top of the oil in tank 2 acts simply as a reservoir of energy, which is instantly available for the operation of the governor. If the load on the wheels suddenly increases, all the various motions above described are reversed.

84. The governor as above described would not give satisfactory regulation. Whenever piston valve 8 is moved out of its normal position by variations in the speed of the balls, it would remain out of position until the balls came back to their normal position. All the while that valve 8 is out of its central position, the main piston and rack will be traveling in one direction or the other. As before stated, the centrifugal balls do not respond instantly to change in speed and the result is that the governor overshoots the mark. In the Lombard governor this defect is overcome by an attachment that centralizes valve 8 before the time that it would be centralized by the balls coming back to the normal position. This auxiliary movement of the valve stem is controlled by the gate movement through the lever 20 attached to rack 12. Lever 20, through dashpot 21, operates a small rack and pinion 22 by means of which the valve stem of valve 8 is raised or lowered independently of the movement due to the balls, and the movements of valve 8 are thus regulated so that by the time the balls resume their normal position, the gate opening is such that the wheel runs at the proper speed and there is no hunting or seesawing. In order to allow the speed of the wheel to be regulated from a distant point, for example, from the switchboard, a small motor 23 is geared to the valve stem of valve 8; this motor is reversible so that the stem can be raised or lowered at will when it is necessary to adjust the speed. This attachment is particularly convenient when alternators are to be synchronized.

REQUIREMENTS FOR GENERAL POWER-STATION BUILDINGS

85. In conclusion, it may be stated that the design of power-station buildings and the arrangement of apparatus therein admit of wide variation. No set rules can well be laid down to govern all classes of work. The capacity of the station, the types of equipment selected, and the environment of the plant will all have a bearing thereon.

There are, however, certain leading points that should always be kept in mind, and these may be summed up as follows:

(*a*) Substantial foundations.

(*b*) A fireproof building.

(*c*) Ample working space around all apparatus.

(*d*) Such a convenient arrangement of apparatus as will require the minimum of labor for its proper attendance.

(*e*) Doorways of ample width and height to admit of taking in and getting out the largest pieces of equipment.

(*f*) In front of boilers, ample space to withdraw and replace tubes.

(*g*) Around engines, ample space to take out pistons, remove and replace shafts, wheels, generators, etc.

(*h*) In engine rooms, a traveling crane or other suitable means for handling heavy parts with safety and minimum of labor.

(*i*) Ample facilities for light and good ventilation.

(*j*) Suitable lavatory accommodations for employes, according to sanitary rules.

(*k*) Such design of buildings and placing of apparatus as will admit of future extensions without destroying existing plant or interfering with its operation.

(*l*) Avoid architectural monstrosities, to the end that the character of the finished building may show the purpose for which it was designed.

TELEGRAPH SYSTEMS

(PART 1)

TELEGRAPHY

NOTE.—While it is essential for an electrical engineer to understand the principles involved in telegraph systems and apparatus, it is not necessary for him to know about all the systems and apparatus used, nor their practical manipulation, adjustment, and care; this work belongs to the telegraph operator and telegraph engineer. Therefore, only the principles of the systems and apparatus most extensively used in telegraph systems will be considered.

1. Electric telegraphy is the art, science, or process of transmitting intelligible signals or signs between distant points by means of electric impulses apparently moving between the two points. Messages may be transmitted in this manner by audible or visible signals, both methods being largely used. A simple Morse telegraph system, patented by an American, Samuel F. B. Morse, in 1837, consists of a device, called a *key*, for opening and closing an electric circuit, a *receiving device*, by means of which electric current impulses may be recorded or heard, a *line wire* connecting the two stations, and a *battery* for supplying the electric current required.

The **key** is an instrument for manually opening and closing the circuit in order that various combinations of long and short current impulses may be produced in the circuit. For receiving audible signals an instrument called a **sounder** is extensively used. It consists of an electromagnet and a pivoted armature adapted to give forth sounds corresponding to the currents produced in its magnet coils.

In the simplest form of the Morse telegraph system, there is usually one key and one sounder at each station and there may be one battery at each station, or merely one battery for all stations. The apparatus may be arranged in one of two ways, known as the *Morse closed-circuit system* and the *Morse open-circuit system*.

MORSE CLOSED-CIRCUIT SYSTEM

2. The arrangement of the line and apparatus shown in Fig. 1 constitutes the simplest form of the **Morse closed-circuit system.** It is called the closed-circuit system because when no messages are being sent all keys are closed and the battery is connected in the line circuit, causing current to be normally flowing through the line circuit. *L* represents the line wire connecting the two stations *W* and

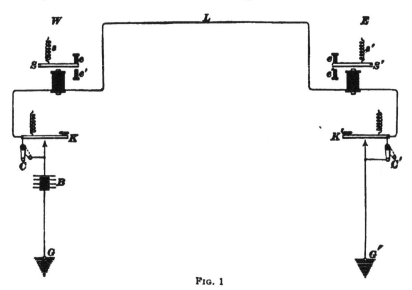

Fig. 1

E; *K*, *K'* are keys; *B* is a battery; *S* and *S'* are sounders for receiving a message. *G*, *G'* are points where a wire is connected with the earth, or *grounded*, as it is usually termed.

The circuit is traced as follows: When both keys *K*, *K'* are closed, the current starts from the plus pole of the battery *B* and passes through key *K*–sounder *S*–line wire–sounder *S'*–key *K'*–*G'*, through the earth to *G* and the minus pole of battery B. The earth is commonly used instead of a return wire, and may, for all practical purposes be considered as a conductor of very small resistance, for, although it is comparatively a poor conductor, its practically unlimited cross-sectional area renders its resistance usually negligible in comparison with that of a long line wire.

When both keys are closed, a continuous current will flow around the circuit, so that the electromagnets of the sounders will attract their armatures. If, now, one key is opened the current will be interrupted; the electromagnets will therefore release their armatures and allow them to be drawn upwards by the springs s, s'. If the key is closed again, both armatures will be drawn down as before. The motions of the armatures are limited by the stops e, e'. The downward movement of the armature causes a sound distinguishable from that made by the upward movement; these movements are called the *down* and *up* strokes, and by certain combinations of these strokes messages are received by sound.

When the line is not in use the circuit is left closed, for which purpose small switches C, C' are provided for every key. Closing one of these switches accomplishes exactly the same result as depressing the key. It saves the operator the inconvenience of having to hold down his key when he wishes to leave the circuit at his station closed for some time. In Fig. 1 the switches C, C' are in the positions to enable the operator at station W to send to station E. The switch at the receiving station must be closed, and that at the sending station open. When the line is idle, both switches must be closed. The necessity for these rules concerning the use of the switches will be appreciated by supposing that the operator at E has left his switch open. It will then be impossible for the operator at W to control the sounder at E, for, no matter whether the circuit is open or closed at W, it is open at E, and therefore no current can pass through the sounder at E.

THE RELAY CIRCUIT

3. When the telegraph line is more than about 30 miles in length, it becomes difficult to render the current in it strong enough to satisfactorily operate the somewhat heavy armatures of the sounders. An instrument called a **relay** is then inserted in the line circuit in place of the sounder. The relay has a very light armature and usually its coils are wound · with a greater number of turns of finer wire than a sounder. The armature carries a contact point that opens and closes a local circuit containing a sounder and a local battery.

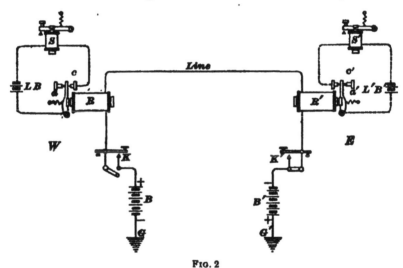

FIG. 2

4. The arrangement in which relays are employed at two telegraph offices W and E is illustrated in Fig. 2. In this figure, R, R' are the relays; K, K', the keys; B, B', the main-line batteries; LB, $L'B'$, the local batteries; S, S', the sounders; and G, G', the ground connecting plates. B, B' are called the *main-line batteries* because they are connected directly in the main or line circuit, while LB, $L'B'$ are connected with the sounders in circuits that do not go outside of the office and hence are called *local batteries*, and the circuits containing them are called *local circuits*. A current starting from the battery B passes through the key K, relay R, the line, relay R', key K', battery B' to the earth through the ground plate G', and returns through the earth and the ground plate G to the battery B. It should be noticed that the batteries B, B' are in series with each other, B having its negative pole to earth and its positive pole to the line, while B' has its positive pole to earth and its negative pole to the line. When the circuit is closed, both relays are energized and attract their armatures a, a'; and when the circuit is opened, both relays lose their magnetism and release their armatures. The armatures a, a' therefore make and break the two local circuits at contacts c, c', and thus act as keys in the local circuits, each of which contains a sounder and a battery.

As many cells of battery as are necessary may be used in the local circuit at $L\,B$, so that the sounder may be made to produce a loud sound, even though the current in the line wire is exceedingly feeble. One cell is often sufficient for this purpose, but it is customary to use two cells. Except in large offices, where dynamos or storage batteries are used, all current for both main-line and local circuits is usually obtained from gravity, or crowfoot, cells.

5. When the telegraph operator at one office desires to send a message to the other, he interrupts the flow of the current by opening the switch of his key. This causes the relays to lose their magnetism, release their armatures, open both the local sounder circuits, and cause the sounders to click. Now, if he operates his key by closing and opening the circuit so as to form the characters representing the letters of the alphabet in the order in which they occur in a message, the armatures of his own and the distant relays, as well as those of the sounders controlled by them, will respond to every make and break in the circuit caused by operating the key, and, consequently, the message may be read by ear from the clicks made by the sounders and the receiving operator writes it down as fast as it comes to him. Since the sending operator's sounder also responds and gives out the message, the receiving operator may evidently interrupt him at any time by opening the line circuit at his key, and thus stop the flow of current through the relays. This causes the sending operator to realize that the circuit has been broken, because his own sounder no longer corresponds to the movements of his own key. He then closes his switch to give the original receiving operator an opportunity to communicate with him.

The Morse closed-circuit system is used all over the United States, Mexico, and Canada, except for submarine-cable telegraphy.

THE LINE CIRCUIT

6. Distributed Electrostatic Capacity. — Underground wires, submarine cables, and air lines form condensers, and may be likened to a Leyden jar, as follows: The wire or cable corresponds to the inside conductor, the insulation or atmosphere corresponds to the glass, and the earth corresponds to the outside conductor. If the circuit is metallic, one wire forms one plate of the condenser, the air between the two wires is the dielectric, and the other wire forms the other plate. The capacity of a line is distributed throughout its entire length, and is therefore termed **distributed capacity**. Each element or short piece of the line wire may be considered as forming one plate of a condenser, the other plate of which is formed by corresponding portions of surrounding conductors and the ground. The line circuit may therefore be considered as an infinite number of small condenser plates.

When an electromotive force (E. M. F.) is applied to a line wire, a current will be sent over the line, but a portion is consumed along the line in charging it. Although the portion of the line nearest the battery will be charged first, nevertheless the whole line, if not too long, will be charged in a very short time.

However, it must not be imagined that this slowness of the distant part of the line to take its charge is due to the speed at which an electric wave may travel along a conductor, for this speed is practically equal to that of light,

186,000 miles per second, and, on the longest line obtainable, the time necessary for an electric impulse to flow through it is almost too small to measure. It should rather be looked at in the following light: The amount of electricity, in coulombs, that will flow through a conductor depends on the number of amperes flowing and on the length of time the current continues to flow. The charge of a condenser may be measured in coulombs, 1 coulomb being that amount of electricity represented by a flow of 1 ampere for 1 second. Obviously, here is a time element that is not dependent on the actual velocity of electricity. If 1 ampere flows into a condenser for $\frac{1}{2}$ second, the charge assumed by the condenser during that time will be $\frac{1}{2}$ coulomb, and in $\frac{1}{4}$ second will be $\frac{1}{4}$ coulomb. Similarly, the amount of electrical energy that can flow through a conductor depends on the strength of the current, the voltage, and the time of the flow.

If an E. M. F. is applied to a line only a brief time, enough electricity may not flow into the line to charge the whole line; hence, the near portion of the line is charged first, because it offers less resistance to the charging current than the more distant portion of the line. The time during which the E. M. F. is applied may be so short that the distant end of the line is not appreciably charged, in which case no appreciable current will flow through apparatus connected between the distant end of the line and ground.

7. The C R Law.—From the foregoing, it may be seen that the length of time necessary for an impulse of current to reach the distant end of a line depends not only on its distributed capacity C, but also on its resistance R. It has been proved by extensive experiments in telegraphy that the length of time required for a current to reach its maximum strength at the distant end of a line varies directly as the product of its capacity C and resistance R. Since both the capacity and the resistance are proportional to the length of the line, or cable, it follows that the product CR, which is called the **time constant** of the line, or cable, increases as the square of the length of the line, or cable.

INTERMEDIATE OFFICES

8. A number of intermediate telegraph offices may be connected in the same circuit with the two terminal offices. Fig. 3 shows one intermediate office I connected in the line between the two terminal offices W, E, with one-third of the whole number of main-line cells at each office. The keys, relays, and also the intermediate batteries, wherever used, are connected in series in the same line circuit.

All the cells may be located at one terminal station, as in Fig. 1, or one-half the number may be at each end station, as in Fig. 2, or the cells may be distributed, some being placed at each station, as shown in Fig. 3. Where several sets of cells are used, the cells in each set must not only be connected in

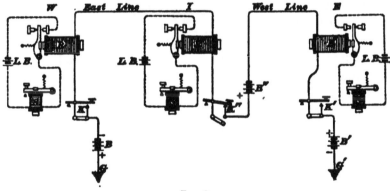

FIG. 3

series with one another, but the various sets must all be connected in series in the circuit and not opposing one another.

As many as thirty or forty intermediate offices are sometimes connected in a single circuit, but twenty instruments are probably as many as should be placed in a single circuit to work advantageously. Of course, only one of the operators can be sending at one time, but all the others may receive the message. The message may be of interest to only one or two offices out of the thirty or forty on the line, but all the other offices have to remain idle until the one sending is through, or else interrupt him if the other business is so much more important that it is allowable to do so.

9. Intermediate Batteries.—It is not very often necessary to connect batteries in the line at small intermediate stations. The best arrangement is to have an equal number of cells at each terminal station. When one terminal station is large and well equipped with dynamos, which are now rapidly coming into use for supplying the current for telegraph lines, and the other station is not so well equipped, it may be advantageous to let the former station supply all the current.

MORSE OPEN-CIRCUIT SYSTEM

10. In Europe, what is known as the **Morse open-circuit system** is used. This system, with two terminal offices and one intermediate office on one line, is shown in Fig. 4. *R*, *R'*, and *R''* are the relays; *S*, *S'*, and *S''*, the sounders; *B*, *B'*, and *B''*, the main-line batteries; *K*, *K'*, and *K''*

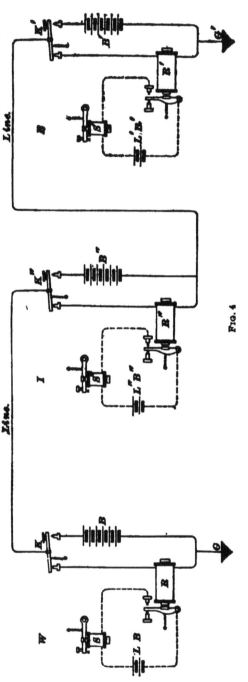

FIG. 4

the operator's sending keys; and LB, $L'B'$, and $L''B''$ the local batteries for operating the local sounders at the W, E, and I offices, respectively. It should be noted that the like poles of the batteries B, B', and B'' are connected to the front contacts of the keys, for the sake of uniformity. This is not necessary, however. When no message is being sent all the batteries are on open circuit, although all the relays are connected in series in the circuit. Thus, normally, no current flows through the line or through the local sounder circuits, and the batteries are all on open circuit, from which fact it derives its name. When a message is to be sent, the sending operator closes his key, the battery at his station is introduced into the line circuit, and his relay is cut out. Thus, current is sent over the line operating all but the home, or sender's, relay.

11. Comparison of the Two Systems.—In the closed-circuit system the whole battery may be located at any one station or divided among any number of stations. Since the current is flowing even when not sending, it does not decrease in strength when sending commences. However, battery material is being steadily consumed whether the line is in use or not. A continuous current seems to increase flaws in cables, and, for this reason, the closed-circuit system is not used for submarine telegraphy.

With the open-circuit system, consumption of battery materials takes place only when the line is in use. However, it is necessary to have a sufficient number of cells at each office to operate the whole system, and a readjustment of relays may be necessary whenever another station commences to send, because it is difficult to keep all the batteries in the same condition.

———

TELEGRAPH INSTRUMENTS

12. The Key.—A key very widely used is shown in Fig. 5, and consists of a steel lever l pivoted in trunnion screws c, mounted in standards projecting upwards from the brass base plate m. Locknuts c' serve to bind the trunnion screws in any position to which they have been adjusted. A coiled spring u, the compression in which may be adjusted by the screw y and locknut y', serves to keep the forward end of the lever in a raised position. The upward movement of the forward end of the lever is regulated by the screw j and its locknut j'. As the knob x is pressed down, a platinum contact point v carried on the under side of the lever makes contact with a point of similar material carried on, but insulated from, the base m. This lower contact point is in metallic connection by means of strip s with the binding post d, which is

Fig. 5

also insulated from the base plate m. The other binding post d' is connected directly to the base plate. These binding posts d, d' form the terminals of the key, and the path through the instrument may be traced as follows: From post d, by strip s, to the lower contact point; then, when the key is depressed, to the upper contact point v; thence by the trunnions and spring u to the base plate m, and to the other binding post d'. The switch handle z is connected with a metallic lever pivoted directly on the base m, and when pressed toward the key lever makes contact with an extension of the strip s, thus short-circuiting the key. This switch will be easily recognized

Fig. 6

as performing the same functions as the switch C, Fig. 1.

The contacts on the key are made of platinum, because of the ability of that metal to resist the corroding and fusing action of the electric arc that is always formed at the break. It may be well to state that the down stroke of the key is often called the *make* and the up stroke the *break*, referring, of course, to the making and breaking of the circuit.

13. The Sounder.—The sounder is shown in Fig. 6. It consists of the electromagnet having the coils *m, m'* and the armature *a* of soft iron, the latter mounted on a brass or aluminum lever *l* pivoted between the trunnion screws *k*. The armature is normally held in its upper position by means of the coiled spring *s* bearing down on the short end of the lever *l*, the compression of the spring being regulated by the thumbscrew *h*. The down stroke of the lever is limited by the screw *g* striking against the anvil *n*, and the up stroke by the lever striking against the screw *f*. The play of the armature can therefore be adjusted by means of the screws *f, g*, and after the proper adjustment is obtained it can be made permanent by the locknuts *f', g'*. The binding posts *b, b* form the terminals of the circuit through the coils, the current passing through them in series. The sounds given out by the sounder may be augmented by mounting the instrument on a sounding board. The base plate and the board *c* on which the instrument is mounted are usually constructed with this idea in view. The coils are usually wound to a resistance of 4 ohms with No. 24 B. & S. copper wire when the sounder is to be used with a relay, and to a resistance of about 20 ohms with No. 25 B. & S. wire when the sounder is to be used in a line circuit, in which case it is called a *main-line sounder*.

14. An ink-recording register is an instrument that may be used in place of a sounder when it is desired to automatically record the signals. It is used quite extensively in small offices, district-telegraph systems, and wherever a permanent record is desirable. One is shown in Fig. 7. As much of the apparatus as possible, including a clockwork to draw along the paper tape, is placed inside a suitable case intended to be dust-proof.

When a current flowing through the magnet coils causes the armature to be attracted, the armature lever, carrying the paper with it, moves up against the disk *e*, which is kept moistened with ink by an ink roller *n*. When the current ceases, the spring *s* draws the armature lever and paper

away from the disk, or *printing wheel e*, as it is called.   The
ink roller *n* is lightly pressed against the disk *e* by the
spring *c*.   The paper *p p* passes through a guide on the armature lever just under the ink disk *e* and then between two
rollers *a, r*, the rotation of which pulls the paper along.   The
rollers *a, r, n* and the disk *e* are kept rotating by the clockwork as long as signals are coming in over the line.   Whenever necessary, the clock spring is wound up by the handle *H*.

FIG. 7

While receiving a message, the armature lever causes the
paper to be alternately pressed up against and withdrawn
from the disk *e*.   By this operation, a long or short mark is
made on the paper, according to the duration of the contact,
between the disk and the paper.

When used in a local circuit, the magnet coils of an ink
register are wound to have a resistance of about 4 ohms.

Registers are usually provided with an automatic self-
starting and self-stopping device; that is, when a message

is started the clockwork starts feeding the paper tape and continues to keep the paper moving until the signals have ceased for a certain predetermined length of time, as at the end of a message.

15. The Relay.—The relay consists of an electromagnet that, by its action on an armature, opens and closes the circuit of a local battery powerful enough to operate a sounder or register. The magnet is generally wound with a large number of turns of insulated wire in order to enable the feeble line current usually employed to produce sufficient magnetizing force to cause the magnet to attract its armature.

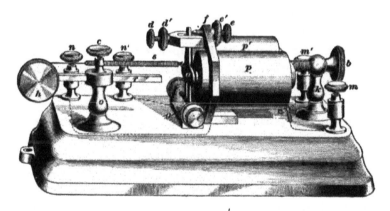

FIG. 8

A main-line relay is shown in Fig. 8. The electromagnet consists of two soft-iron cores on which are the coils p, p', the cores being connected together at the rear by a soft-iron yoke piece. To this yoke piece is attached one end of the screw b, by means of which the cores can be moved backwards or forwards through the coils and the supporting frame f, which is securely fastened to the wooden base. This screw b is supported by the pillar k. This piece f also carries the adjusting stop-screws e, d and their locking nuts e', d'. The armature a is pivoted between trunnion screws, supported by the brass piece g, which is fastened to the wooden base, but insulated by the wooden base from the piece f. To the armature a is attached the retracting spring s, and to the spring is fastened one end of a piece of silk cord, the other end of which is fastened and wrapped around the adjusting screw h. This screw h passes through one end of the rod r, which slides easily through the pillar o, but is secured in position by the setscrew c. By turning the screw h, the thread is wound or unwound, increasing or decreasing, respectively, the pull of the spring s on the armature. When the thread is all wound up and the end of the spring reaches the screw, the screw and the rod r should be moved out away from the armature. The front stop-screw e should be so fixed that when the armature is against it, there is at least the thickness of a piece of ordinary writing paper between the iron armature and iron cores. The back stop d should be adjusted so that the armature shall not have over $\frac{1}{32}$ inch play. The tip of the back stop-screw contains a piece of insulating material, such as hard rubber or fiber, so that the armature cannot make a metallic contact with the screw d or frame f.

The binding posts m, m' are the ones to which the main lines are connected, and, for this reason, are called the **main-line** binding posts. One end of the coil p is connected with the binding post m, the other end is connected to one end of the coil p', while the other end of the coil p' is connected with the binding post m'. The coils p, p' are enclosed and protected by polished hard-rubber casings. The binding posts n, n' are called the *local binding posts*, because they are connected with the local battery and sounder circuit. The binding post n is connected through the brass piece f to the screw e, the other post n' is connected with the metal piece g and through the trunnion screws with the armature a. Consequently, when the magnet draws the armature against the stop-screw e, the local circuit is closed at that point. The screw e and the end of the armature that comes in contact with it are tipped with a piece of platinum. The coils p, p' are usually connected in series with each other, with the windings in such a direction around the iron cores that the front end of one iron core has north polarity and the front end of the other iron core has south polarity.

Main-line relays may be wound to have any resistance from 37½ to 300 ohms, but 150-ohm relays are the most extensively used. A **pony relay** is somewhat smaller than a main-line relay, has fewer adjustments, varies from 20 to 100 ohms in resistance, and is used on shorter lines.

WINDING FOR SOUNDERS AND RELAYS

16. The resistance of a line wire increases directly as its length. With a given line circuit, the only way to increase the current is to increase the difference of potential at the terminals of the circuit. But it is not advisable or practical in telegraph circuits to use too high an E. M. F., so that it is not always possible, with the highest feasible E. M. F., to get through a long, and, therefore, high-resistance line a current strong enough to operate a sounder properly. To decrease the line resistance sufficiently would require such a large line wire that its cost would prohibit its use entirely. But, by putting the sounder, with a separate battery, in a local circuit whose continuity is controlled by a relay armature, the desired results can be obtained, because only sufficient pull need be exerted on the relay armature to bring two contact points together, and, with the sounder in a local circuit of low resistance, there is no difficulty whatever in securing a current large enough, even with only a few cells of battery. By having sufficient turns of wire on the relay, a very small current will be sufficient to cause the magnet to attract the armature and so close the local circuit containing the sounder and local battery.

17. Coils Designated by Their Resistance.—The number of turns in a coil, since this varies as the length of the wire, has a definite relation to its resistance, and, therefore, the resistance of a coil may be taken as a measure of the number of turns of wire it contains. Now, it is easy to measure the resistance of a finished coil, but it is not so easy to determine the number of turns or the length of wire used. On account, therefore, of this practical convenience, and not because the resistance itself is a desirable quantity, it is customary to speak of an electromagnet as having a certain

resistance instead of a certain number of turns, and, therefore, to designate it by its resistance. Thus, we speak of a
150-ohm relay and not of a relay of 8,640 turns, although this
latter would be a more direct way of indicating the value of
the relay, because the more turns there are, the smaller need
be the magnetizing current. In order to put more turns in
exactly the same space, a smaller wire, whose resistance per
unit length is greater, must of course be used.

RESISTANCES OF RELAYS IN A CIRCUIT

18. All telegraph electromagnets connected in series in
the same circuit, should be similar in construction and have
the same resistance. This will cause all the electromagnets
so connected to work equally well. Thus, a 20-ohm main-line
sounder should not be used in the same circuit with a 20-ohm
relay, because the sounder, even if it has the same number of
turns, requires more ampere-turns than the relay since its
armature is heavier, farther from the cores, and must be
moved more vigorously in order to give a loud click.

Two electromagnets of the same size and construction
require the same number of ampere-turns in their coils to
operate them equally well. A 150-ohm relay has about 8,640
turns and requires at least .02 ampere, or 8,640 × .02 = 172.8
ampere-turns. Now, if this relay is connected in series with
a 15-ohm relay and the current is still .02 ampere, then the
15-ohm relay, having, say, about 2,750 turns, will have only
2,750 × .02 = 55 ampere-turns, whereas it requires at least
172 ampere-turns to pull the same armature the same distance.
Hence, it is unreasonable to expect the 15-ohm relay to do
the same work with the same current as that required by a
similarly constructed relay having more turns, that is, a higher
resistance. If the 15-ohm relay is put in another circuit
where it can get current enough (approximately $\dfrac{173}{2,750} = .063$
ampere) it will work all right and consume no more power
$[I^2 R = (.02)^2 \times 150 = (.063)^2 \times 15]$, because a proportionately smaller E. M. F. is required to force .063 ampere
through 15 ohms than .02 ampere through 150 ohms.

For clear signals and quick action, the combined resistances of all the relays on the same line should not exceed that of the line and batteries combined. Beyond that proportion, the time constant of the circuit increases so rapidly as more relays are added as to appreciably reduce the maximum speed at which it is possible to receive clear signals over ordinary line circuits.

The longer the line, the greater will be its resistance, if constructed of the same size and kind of wire, and the smaller will be the current with the same E. M. F., and consequently the greater must be the number of turns and resistance of the relays, if the E. M. F. cannot be increased sufficiently.

SOURCES OF CURRENT

19. The primary batteries most extensively used in the United States for operating telegraph systems consist of gravity Daniell cells, usually connected in series. For operating sounders in local circuits usually two cells are used, but sometimes only one or as many as three may be required. For a one line circuit several hundred cells may be required.

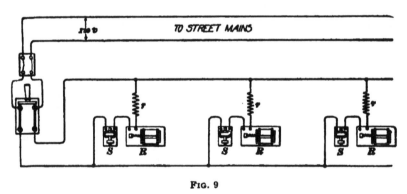

TO STREET MAINS

FIG. 9

In very many places, especially in large offices, dynamo or street-lighting circuits are rapidly replacing gravity batteries for operating all local circuits. One of the best arrangements when a lighting circuit is used is shown in Fig. 9. The sounders *S* have a resistance of 200 ohms and the non-inductively wound German-silver wire coils *r* a resistance of

4,000 ohms. Assuming that the sounder is closed 60 per cent. of the time and that current at 110 volts costs $1\frac{1}{2}$ cents per ampere-hour ($\frac{3}{4}$ cent per 16-candlepower lamp per hour), the cost for operating one sounder for 1 day (10 hours) is a trifle under $\frac{1}{4}$ cent, which amounts to less per year than the cost of maintaining two gravity cells, to say nothing of the saving in attendance, dirt, and especially space, the latter being a very important item in city branch offices where rents are high. Although there is some loss of energy in the resistance r, it reduces the time constant of the circuit and hence makes the sounder act more promptly. It is, moreover, necessary in order to obtain a small current, for it would not be economical to use a larger current and fewer turns on the sounder, nor practical to put all the required resistance on the sounder coils, because too fine wire would be required.

TELEGRAPH CODES

20. **Telegraph codes** consist of combinations of dots, dashes, and spaces that represent letters, numerals, and punctuation marks. The dot is taken as the unit by which the lengths of the dashes and spaces are measured. The following table gives the relative lengths of the different dashes and spaces:

SIGNAL		DURATION OF SIGNAL
Dot	▪	1 unit
Dash	—³	3 units
Long dash (*l*)	—⁵	5 units
Extra long dash (*cipher*)	—⁷	7 units
Space between parts of a letter	—¹—	1 unit
Space in spaced letters	—²—	2 units
Space between letters	—³—	3 units
Space between words	—⁶—	6 units

It will be noticed that there are four lengths of spaces and three of dashes, or four including the dot. Theoretically, the long dash (*l*) should be twice as long as the dash and the extra long dash (*0*, cipher) should be one-half longer than the long dash (*l*), that is, 9 units in length when *l* is made 6 units. However, the long dash (*l*) is seldom made longer than 5 units and the cipher seldom longer than 7 units. Furthermore, in practice, the *l* and *0* (cipher) are frequently made the same; occurring alone or in words, the long dash would be read as *l*, but when found among figures, it would be translated as *0* (cipher).

21. Several telegraph codes are now in use. The system of combining dots, dashes, and spaces to represent the letters, numerals, and punctuation marks that was devised by Morse, but rearranged by Alfred Vail, is known as the **American Morse code,** or simply as the **Morse.** The **Phillips punctuation code** has superseded it for punctuation, however, and is much more complete and systematic. Except for submarine telegraphy, the Morse code for letters and numerals, and the Phillips code for punctuation are used throughout the United States and Canada. A modification of the Morse code, called the **Continental,** is used all over the world for submarine telegraphy, and for land telegraphy in almost every country except the United States, Canada, and parts of Australia. On account of its extensive use, it is coming to be known as the *universal* code. The Continental is much superior for signaling through long submarine cables, and, owing to the fact that it has no spaced letters that are apt to be taken for double letters, it is freer from errors of transmission. For instance, with the Morse code, it is very easy for *ee* to be taken for *o*. On a siphon submarine-cable recorder, it would be practically impossible to avoid such errors. The American, or Morse code, owing to the fact that there are fewer dashes in it, is about 5 per cent. more rapid than the Continental. These codes are as follows:

ALPHABETS

LETTERS	MORSE	CONTINENTAL
A	▪ ▬	▪ ▬
B	▬ ▪ ▪ ▪	▬ ▪ ▪ ▪
C	▪ ▪ ▪	▬ ▪ ▬ ▪
D	▬ ▪ ▪	▬ ▪ ▪
E	▪	▪
F	▪ ▬ ▪	▪ ▪ ▬ ▪
G	▬ ▬ ▪	▬ ▬ ▪
H	▪ ▪ ▪ ▪	▪ ▪ ▪ ▪
I	▪ ▪	▪ ▪
J	▬ ▪ ▬ ▪	▪ ▬ ▬ ▬
K	▬ ▪ ▬	▬ ▪ ▬
L	▬	▪ ▬ ▪ ▪
M	▬ ▬	▬ ▬
N	▬ ▪	▬ ▪
O	▪ ▪	▬ ▬ ▬
P	▪ ▪ ▪ ▪ ▪	▪ ▬ ▬ ▪
Q	▪ ▪ ▬ ▪	▬ ▬ ▪ ▬
R	▪ ▪ ▪	▪ ▬ ▪
S	▪ ▪ ▪	▪ ▪ ▪
T	▬	▬
U	▪ ▪ ▬	▪ ▪ ▬
V	▪ ▪ ▪ ▬	▪ ▪ ▪ ▬
W	▪ ▬ ▬	▪ ▬ ▬
X	▪ ▬ ▪ ▬	▬ ▪ ▪ ▬
Y	▪ ▪ ▪ ▪	▬ ▪ ▬ ▬
Z	▪ ▪ ▪ ▪	▬ ▬ ▪ ▪
&	▪ ▪ ▪ ▪	

NUMERALS

FIGURES	MORSE	CONTINENTAL
1	▪ ▬ ▬ ▪	▪ ▬ ▬ ▬ ▬
2	▪ ▪ ▬ ▬	▪ ▪ ▬ ▬ ▬
3	▪ ▪ ▪ ▬	▪ ▪ ▪ ▬ ▬
4	▪ ▪ ▪ ▪ ▬	▪ ▪ ▪ ▪ ▬
5	▬ ▬ ▬	▪ ▪ ▪ ▪ ▪
6	▪ ▪ ▪ ▪ ▪ ▪	▬ ▪ ▪ ▪ ▪
7	▬ ▬ ▪ ▪	▬ ▬ ▪ ▪ ▪
8	▬ ▪ ▪ ▪ ▪	▬ ▬ ▬ ▪ ▪
9	▬ ▪ ▪ ▬	▬ ▬ ▬ ▬ ▪
0	▬▬▬	▬ ▬ ▬ ▬ ▬ or ▬

PUNCTUATIONS, ETC.

	MORSE	CONTINENTAL	PHILLIPS
. Period			
:: Colon			*KO*
:| Colon dash			*KX*
:, Semicolon			*SI*
,, Comma			
?, Interrogation			
~ Exclamation			
I I Fraction line			
— Dash			*DX*
, , Hyphen			*HX*
, Apostrophe			*QX*
$ Dollars			*SX*
c Cents			*C*
£ Pound sterling			*PX*
/ Shilling mark			
d Pence			*D*
Capitalized letter			*CX*
: " Colon followed by quotation			*KQ*
. Decimal point			
¶ Paragraph			*PN*
Italics or underline			*UX*
( ) Parenthesis			
[ ] Brackets			*BX*
" " Quotation marks			*QN*
" " ' ' Quotation within a quotation			*QX*

CIRCUIT ACCESSORIES

PROTECTING DEVICES

FUSES

22. Fuses are pieces of soft wire that melt and open the line circuit, if the current exceeds a certain value. For telegraph circuits, it is desirable that fuses should be melted by about 1 ampere. The larger the current above the limiting value, the quicker the fuse will melt. When a simple fuse melts, it merely opens the line circuit and does not usually ground it; consequently, a fuse cannot be considered as a protection against lightning, though it is a fairly satisfactory protection against crosses.

Enclosed fuses are now used extensively by both telegraph and telephone companies to protect the wires in cables and station apparatus from excessive currents caused by crosses on open overhead wires. This fuse consists of an enameled wood, paper, or fiber tube, about 4 inches long and ½ inch in diameter, with metal terminals secured to the ends. The tube prevents the scattering of the fuse that is inside, when it melts, and, being almost air-tight, protects the fuse from currents of air and thus causes it to operate more uniformly. The fuse is made long so that at least 2,000 volts at its terminals will be required before an arc can be maintained between them at or after the melting of the fuse. Hence, after melting, the passage of lightning discharges is prevented, especially when used in connection with a plate arrester having insulating material, only 5 or 10 mils in thickness, between the line and the ground. Excellent enclosed fuses, mounted on porcelain bases and made by the D. and W. Fuse Company, are now used by telegraph and telephone companies.

LIGHTNING ARRESTERS

23. A lightning arrester is a device designed to protect telegraph offices and their instruments from injury during lightning storms, due to atmospheric electricity, which, when it charges or strikes the line wires, follows them into the offices. If unprotected, the fine-wire coils of instruments would often be burned out and the operators might be injured.

The **plate, or static,** arrester consists of a plate or disk of metal or carbon separated and insulated from another similar plate by 5 to 10 mils of insulating material, such as mica, silk, or paraffined paper. One plate is connected to the line and the other to the ground. The lightning discharge jumps from the line plate through the insulating material to the grounded plate and then to the earth. Mica is probably the best insulating material for this purpose, as it does not carbonize. Where paper is used it should be renewed after every lightning storm. This type of arrester is extensively used in the United States by both telegraph and telephone companies and is built in a great variety of forms.

24. Action of Lightning Arresters.—The resistance offered by the insulation between the two plates is the same for alternating currents of high or low frequency as it is for steady direct currents. But lightning discharges are oscillatory in character, that is, the current surges back and forth thousands of times per second, and a coil of wire, especially when wound on an iron core, has an apparent resistance that is enormously greater for such an oscillatory current than for a steady direct one. The excess resistance that a coil of wire offers to an alternating current over a direct current is due to that property of the coil called its *self-induction.* For a given coil and a given intensity of the magnetic flux, the inductance L is constant, but the apparent resistance opposing the current increases rapidly as the number of alternations of the current per second increases; that is, the higher the frequency of alternation, the greater will be the so-called apparent resistance of the coil of wire.

To a steady direct current, the resistance of a given circuit is found by Ohm's law to be $R = \dfrac{E}{I}$. But when E and I are alternating in character, the relation between E and I will have so changed that the quotient $\dfrac{E}{I}$ will no longer give the same value for the resistance as found above, but will give some other value, which we will call Z. It is a well-recognized fact that for a simple alternating current the value of Z may be found from the following formula:

$$Z = \sqrt{R^2 + (2\pi n L)^2} \qquad (1)$$

in which R = ordinary resistance that circuit will offer to a
steady direct current, i. e., the ohmic resist-
ance;

L = inductance of the circuit;

π = 3.1416;

n = the frequency.

The frequency is the number of complete cycles per second, or twice the number of alternations per second. This quantity $\sqrt{R^2 + (2\pi n L)^2}$ is called the *impedance* of the circuit whose ohmic resistance is R and whose inductance is L. The value of this expression evidently increases if any of the quantities $R, n,$ or L increases, and, conversely, decreases if any of them decreases. For a lightning discharge, n is very large, but for an air gap, even if the air space is replaced with mica or any insulating material, L is zero. Consequently, the impedance of the air gap is always about equal to R, no matter how large n is, because $2\pi n L$ is zero when L is zero. Therefore, $\sqrt{R^2 + (2\pi n L)^2} = R$ when $L = 0$. But for the coils on the instrument, L has an appreciable value; it may amount to several henrys or more. Consequently, when n has a very large value and L is not too small, the value of $\sqrt{R^2 + (2\pi n L)^2}$ will be very large. As a matter of fact, n is large enough in lightning discharges to make the value of R^2 insignificant compared to $(2\pi n L)^2$. Therefore, for a lightning discharge, the impedance of the air gap remains equal to R, because L is zero, as already stated; but the impedance of the coils of wire on the instruments increases so much in value that the air gap becomes, for a lightning discharge, a path of low resistance in comparison with it, and since a current will always take the easiest path, the discharge will jump the air gap in its effort to reach the ground in preference to going through the coils on the instruments.

25. If the lightning were compelled to go through the coils to earth, it would invariably burn out the fine wire in the coils and also ground the coil to the iron core. It would thus ruin the coil, and, in its effort to reach the ground, would probably do much damage. Direct or low-frequency currents will not jump the air gaps between any of the plates, because the difference of potential is not usually great enough, and

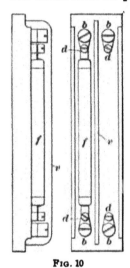

FIG. 10

because to them the coils offer an easier path. The E. M. F.'s of lighting and power circuits ordinarily in use, and against which the telegraph wire may come into contact, are not usually high enough to start an arc across the air gap. However, some lightning arresters used in telephone and telegraph work will ground the line if the potential exceeds 350 volts.

26. Postal Telegraph Fuse and Arrester.—Fig. 10 shows a style of fuse block in present use by the Postal Telegraph Cable Company. The block itself is of porcelain. The wires are brought to binding posts *b, b* and the enclosed fuses *f* attached to the binding screws *d, d*, which are a part of the same casting as that to which *b, b* are attached. The partition *v* is a strip of fiber fitting into small sockets cast in the porcelain and is removable.

Fig. 11 shows the lightning arrester used by this company; this also has a porcelain block. The line wires are brought to binding posts, or screws, *b, b* fastened to the metal straps *e, f;*

h, h are connected together at the back of the porcelain block and also to the central pin that holds the brass plate *p* in place. This plate is broken away in the illustration to show the line strap *e*. The side view shows several parts of the arrester in cross-section. Behind the brass plate *p* is placed a carbon plate *d*, and between that and the line straps *e, f* is placed a piece of perforated mica that insulates the carbon plates from the line straps and also leaves an unobstructed air gap between them. The whole is fastened by nuts that screw on to the pin connecting with the ground through *h, h*. Thus, a charge of lightning entering on either line strap jumps the air gap in the perforated mica and thence goes to

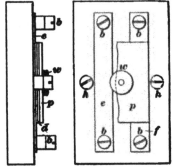

FIG. 11

the ground through the carbon and brass plates. Two ground-wire posts are furnished in order that a ground wire may be attached on either side as desired.

27. The Argus protector is a combination lightning arrester and fuse now extensively used. As shown in Fig. 12, it consists of a fuse secured between two binding posts

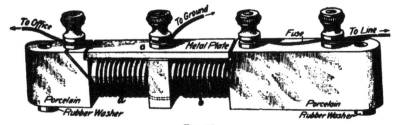

FIG. 12

mounted on a porcelain base and a very uniform spiral of bare copper wire *a b* wound on a porcelain core. A grounded metal plate *c* that is insulated and separated by an air gap from the spiral of bare wire is inclined so that the air gap at the line end is $\frac{1}{8}$ inch and at the office end $\frac{1}{32}$ inch. A lightning charge is supposed to be broken into a number of

discharges passing from the various convolutions of the
spiral to the grounded plate. Each discharge is generally
so small as to be harmless, and at no one point is there a
discharge of sufficient intensity to produce a burr, as frequently occurs on simple metal-plate arresters. Burrs on
metal-plate arresters frequently allow the direct current to
form an arc and then produce a permanent ground that must
be removed by filing. The Argus lightning arrester is said
to remain in condition to protect the circuit at all times from
lightning and requires little or no attention. The spiral has
very little resistance for a direct current, but offers considerable opposition to an oscillatory lightning discharge.
Furthermore, this opposition increases the nearer to the
office binding post that the lightning discharge is able to
penetrate, but as the air gap decreases in length its opposition to the discharge to ground decreases. Consequently,
the opposition toward the office circuit increases with each
turn, while that to the ground decreases.

SWITCHES

28. The **plug switch,** a portion of one of which is
shown in Fig. 13, is a form much used. It consists of
alternate brass plates
P, P_1, and rows of
brass buttons or
disks B, B_2; the disks
in each horizontal
row are connected
together. Thus, any
wire from the hori-
zontal side may be
connected to any on
the vertical side by
inserting a plug, such

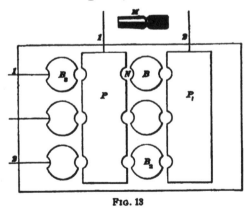

FIG. 13

as is shown at M, into the proper aperture, such as at N.
The lower part of the plug M is of brass, and should be
slotted to insure elasticity, while the upper part is of hard

rubber. The circuits in a large station are all appropriately numbered, so that wires *1* and *1*, for instance, as well as *2* and *2*, may be connected together by inserting plugs at the junction of $B_{\mathbf{.}}$ and P, and at the junction of $B_{\mathbf{.}}$ and P_1, respectively.

_____

PLUG SWITCH FOR TWO LINES

29. In Fig. 14 is shown the arrangement of apparatus at an intermediate office or way station adapted to make the necessary changes in connections between two through lines, the ground, and the instruments an easy matter. The switch is fastened to the office wall in a vertical position. The binding post *u* is connected to the disks *a* and *b* by

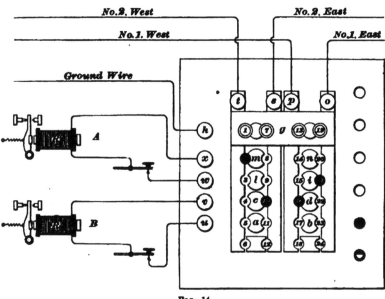

Fig. 14

wires behind the baseboard; similarly, *v* is connected to *c* and *d*, *w* to *l* and *i*, and *x* to *m* and *n*; *h* is connected to the plate *g*, which, being separated and insulated from the vertical strips by a thin piece of mica or paraffined paper, forms the ground plate of a lightning arrester. Any vertical strip and the line wire connected to it may be grounded by inserting a plug in the proper hole in the top horizontal row. Two

47—21

sets of instruments located in the way office and the two through main lines, ordinarily called *No. 1* and *No. 2* lines, are connected to binding posts on the switch as shown.

30. Instruments Cut Into Both Line Circuits. To loop the set *A* in line *No. 1*, that is, to connect the set *A* in series with the line wires *No. 1 East* and *No. 1 West*, and also to connect the set *B* in series with the wires *No. 2 East* and *No. 2 West*, insert plugs in holes *21, 14, 11,* and *4*. Current from the line *No. 1 East*, for instance, will pass down the strip *o* through the plug in hole *21* to the disk *i*, to binding post *w*, through the key-and-relay set *A* to the binding post *x* and disk *n*, through the plug in hole *14* to the strip *p* and out to line *No. 1 West*. The circuit of line *2* through set *B* may be traced in a similar manner.

31. Cutting Off One Side of a Line.—To cut off one side of one line (say *No. 2 East*) leaving the office set in the other side of the line (say set *A* in *No. 2 West*), insert a plug in hole *7* and plugs in holes *2* and *9* (or *3* and *8*). *No. 2 East* is now connected directly to ground and *No. 2 West* is connected through the set *A* to ground.

32. Cross-Connections.—To cross-connect the line wire *No. 1 East* to *No. 2 West*, and *No. 2 East* to *No. 1 West*, with or without an office set in each circuit, is easily done. In the figure, the plugs are shown in holes *21* and *2* in order to connect line wire *No. 1 East* and *No. 2 West* with the office set *A* cut in, and in holes *10* and *16* in order to connect line wire *No. 2 East* to *No. 1 West* without an office set. The set *A* may be cut out by simply shifting the plug from hole *21* to *20*, and set *B* may be cut into circuit with line wires *No. 2 East* and *No. 1 West* by shifting the plug from hole *16* to *17*.

33. Two East-Line Wires Connected Together. Connecting together two east or two west lines is called *looping;* the loop may or may not include an office set. For instance, if it is desirable to connect the set *A* in the loop between the line wires *No. 1 East* and *No. 2 East*, insert plugs in holes *20* and *9*. If *No. 1 West* and *No. 2 West* are to be looped directly together without an office set, insert plugs in holes *16* and *4*. Other combinations may be made between the various line wires and the office sets, but the above sufficiently illustrates the manner of using this form of switch.

The plug switch described for two main lines that enter a way station may be extended to accommodate almost any number of lines and instruments.

POSTAL TELEGRAPH SWITCH FOR SMALL OFFICES

34. Fig. 15 shows the form of a so-called cut-out board used by the Postal Telegraph Cable Company, in small offices where but one or two wires enter. On the extreme left are the line fuses and arrester and a so-called spring jack, next are the instrument fuses and arrester, and on the extreme right are the battery fuses and arresters (where a battery is used). As will be readily seen, the main-line circuit passes from line W through one fuse and one strap of the arrester to one side of the spring jack, thence to the opposite side of spring jack to remaining arrester strap and line fuse to line E.

A side view of the board is shown at the extreme left. Normally, the screws a, b are connected together through the brass contact pieces n, c, which are firmly pressed together by the spring s. When a wedge is inserted in the spring jack, one side of the wedge is connected to n and the other side to c, which is raised by the insertion of the wedge. Similar jacks and wedges are used by the Western Union Telegraph Company. The local connections pass from the instrument through fuses and arrester to the looping wedge I. When I is inserted in a spring jack, the instrument is cut into the line; and if a battery is required, it is looped in by placing the wedge B in the jack on top of the wedge I. Wedge G has both sides grounded on the ground connection of the arrester, so that if it is desired to ground the line in one direction, this plug is inserted in the spring jack on the top of the instrument and battery wedges, and underneath them to ground in the other direction.

This makes a very neat and simple arrangement for a small office, and being assembled on one board may be removed by

simply taking out the four large screws that hold it to the wall.
In the side view will be seen the insulating knobs that keep

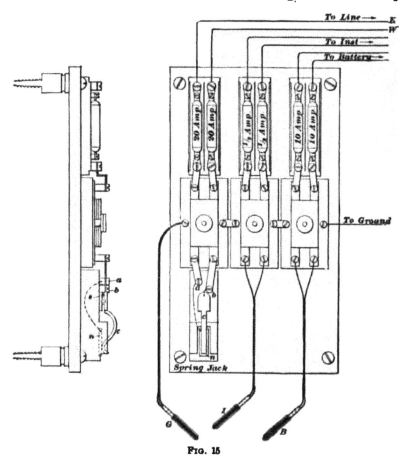

Fig. 15

the whole installation at the proper distance from the wall.
The fuses used are 20-ampere, ½-ampere, and 10-ampere in
line, instrument, and battery circuits, respectively.

POSTAL TELEGRAPH INTERMEDIATE SWITCHBOARD

35. Fig. 16 shows the style of switchboard, protecting
fuses, and arresters used by the Postal Telegraph Cable
Company, at its way offices where there are three or four
wires. All the apparatus is assembled upon one back board
and the local connections made up before the board is

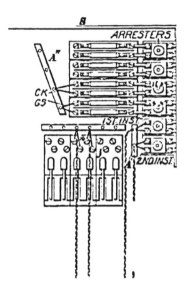

SWITCHING BOARD

CONNECTING PLUG

2ND.
INST.

FIG

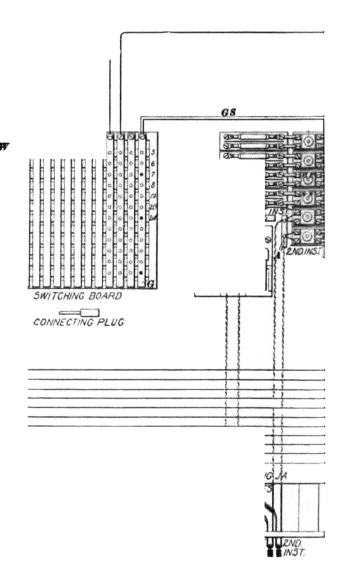

W

SWITCHING BOARD

CONNECTING PLUG

GS

2ND.INST.

RND.
INST.

Fic

shipped, so that it is a simple matter to put it in position and make the necessary connections to the line wires and apparatus. At the extreme right of the figure is shown a bank of arresters and fuse blocks. The bases are porcelain of the standard Postal type shown in Figs. 10 and 11, except that the porcelain block is narrower and has its ground at the bottom between the line posts; this ground post is made much shorter than the line posts, so that the line wires may readily pass over without touching it. This particular board being designed for twelve wires, the general appearance of the complete installation was improved by mounting ten of these blocks vertically and the remaining two horizontally, together with the four sets required for local instrument and broker loops. Loops are merely one or two wires used to connect a switchboard with telegraph instruments located in any part of the same building or city.

At the left of the figure is shown the switching board, which is extremely simple, and next the spring jacks for the loops and instruments connected with the board. All wires going in one direction are brought to vertical brass strips, fastened to one face of a wooden frame; and the lines going in the opposite direction are connected to horizontal brass strips, fastened to the opposite side, or face, of the wooden frame, which insulates them from one another. Where these vertical and horizontal strips intersect, a hole is bored in each so that a connecting plug inserted in these holes connects the two strips together electrically, so that the wire entering from the east on a vertical strip may leave going west by connecting the vertical strip to a horizontal strip by means of a connecting plug.

The board illustrated is, at present, only carrying seven wires, numbered as shown. The extreme right-hand vertical strip and the bottom horizontal strip are connected to the ground wire, giving a ground east and west for testing purposes. The board, as connected up, shows wires 5, 6, 19, and 20 connected through it east and west, while 7 east is patched, that is, connected to 8 west, 7 west being grounded. 24 is grounded west, as this office is its eastern terminal.

When it is desirable to table an instrument, that is, to connect a relay set in a line circuit, a switch plug like P is inserted in the proper hole in the switching board in place of a simple connecting plug. A switch plug is made in two sections insulated from each other and when placed in a hole in the switching board one section makes connection with the rear strip and the other with the front strip. Flexible wires connect the two sections of the switch plug to the spring-jack terminals s, j and the instruments are tabled by inserting a wedge in the spring jack. Where the cords pass through the board to the back side, they pass through insulating strips of fiber A, A', A''. The ground wire is connected to a brass strip GS that connects together all the ground connections of the lightning arresters and the grounded strips of the switching board. The connections between arresters and fuse blocks are also made by brass strips. All the other connecting wires are neatly cabled and this cable attached to the face of the assembling board, making a neat appearance. It will be noticed that an incoming line passes through a fuse, thence to one strap of a lightning arrester, to one side of switching board, then through switch plug, spring jack, wedge, and cord, to one strap of arrester and fuse, that protects the instrument, through the instrument back to fuse, arrester, cord, wedge, spring jack, switch plug, other side of switching board, arrester, and fuse to the line running in the opposite direction. For the broker loops, like those marked HF and SN, two sets of fuses and but one arrester are used. This is to give fuse protection no matter from which direction the dangerous current may happen to come. The fuses between line and switching board are 20 amperes of the enclosed D. and W. type, and those between switching board and instruments are of the same type, but of $\frac{1}{2}$ ampere capacity.

Any wire can be readily traced from the cable through the switchboard back to the cable again by referring to the diagram. Suppose, for instance, that it is desired to put 7 east to 8 west, and table an instrument on this combination. All that is necessary is to open 7 east and west, that is, remove the connecting plugs, insert switch plug where 7 east and *8* west intersect, and to insert an instrument wedge in the spring jack to which the switch plug used is connected.

MAIN-OFFICE SWITCHBOARDS

36. At terminal stations, provision must be· made, not only for interconnecting, in various ways, the lines and office sets, but also for connecting the main-line and intermediate batteries in the circuit with either or both the above. At large telegraph centers, where thousands of wires terminate, and all kinds of office sets, such as repeaters, duplex, quadruplex, simple relays and keys, and others, are in use, requiring batteries of various potentials, the connections are apt to appear very complicated. Main-office switchboards resemble those used at intermediate offices, but they are usually larger and have one or two spring jacks in series with each vertical strap to give increased facilities for connections between lines, instruments, and dynamos.

In all large offices dynamos, or direct-current street-lighting circuits, usually furnish all the current required in place of primary batteries, as formerly used and now used in small offices. It should be borne in mind that, although batteries are very often shown in illustrations of telegraph switchboards and systems, dynamos of suitable voltage may usually be substituted, but a protecting resistance, such as an incandescent lamp, is usually necessary in each circuit supplied. The disks in each row leading to dynamos are not joined directly together behind the board, but are first connected to incandescent lamps or non-inductive resistance coils, and the other terminal of the lamp or coil is then joined to the dynamo lead or bus-bar.

TELEGRAPH REPEATERS

37. The current at the end of a line several hundred miles long is very feeble because of leakage and the high line resistance; and, moreover, the electrostatic capacity of a long line diminishes the speed of signaling. But if a very small current arrives at the distant end of the wire, it may be made to close a new circuit containing another main battery, and thus repeat the message. These repeating devices are required at about each 500 miles of line, depending on the resistance, insulation, and condition of the weather.

38. **Button repeaters** are those requiring the manipulation of a switch, once called a *button*, by the hand of an attendant operator in order to change from repeating in one direction to repeating in the opposite direction. With such repeaters an operator must listen to what is passing, and be ready at any moment to turn the switch in order to reverse the direction in which messages may be sent, and so allow the operator at the receiving end to send, and vice versa. A button repeater, since it requires the constant attendance of an operator, is generally used for temporary purposes only. The *Wood button repeater* is about the only one worth considering here.

THE WOOD BUTTON REPEATER

39. The arrangement of the **Wood button repeater** is shown in Fig. 17. M is a switch so arranged that the lever k, which is pivoted at the center, is always in contact with one or both of the brass pieces c, d; o, o_1, o_2, and o_3 are binding posts, each joined to the respective brass pieces a, b, d, and c; g is a ground switch connecting the lever k of the switch M with the ground at G; W is the western and E the eastern main line; R, R_1 are the western and eastern relays; S, S_1 the western and eastern repeating sounders; and B, B_1 the western and eastern main-line batteries. B, B_1 must be arranged in series, the plus pole of one being connected to the minus pole of the other, for this arrangement allows the line to be connected straight across, as will be explained later. V, V_1 are the local batteries.

40. **Operation.**—If the ground switch g is closed and lever k is placed from d to a, the western circuit will repeat into the eastern. Suppose that the keys at both the western and eastern stations are closed. A current proceeds from the plus pole of battery B, through the relay R to the western station, where the line is grounded, then back through the earth to G, through levers l and k, to d–o,–2 to the minus pole of battery B. This current causes relay R to close its local circuit, which causes sounder S to close the circuit of battery B_1. This circuit may be traced as follows: From the plus pole of battery B_1 through 5–i–j–3–4–o, levers k and l to the earth at G. Then, through the earth to the eastern station and back through the eastern relay R_1. This repeats the message and also operates the sounder S_1. No circuit

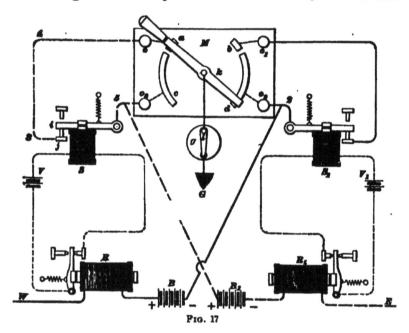

FIG. 17

is closed, however, by the armature of the sounder S_1, as there is a break between b and k. It therefore acts merely as a reading sounder.

If the two sounders do not work in unison, the attendant operator understands that the lever k must be instantly turned, as it indicates that the person receiving has broken and desires to become the sender.

If the lever k be placed from c to b while switch g is closed, the eastern circuit will repeat into the western, and the circuits may be traced as before, beginning, however, with eastern line and battery B_1. The single ground at G will serve for both eastern and western circuits.

If the lever k connects c and d and switch g is opened, the eastern and western circuits are connected straight across. This circuit is as follows: From the plus pole of battery B, through relay R to the western station, through the earth to the eastern station, then to relay R_1, and to the minus pole of battery B_1. Batteries B and B_1 are connected as one battery in series through 5–o_2–c–k–d–o_2–2. If k connects c and d, and switch g is closed, two independent circuits are formed, namely, B–R–W–G–l–k–d–o_2–2 and B_1–5–o_2–c–k–l–G–E–R_1.

MILLIKEN AUTOMATIC REPEATER

41. An **automatic repeater** is one that will automatically repeat in either direction without the necessity of turning a switch. An operator, however, is always needed to adjust the armatures of the relays and sounders, and to care

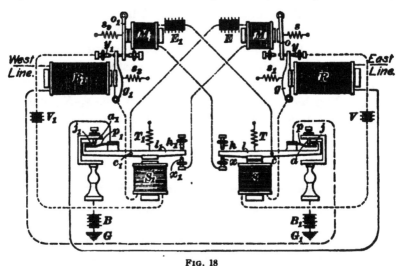

for the batteries, but, of course, his time may be largely devoted to other duties.

42. The **Milliken repeater**, although one of the earliest automatic repeaters used in telegraphy, is still regarded in the United States as the standard repeater; it is shown in Fig. 18. R, R_1 are main-line relays mounted on metal standards that hold them rigidly in place with respect to the extra magnets M, M_1. The levers of the relays and extra magnets are pivoted, as shown in the figure; the springs s, s_2 are so much stronger than s_1, s_3 that the levers g, g_1 are pressed against the contacts y, y_1 when there is no current in either R or M, neither in R_1 nor M_1, respectively. The telegraph instruments T, T_1 are called *transmitters*. When current flows through the electromagnet S, the armature lever l is attracted, causing the insulated spring, or *tongue* p, as it is called, to come into contact with the stop a slightly before the other end of the lever l touches x. When the current stops flowing through the coil of the transmitter, the lever is released, as shown at T_1, causing the contact at x_1 to be broken slightly before the contact is broken at a_1. B, B_1 are main-line batteries; V, V_1 local batteries; and E, E_1 so-called extra local batteries.

Normally, all circuits are closed. The western main-line circuit may be traced from the western office through R_1–p–a–B_1–G_1 and the ground back to the western office. The eastern main-line circuit is from the eastern office through R–p_1–a_1–B–G and through the ground back to the eastern office. The local circuit of R_1 includes V_1–y_1–g_1–S_1, and the local circuit of R includes V–y–g–S. The extra local circuits, including the magnets M, M_1, are, respectively, M–x_1–l_1–c_1–E and M_1–x–l–c–E_1.

43. Operation.—Suppose that all circuits are in their normal condition, that is, closed. If, now, the western key is opened, the relay R_1 will lose its magnetism, but the magnet M_1 retains its magnetism; hence, the armature g_1 is released by the relay magnet and is not held by the spring s_3; therefore, it breaks the local circuit between g_1 and y_1, causing the lever l_1 of the transmitter T_1 to first break at x_1 the extra local circuit containing M, and then to break the eastern main-line circuit between p_1 and a_1. Thus, M is first demagnetized and the spring s presses the lever o against the lever g, so that when a moment later R is also demagnetized by the opening of the circuit between a_1 and p_1, the lever g is still held against y, since the spring s is adjusted to overcome the pull of the spring s_1. Thus, the opening of the circuit containing the electromagnet S of the transmitter T is prevented. The opening of this circuit, when the western circuit is repeating into the eastern circuit, would be fatal to the successful operation of the repeater. But when the western key is opened, the eastern circuit may be opened without opening the western circuit at the repeating station. In the figure, the instruments are shown in their proper position when the western key is open.

The chief function of an automatic repeater is to automatically prevent the opening or breaking of the sending circuit at the repeater station. For instance, the transmitter that controls the western circuit must not open the western circuit at the repeating station when the western circuit is repeating into the eastern circuit.

As each circuit in any kind of repeater is closed, a short delay occurs in the transmission of a message, for each armature moves over a short distance before the circuit is complete. This shortens the dots and dashes in proportion to the number of contacts to be closed, and thus the dots are sometimes wholly lost. Therefore, in operating such a circuit, the dots and the dashes should be made longer, or, as operators term it, the "sending should be heavy."

There are a number of automatic repeaters used more or less, but it is beyond the scope of this Course to describe them.

MULTIPLEX TELEGRAPHY

44. Thus far, methods for transmitting only single messages over a line have been discussed. Such systems are frequently called *simplex* to distinguish them from *multiplex* systems. **Multiplex telegraphy** is the transmission of two or more messages over the same wire at the same time. It is quite obvious that, if instruments can be arranged so that two simultaneous messages can be sent through the same wire, the work of the system is equal to that of two lines. If four messages can be sent simultaneously over the same line, the system is equivalent to a four-wire system. In these two cases a good ground return is assumed. If, then, one line can be made to do the work of four lines, the expense of erection and maintenance of three lines is avoided.

The transmission of two telegraphic messages simultaneously in opposite directions over the same wire is called *duplex telegraphy*. On this system there is one sending and one receiving operator at each end or office, four operators in all. The transmission of two telegraphic messages simultaneously in the same direction over the same wire is called *diplex telegraphy*. On this system there are two sending operators at one end and two receiving operators at the other end, four operators in all. The simultaneous transmission of four independent messages, two in one direction and two in the other, is termed *quadruplex telegraphy*. On this system there are two sending and two receiving operators at each end, eight operators in all.

DUPLEX TELEGRAPHY

DIFFERENTIAL DUPLEX

45. There are three systems of duplex telegraphy—the *differential*, *polar*, and *bridge* duplex.

The **differential duplex** is also known as the Stearns duplex. Its essential feature is a differentially wound relay, that is, a relay with two windings, through which currents may circulate in the same or opposite directions around the iron core. The principle of the differential duplex may be explained by the aid of Fig. 19. R, R_1 represent ordinary

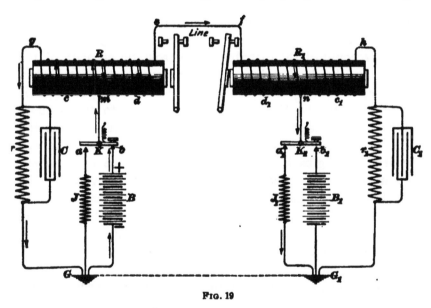

Fig. 19

relays differentially wound, and are called *differential neutral relays*, or simply *neutral relays;* the winding on each relay has a connection made at its middle point, so that the two coils c, d into which the whole winding is divided, have an equal number of turns and an equal resistance. K, K_1 are keys which, when depressed, connect exactly similar batteries B, B_1 to the circuit; in their normal position they rest against their back stops a, a_1, thus connecting the resistances J, J_1 to the circuit. Each of these resistances is adjusted so that it is exactly equal to the internal resistance of the adjacent battery. The resistance r is equal to the resistance of the circuit e-f-n plus the joint resistance of n-h-r_1-G_1 and n-a_1-J_1-G_1, which are in parallel. As r_1 is very large compared to J_1, the joint resistance of r_1 and J_1 in parallel is practically equivalent to J_1 alone. Hence, r is very nearly equal to the resistance of the circuit e-f-n-a_1-J_1-G_1. Similarly, r_1 is considered to be equal to the resistance of the path f-e-m-a-J-G. The ground resistance is invariably neglected.

The resistance r is adjusted so that when the key K is depressed the current from the battery B will divide at m into two equal parts—one part flows through the coil d–line–coil d_1-n-K_1-a_1-J_1-G_1-G-B; the other part through coil c-g-r-G-B. Only a very small current flows from n through c_1-h-r_1-G_1.

On account of the distributed electrostatic capacity of the line wire, the line receives a charge from the battery B when the key K in Fig. 19 is depressed, for the plus pole of B is connected to the line and the minus pole to the ground. When contact is broken at b the two conductors—the line and the earth—are connected by the contact at a, and the discharge coming back through relay R would give a *false signal*, as it is called, if means were not taken to prevent it. To neutralize this effect on the relay, the condensers C, C_1 are used. Their capacities are adjusted to equal the electrostatic capacity of the line circuit. Consequently, r, C and r_1, C_1 have a resistance and capacity equal to that of the line. Such a combination of a resistance and a condenser is called an *artificial line*, since its purpose is to form a branch to earth having the same electrical properties as the line wire. In practical systems, both the resistance and capacity of the artificial lines are adjustable. When both the resistance and capacity of an artificial line are properly adjusted, so that the operation of the home key produces no effect on the home relay, the duplex set is said to be balanced.

A resistance J equal to the internal resistance of the battery B, must be inserted between the ground plate G and the rear contact a of the key. This will give a path of equal resistance from m to the ground G whether the key K rests on the front or rear contact. J_1 is a similar resistance equal to the internal resistance of B_1. If such resistances are not used, the home relay would be more strongly magnetized when the home key is open than when closed, because the current through the line coil d of the home relay would not be quite twice as great when the home key K is closed as when open. The unequal magnetization of the relay would produce in the signals an inequality that it is very desirable to avoid.

46. One Key Closed.—Suppose that K_1 rests on the back contact a_1 and that K is pressed against the front contact b, or **closed**, as it is called. Current then flows from the positive pole of B, charging both the line and condenser C, and, when it reaches its maximum value, flows steadily through b to m, where it divides equally, one half flowing through c and r back to the battery B, and the other half flowing through d–e–f–d_1–n–a_1–J_1–G_1–the ground-plate G, and back to the battery B. There is also a closed circuit from n through c_1 and r_1 to G_1, but the resistance of this path is so very large, compared to that of the path through a_1 and J_1 to G_1, that it need hardly be considered. Moreover, even if there is an appreciable current in the artificial-line coil c_1, it flows in the proper direction in this case through the coil c_1 to help, and not to oppose, the magnetizing influence of the current through the line coil d_1. Thus, the closing of the key K will not magnetize the relay R, because the currents through the two coils c, d are equal and circulate in opposite directions around the iron core, producing, therefore, no resultant magnetizing force.

Not only will the steady or final current strength in both coils be the same, but since the capacities, as well as the resistances, in the two circuits are equal, the currents in both coils of the neutral relay will rise and fall at exactly the same rate. If the current should reach its maximum value, or fall from its maximum value to zero much quicker in one coil than in the other, it would cause a movement, or momentary *kick*, as it is called, of the armature every time the home key is opened or closed. This kick of the armature would cause false signals every time the home key was operated and would seriously interfere with incoming signals and render the method useless, except, perhaps, on very short lines.

By the arrangement shown in this figure, however, the home relay R is not sufficiently affected by the ordinary operation of the home key K to cause the relay to attract and hold its armature. However, the distant relay R_1 is magnetized because the currents through the two coils c_1, d_1 are not equal and opposite in direction, and, furthermore, the current in the coil d_1 is alone strong enough to cause the armature to be attracted. Consequently, the relay R is not magnetized, but R_1 is magnetized when the battery B is connected to the circuit by closing the key K.

47. Both Keys Closed.—If, while the key K is against the front contact b, the key K_1 is closed, the batteries B, B_1 will be in opposition in the circuit $B-b-m-d-e-f-d_1-n-b_1-B_1-G_1-G-B$. These two batteries contain the same number of cells and have the same E. M. F.; consequently, in the line circuit just traced, the current will be zero, since their E. M. F.'s are opposed to one another. With both keys closed, the currents in the artificial-line circuits, that is, in $B-b-m-c-g-r-G-B$ and in $B_1-b_1-n-c_1-h-r_1-G_1-B_1$, are due to the E. M. F. of only one battery in each circuit; hence, these currents will have their normal strength. Consequently, there is no current in the line coils d, d_1, but there is sufficient current in the artificial-line coils c, c_1 to magnetize both relays R, R_1. Thus, when both keys are closed at the same time, both relays will be closed. Although current from the home battery really closes the home relay, nevertheless it is the distant key that controls the opening and closing of the home relay. The home key has no control over the home relay.

48. Thus, it has been shown that the distant relay R_1 is energized and the home relay R unaffected when only the home key K is closed, and that both relays are energized when both keys are closed. From similar reasoning it is

47—22

evident that the relay R is energized and R_1 unaffected when only K_1 is closed. Furthermore, neither relay is magnetized when both keys are open, because both batteries are then cut off. The differential duplex is not now used alone, but the principle is employed in quadruplex systems, and hence it should be clearly understood.

49. Cause and Prevention of False Signals.—If, at the same moment, both keys should be in an intermediate position, touching neither the front nor the rear contact, there would be no current in any of the relay coils. Consequently, both relays would open every time this occurred, causing false signals and confusion, if means were not taken to prevent them. When gravity cells are used, false signals may be avoided by using a continuity-preserving transmitter that is so constructed that when it closes, contact is made with one stop just a moment before the contact with the other stop is broken.

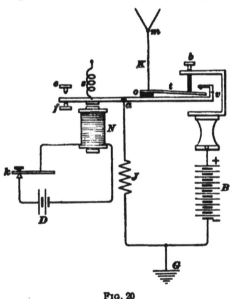

FIG. 20

50. A continuity-preserving transmitter is shown in Fig. 20. It is lettered to resemble the key arrangement in the preceding figure, so it will not be difficult to see how it would replace the key in that figure. When the key k is closed, current from the local battery D energizes the electromagnet N, causing the wire K to be connected through the spring tongue t and contact screw b to $+ B$. The piece of hard rubber o insulates t from the lever a. When k is open the spring s pulls the lever against stop e, causing t to make contact with v just a moment before it is separated from b.

Thus, the circuit from *m* to *G* is always closed, even in the intermediate position of the transmitter. The battery *B* is momentarily short-circuited through the resistance *J* in the intermediate position of the transmitter, but this is not sufficiently hard on either gravity batteries or the transmitter contacts to preclude its use.

51. Differentially Wound Relays.—The idea to be kept in mind in winding differential relays is to so arrange the two windings that the resistance and the number of turns in each winding shall be exactly equal, and that the effect of equal currents in each winding on the movable part of the relay shall be the same in intensity. A way that has proved satisfactory, consists in winding four separate coils, two on each core, the rear coil on one core, together with the forward coil on the other core, forming one-half of the differential winding, and the forward coil on the first core together with the rear coil on the second core forming the other half. This method of winding an ordinary relay is shown in Fig. 21. When the current circulates in the coils in the direction shown by the arrows, the magnetizing forces due to current in the two windings neutralize each other. The ends of the coils are brought out to the binding posts *a*, *b*, *c*, and *d*. Evidently, an excess of current in one winding over that in the other will magnetize the relay. Current in either direction in either winding alone will also magnetize the relay.

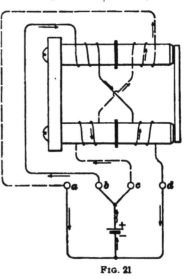

FIG. 21

POLAR DUPLEX

52. Polarized Relay.—The essential feature of the polar duplex is a differentially wound polarized relay. A simple **polarized relay** is one that requires the direction of the current flowing through it to be reversed in order to move the armature from one stop to the other. The mere absence of a current will leave the armature of the relay against whichever stop the last current may have moved it. Dots and dashes are made by currents flowing in one definite direction and spaces by currents flowing in the opposite direction. A battery reversing key must be employed in place of the ordinary make-and-break key.

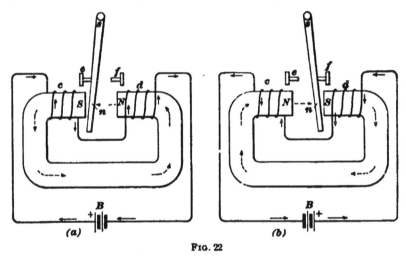

Fig. 22

Fig. 22 represents a bar of soft iron, bent so as to bring the two ends opposite each other. When a current from the battery circulates in the direction of the arrows shown in Fig. 22 (*a*), the current in each coil will magnetize the iron in the same direction, and thus produce magnetic lines of force in the direction of the dotted arrows, and, consequently, a north pole at *N* and a south pole at *S*. If a permanently magnetized piece of steel be suspended, so that its north pole *n* is free to move between the poles of the electromagnet, the south pole *S* of the electromagnet will attract the north pole *n* of the permanent magnet, and the north pole N of the electromagnet will repel the north pole n of the permanent magnet. Consequently, the north pole n of the permanent magnet will move over as near to the south pole S as the stop e will permit. If the battery and, as a result, the direction of current is reversed in the coils, the lines of force in the soft iron and the polarities of the ends of the soft-iron core will be reversed, as shown in Fig. 22 (b). Now the north pole n of the permanent magnet, being attracted by the south pole S and repelled by the north pole N of the electromagnet, will move from the stop e, as shown in Fig. 22 (a), to the stop f, as shown in Fig. 22 (b). If the current be reversed, the permanent magnet will move back to e. Thus, every time the direction of the current is reversed, the permanent magnet, or **armature**, as it is called, will move from one stop to the other.

53. In order to keep the armature permanently and strongly magnetized, the polarized relay has a strong and rather large permanent magnet. A skeleton view of

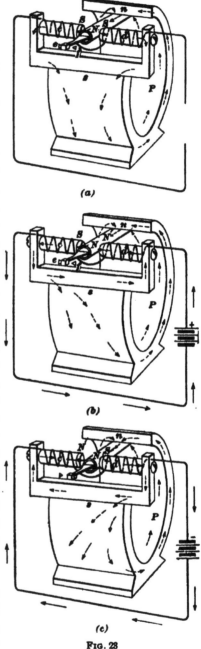

(a)

(b)

(c)

Fig. 23

one form of a polarized relay is shown in Fig. 23. *P* is a curved piece of special magnet steel that is permanently magnetized and not easily weakened, or demagnetized. In the rear end of this permanent magnet, the armature, or **tongue,** as it is also called, is loosely pivoted, and on the front end of the permanent magnet is placed a piece of iron of about the same shape as that shown in Fig. 22. This rectangular piece and the tongue are made of the very best quality of magnetically soft iron. On the ends, or cores, of the soft-iron piece are wound the coils *c, d.* If the rear end of the permanent magnet is a north pole *n,* and the front end a south pole *s,* then, when there is no current flowing in either coil, the soft-iron parts will be magnetized on account of their contact with the permanent magnet, so as to have north and south poles where indicated by the letters *N* and *S,* respectively.

The dotted arrows in Fig. 23 indicate the direction of the lines of force through the various parts of the magnetic circuit. In Fig. 23 (*a*), the lines of force are due entirely to the permanent magnet. If the tongue was exactly half way between the faces of the iron cores, it should, theoretically, remain there because each core would attract it with exactly the same force. But the least deviation in the equality of these two forces, due to the least deviation of the tongue from the exact middle position, will cause the tongue to fly against the stop *e* or *f,* toward whichever one the pull is the greater. In Fig. 23 (*b*) is shown the direction of the lines of force when the positive pole of a battery is connected to the coil *d.* A south pole is produced at the right-hand end of *c* and a north pole at the left-hand end of *d.* In this case *S* attracts and *N* repels the armature that has north (N) polarity. These polarities are reversed when the battery is reversed, as shown in Fig. 23 (*c*), and the armatures move toward *d.*

Polar relays are usually wound differentially so that they can be used in polar duplex and quadruplex systems. A differential polar relay is wound with two coils on each core in the same manner as the differentially wound neutral relay.

To avoid complicated diagrams, it is quite customary to represent differentially wound neutral and polar relays as having only two coils, one on each core, although they invariably have two coils on each core. Polarized relays are made in various forms, but the principle is invariably the same.

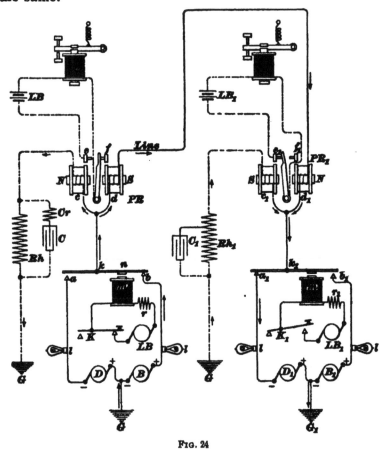

FIG. 24

54. Connections of Polar Duplex.—The connections of the polar duplex when dynamos are used to supply all current are shown in Fig. 24. PR, PR_1 are differentially wound polarized relays, the adjustable resistances Rh, Cr and the condenser C represent the artificial line at the left-hand stations Rh_1, C_1 represents the artificial line at the right-hand station. The artificial-line circuits are arranged in a slightly different manner at the two stations, in order to illustrate two methods used. The resistance Cr, called a *retarding coil*, because it retards the discharge of the con-denser, is placed in series with the condenser C and is adjusted until the artificial line charges and discharges neither slower nor faster than the line.

When a current is flowing through a wire, the difference of potential between two points near together is less than that between two points farther apart. Since the charge that a condenser receives depends on the difference of potential at its terminals, the charge that the condenser C_1 receives may be regulated by connecting the upper terminal of the con-denser to different coils of the rheostat Rh_1. The nearer this connection is made to the line, the greater will be the resistance and potential difference between the terminals of the condenser; and, hence, the greater will be the charge taken by the condenser. The portion of the rheostat Rh_1 between the condenser connection and the line acts also as a retarding coil. This is the later and better arrangement.

55. The dynamo, or so-called walking-beam type of pole changer, which is shown in this figure, consists of a lever, or beam, k to which is fastened the iron armature n, two adjustable stops a, b, an electromagnet m, the current through which is controlled by the operator's key K. LB is a 23- to 40-volt dynamo and r a resistance that allows just sufficient current to flow to properly operate the pole changer. When K is closed it causes the positive pole of dynamo B to be connected through contact b to k and when open the nega-tive pole of dynamo D is connected through contact a to k. The beam k, in moving from one position to the other, momentarily opens the circuit when in its intermediate posi-tion, as is shown in this figure, and the dynamos are never short-circuited.

In circuit with each machine is a non-inductive resistance l, either an incandescent lamp or a non-inductively wound coil of German-silver wire. This resistance serves **two** purposes: it reduces sparking at the contact points, because it limits the strength of the extra current when the pole changer opens the circuit, and it prevents injury to the dynamo due to overheating in case there is a short circuit. For duplex and quadruplex circuits, this resistance varies from about 600 to 800 ohms. All four dynamos, D, B, D_1 and B_1, generate current at the same voltage. The connections at the two ends are identical, except for a slight difference in the artificial lines that has already been explained.

56. Both Keys Open.—When both keys K, K_1 are open, the beams of the pole changers rest on the rear contacts a, a_1, and, since the negative poles of two equal dynamos, one at each end, are, in this open position of the two keys, connected to the line, there will be no current flowing in the line coils d, d_1. However, there will be current in the two artificial-line coils c, c_1 and the direction in which the current flows around the soft-iron cores and the polarization of the armatures will be such as to hold the armatures of both polarized relays against their back stops f, f_1. The current in either artificial-line coil due to the home dynamo may be represented as having the strength of 1 unit.

57. Key K Closed.—If the western operator commences to send by pressing his key K, the positive pole of B will be connected to the relay and line, in place of the negative pole of D. This will reverse the direction of the current through the artificial-line coil c, but its strength will remain the same, namely, 1 unit. The current in the line coils d, d_1 will now have a strength of 2 units, because the two dynamos B, D_1, which generate equal E. M. F.'s, are now connected in series in the line circuit. Hence, a current of 1 unit flows through the artificial-line coil c and a current of 2 units through the line coil d, but the direction of the currents in the two coils, as indicated by the arrows, is such that their magnetizing forces oppose each other, and the result is equivalent to a current of 1 unit flowing only through the line coil d in the direction shown by the arrow at that coil. This will produce a north pole at the left-hand end of the core on which the coil d is wound and a south pole at the right-hand end of the core on which c is wound and, consequently, the tongue, assuming it to have a south pole between the two cores, will remain against the back stop f. Thus, the closing of the key K does not affect the home relay PR as long as K_1 remains on the rear contact a_1.

At the east end, the current in the artificial-line coil c_1 has not changed in strength nor direction; it has a strength of 1 unit.   In the line coil d_1, however, there is now a current of 2 units.   The direction of the currents in the two coils, as indicated by the arrows, is such that their magnetizing forces oppose each other, and the result is equivalent to a current of 1 unit through the line coil d_1 in the direction shown by the arrow in that coil. This will produce a south pole at the left-hand end of the core on which the coil d_1 is wound and a north pole at the right-hand end of the core on which the coil c_1 is wound.   Consequently, the tongue, assuming it to have, as before, a south pole between the two cores, will move against the front stop e_1 and close the local sounder circuit at the eastern office.   Thus, the closing of the key K, when the key K_1 is open, operates the polar relay PR_1 at the distant office, but does not affect the home relay.

58.   Both Keys Closed.—If the key K_1 is now closed, the relay PR will be closed and the relay PR_1 will continue to remain closed until the key K is released.   For when both keys are closed and the levers of the pole changers rest on their front contacts, the two dynamos B, B_1 are connected in opposition in the line circuit.   Consequently, no current flows in either of the line coils d or d_1.   The current in the artificial-line coil c_1 is reversed and will produce a north pole at the right-hand end of the core on which it is wound, and a south pole at the left-hand end of the core on which the coil d_1 is wound.   But this polarity is the same as before and, therefore, the tongue of the polar relay PR_1 remains against the front-stop e_1, being unaffected by the change in the currents caused by closing the home key K_1, although it has reversed the polarity of the home dynamos.

At the western office the current in the artificial-line coil c has not changed in strength nor direction, but there is now no current in the line coil d.   The direction of the current is such that it reverses the polarity of the cores, producing a north pole at the right-hand end of the core on which c is wound and a south pole at the left-hand end of the core on which d is wound.   Consequently, the tongue of the relay PR moves against the front stop e and closes the local sounder circuit.   Thus, the closing of the key K_1 when the key K is closed, closes the relay PR at the distant office, but does not affect the home relay PR_1.

59. Key K_1 Closed.—If the western key K is now released, K_1 remaining closed, the two dynamos D, B_1 will be in series in the line circuit, the current in the line coils of both relays will have a strength of 2 units, and the current through the artificial-line coil c will be reversed in direction, but will have the same strength as before, namely, 1 unit.   The direction of the currents in the coils c, d is such that their magnetizing forces oppose each other, and the result is equivalent to a current of 1 unit flowing from the line through coil d.   This produces a south pole at the left-hand end of the core on which the coil d is wound and a north pole at the right-hand end of the core on which c is wound. Consequently, the polarity is such that the tongue of the relay PR remains against the front stop e when the key K at the western office is released.   Thus, the opening of the key K does not change the polarity of the home relay PR. At the eastern station the effect of a current of 2 units flowing from the dynamo B_1 through the coil d_1 to the line, and a current of 1 unit from the same dynamo flowing through the coil c_1 to the artificial line, is equivalent to a current of 1 unit flowing from the same dynamo B_1 through only the line coil d_1.   This produces a north pole at the left-hand end of the core on which d_1 is wound and a south pole at the right-hand end of the core on which c_1 is wound, thus causing the tongue of the relay to move from the front stop e_1 to the rear stop f_1.   Thus, the opening of the key K, while the key K_1 is closed, produces no effect on the home relay PR, but does open the distant relay PR_1.

It has, therefore, been shown that no matter what may be the position of the distant key, the operation of the home key does not affect the home relay, but that it does properly operate the distant relay.

60. Keys in Intermediate Positions.—It may be well to consider what happens to the relay during the short interval between the opening of the circuit at one contact of the pole changer and the closing of the circuit again at the other contact of the same pole changer. Suppose that the lever of the pole changer k_1 rests on the rear stop a_1, and that the lever of the pole changer k, in moving from the rear to the front contact, remains in an intermediate position, touching neither a nor b. In this position only the dynamo D_1 is in the circuit. It supplies a current of 1 unit to the coil c_1. The artificial line Rh at the western office, the coils c, d, the line, and the coil d_1 are in series with the dynamo D_1. The resistance of this circuit is double that of the line and the two line coils d, d_1, and, consequently, the current in this circuit will have a strength of $\frac{1}{2}$ unit. The current of a strength of $\frac{1}{2}$ unit in coil d_1 will oppose the current of 1 unit in c_1, but the magnetism due to a resultant current of $\frac{1}{2}$ unit flowing from the artificial line through coil c_1, produces a south pole at the right-hand end of c_1 and, therefore, tends to hold the tongue of the relay PR_1 against the back stop f_1, where it is already. Thus, the relay PR_1 will not be affected until the lever k of the pole changer touches the front stop b. A current of $\frac{1}{2}$ unit flows in the same direction through both coils c, d, so that their magnetizing forces help each other and produce a resultant magnetism of the same polarity and strength as when the lever k of the pole changer rested on the rear stop a. Consequently, the home relay is not affected, and the tongue remains stationary against the rear stop f.

Suppose that the levers of both pole changers are in an intermediate position at the same instant. Evidently all four dynamos are cut off and there is no current in any part of the system. But, now, the magnetism produced in the soft iron by the permanent steel magnet will hold the tongues on whichever side they happen to be, thus preventing any false signals.

61. Pole Changers. — The **walking-beam pole changer,** or one similar to it in principle, is invariably used in polar duplex and quadruplex systems operated by dynamos. The pole changer used with dynamos always opens the circuit connected to one pole of one dynamo before it connects the circuit with the opposite pole of the other dynamo. This is done to avoid an unnecessary drain on the dynamos and the formation of bad arcs at the contact points

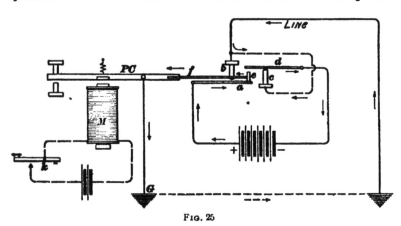

Fig. 25

of the pole changer, which would be the case if the beam of the pole changer could touch one contact stop before leaving the other.

62. Where gravity batteries are used for operating a polar duplex or quadruplex system, a so-called **continuity-preserving pole changer** is preferred because it is less liable to cause the production of false signals. One form of continuity-preserving pole changer is shown in Fig. 25. The contact screw b is placed behind the spring lever f and never touches it. The figure shows the position of the contacts and direction of the current in the various parts of the circuit when the key k is open.

When the key *k* is closed, the magnet *M* draws down the lever *PC* of the pole changer, causing the piece *e* to push *d* away from the contact screw *c*, and allowing *a* to rest against the contact screw *b*. In this position the direction of the current in the line and ground is opposite to that shown in the figure, as may be determined by tracing the path of the current for this position. The springs *d*, *a* may be made long and flexible enough and so arranged that, as the lever *t* moves up, the spring *a* first touches the screw *b*, then *e* touches *d*, then *e* parts from *a*, and finally *e* pushes *d* from *c*. This pole changer momentarily short-circuits the battery, because *e* touches *d* before it leaves *a*, and hence an uninterrupted path from the line to the ground is preserved. This short-circuiting does not injure a gravity battery on account of its rather high internal resistance, nor is the arc produced at the contacts sufficiently intense to render the method impracticable.

63. To Balance Polar Duplex.—The polar relays are first adjusted, with all current cut off, until their armatures will remain against either stop, or move with equal force from the middle position toward one side or the other. The artificial line at the left-hand station is adjusted as follows: The wire at the right-hand station is disconnected from the dynamos at lever k_1 and grounded; then *R h*, *C r*, and *C* are adjusted until the armature of *PR* will remain against either stop, or move with equal force toward one side or the other from the middle position, whether the home key makes contact with *a* or *b*. If the adjustment has been properly made, incoming signals will be equally good whether the home key is open or closed, and also when dots are being rapidly made with the home key. The other end is adjusted in the same manner.

BRIDGE DUPLEX SYSTEM

64. The **bridge duplex system** shown in Fig. 26 is similar in its action to a Wheatstone bridge. *S* is a rheostat so arranged that, as the lever is turned upwards, resistance is taken out of the *ac* arm of the bridge and is added to the

a d arm, and vice versa if the lever is moved in the other
direction. The four arms of the bridge are *a d, a c, d G*₁, and
from *c* through the line and apparatus at the other station
to the grounds *G′* and *G*ₛ. Hence, the resistance of the
artificial line at each end must be equal to the resistance of
the line wire plus the resistance from the distant end of the
line to the ground through the apparatus at the distant station, assuming, as is usually the case, that the resistance
of *ac* is equal to that of *a d*. In any case, the following

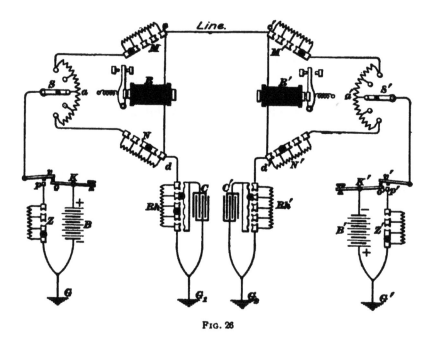

Fɪɢ. 26

proportion must be satisfied: Resistance of *a c* : resistance
of *a d* = line resistance + resistance from *c′* through all paths
at right-hand station to grounds *G′* and *G*ₛ : resistance of the
artificial line *d G*₁.

M, M′, N, and *N′* are adjustable resistances and *K* represents
a continuity-preserving transmitter. When the key is pressed
down, the lever *o* lifts the lever *v* off the contact point *p*,
momentarily short-circuiting the battery in order to avoid
opening the circuit between the ground *G* and the line. The

adjustable resistance Rh and condenser C constitute an artificial line, and Z is a resistance that is adjusted to equal the internal resistance of the battery B. The apparatus and connections at the two stations are similar.

65. If ac bears the same relation to ad that the circuit from c through the line and apparatus at the distant station to ground bears to dG_1, then the relay R, which in this case corresponds to the galvanometer in a Wheatstone bridge, will not be affected by the outgoing current from the battery B for the same reason that the galvanometer in the Wheatstone bridge will not be deflected when the bridge is balanced. If the key K' at the distant station is pressed down and K is up, that is, open, some current will pass along the line and at the point c will divide, a part of it passing through and operating the relay R. The position of the key K will in nowise affect the operation of the relay R, because it does not alter the resistance of the circuit. Thus, the relay at one station will be operated only by the key at the distant station.

Adjustment of resistances is made in ac and ad, first by the resistance boxes M, N, and finally by the rheostat S. If the resistance from c through the line and apparatus at the distant station to the ground is 4,000 ohms, then a resistance of 1,000 ohms in ac, 2,000 in Rh, and 500 in ad will properly balance the bridge. The connection between the condenser C and the resistance box Rh is adjusted until the artificial line charges and discharges in the same manner as the line, so that no momentary kick would be made by the relay.

The bridge principle is used whenever submarine cables are duplexed and very sensitive receiving devices are required. But, while the principle is the same, the apparatus used is quite different from that shown in this figure. The bridge duplex, as applied to submarine cables, will be explained later.

DIPLEX

66. The diplex is a system of telegraphy by which two messages may be simultaneously transmitted in the same direction over one wire. The form described here, although seldom or never used alone, should be thoroughly understood, for it is an essential feature of the quadruplex systems.

The principle of the diplex may be readily understood by the help of Fig. 27, in which PR is a polarized relay, whose operation depends only on the direction of the current; NR, a neutral relay, whose operation depends on an increase in the strength of the current and not on the direction of the current; PC, a pole changer; and T, a transmitter. The

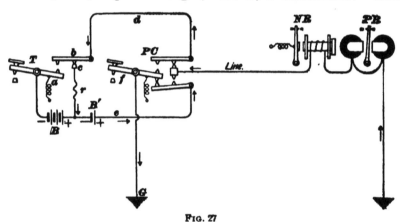

FIG. 27

transmitter is so connected that when the key is open, only one cell B' is connected between the wires d, e. When the key is depressed, the lever a first touches the lever b, thereby short-circuiting, momentarily, the battery B, which consists of three cells, before it lifts b off c. When the lever a has lifted b off c, the two batteries B, B' are connected in series, making one battery of four cells across the two wires d, e. Hence, the number of cells in the circuit has been increased from one to four; with the same resistance in the circuit, the strength of the current will be four times as great as before. If the weaker current has the strength of 1 unit, then the stronger current will have a strength of 4 units; that is, the ratio of the two currents is 1 to 4. In order to keep the resistance of the circuit the same whether B is cut in or out, it is necessary to insert the resistance r, which is equal to the internal resistance of the battery B, in the circuit when the battery B is cut out.

67. When the key f of the pole changer PC is open, that is, up, the line is connected to the wire d, and the ground G to the wire e. When the key is depressed, these connections are reversed. Hence, the pole changer, when operated, reverses the polarity of whatever battery happens to be connected by the transmitter T across the two wires d, e. The operation of the transmitters varies the current from 1 to 4 units, or vice versa, and the pole changer merely reverses the direction of this current through the line whether it be 1 or 4 units. Thus, the transmitter and the pole changer do their work independently of one another.

The action of these two instruments when they are combined in this manner should be thoroughly understood. There are four possible positions of these two keys. If it is not clearly understood why the operation of the pole changer does not affect the strength of the current, and the operation of the transmitter does not affect the direction of the current in the line, draw on a separate piece of paper the three other possible positions of the two keys and note the strength and direction of the current in the line in each case. The tongue, or armature, of the polarized relay will move whenever the direction of the current is reversed, no matter whether the strength of the current is 1 unit or 4 units. The reversal of the 4-unit current will perhaps make the polarized relay operate more vigorously than will the reversal of the 1-unit current, but the 1-unit current will operate it and the intensity of the click of the sounder that is controlled by the polar relay will be the same in either case.

The neutral relay, however, will tend to attract its armature, no matter in which direction the current flows through it, and if the current is only strong enough to overcome the retractile spring, the relay will close its local circuit. The spring is adjusted so that the magnetism produced by the

1-unit current will not be strong enough to overcome it, but the magnetism produced by the 4-unit current will readily overcome the spring and close the local circuit. Hence, the message sent by the operator at the transmitter T is received by the operator at the neutral relay NR, and the message sent by the operator at the pole changer PC is received by the operator at the polarized relay PR. Furthermore, these two messages do not interfere with each other when the apparatus is properly adjusted.

68. Elimination of False Signals. — If the pole changer reverses the direction of the current while the 4-unit current is flowing, in which case the neutral relay is closed, the neutral relay tends to release its armature at the instant of reversal, because when the whole battery is reversed, and, consequently, the direction of the current through the neutral relay is reversed, the magnetism of the neutral relay falls to zero and then increases in the opposite direction. If the interval of no current in the neutral relay, which lasts while the battery is momentarily short-circuited, is sufficiently prolonged, a mutilation of the signal, or a false signal, will be produced that will seriously interfere with the successful operation of the system. However, by adjusting the pole changer so that the interval of no current in the line and relay is as short as possible, and, furthermore, by using a so-called repeating sounder or relay that is closed on the back stop of the neutral relay, and an ordinary sounder that is closed in turn, on the back stop of a pony relay or repeating sounder (for which reason it is called a *repeating sounder*), the tendency to produce false signals can be overcome. This arrangement will give the signals properly, provided the interval of no current in the relay, although it may be long enough to allow the armature to break contact with the regular front stop, is still too short to allow the armature to cross the gap and make contact with the back stop. The second sounder is frequently called the *reading sounder*. This arrangement is used on the neutral-relay side of many quadruplex sets.

TELEGRAPH SYSTEMS
(PART 2)

MULTIPLEX TELEGRAPHY—(Continued)

QUADRUPLEX SYSTEMS

1. The principle of all **quadruplex systems,** in which two messages are sent in each direction simultaneously over one line wire, is about the same. In the Stearns duplex, the differential relay responds only to signals sent from the distant office; the connection and disconnection of the home battery does not affect the home relay, because it is differentially wound. In the polar duplex, the polar relay at the home office responds to the reversals of the distant battery but not to the reversals of the home battery, because the polar relay is differentially wound. In the diplex system one message is transmitted by increasing and decreasing the strength of the current, independent of its direction; while another message is being sent by reversing the direction of the current, independent of its strength. The principles on which most quadruplex systems depend are merely combinations of those involved in the Stearns duplex, the polar duplex, and the diplex.

BATTERY QUADRUPLEX SYSTEM

2. The connections for a quadruplex system operated by primary batteries is shown in Fig. 1. In order to have a clear diagram to use in explaining the system, the apparatus

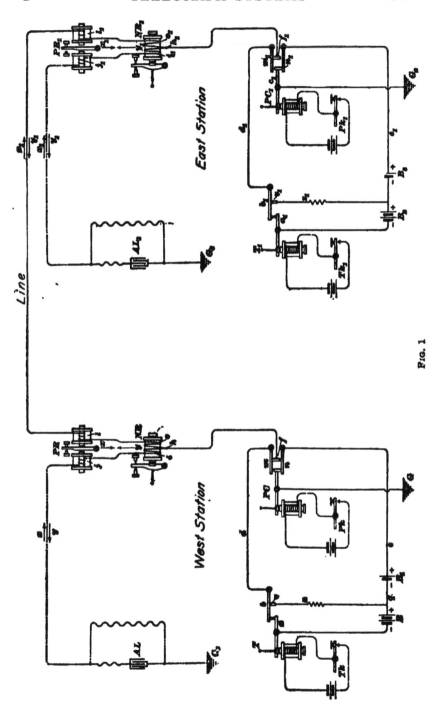

FIG. 1

in this figure has been reduced to as simple a form as possible; and all local-sounder connections have been omitted. The arrangement of the apparatus at the two ends is exactly similar, and the four relays are differentially wound. The artificial line $A L$ is so adjusted that the resistance from h through $A L$ and G_1 to G equals the resistance from h through the line and to the ground at the east station back to G; $A L_1$ is similarly adjusted. The resistance of the ground return can usually be neglected without appreciable error. The battery B has twice the E. M. F. of B_1, as is indicated, by giving B twice as many cells as B_1; hence, if B_1 has an E. M. F. of 100 volts, and B an E. M. F. of 200 volts, then, when B and B_1 are connected in series between d and e, the E. M. F. will be 300 volts; hence, if the strength of the current through a certain resistance, due to B_1 alone, is represented as 1 unit, the current through the same resistance due to both B and B_1 in series will be 3 units. B_1 alone is called the *short-end battery*, but B and B_1 together constitute the *long-end battery*. The same terms naturally apply to similar batteries at the other end.

3. R will be used to represent the resistance from the point h through the line to the ground G_2, at the east station, or from h_1 through the line to the ground G, at the west station. It is also equal to the resistance from the point h through the artificial-line circuit at the west station to the ground G_1, or from h_1 through the artificial-line circuit at the east station to the ground G_2. The resistance of the earth return is in each case neglected. The letters in parentheses in the following explanation refer to the direction of the current and to the branch carrying the largest current. Thus, $\frac{100}{R}$ $(x_1 A L_1)$ means that an effective, or excess, current of $\frac{100}{R}$ amperes is flowing in the direction of the arrow x_1 through the artificial-line coils of the relays at the east station.

An effective current of the strength $\frac{100}{R}$, which we may call 1 unit, is not strong enough to close the neutral relays when their springs are properly adjusted; and the polar relays are so connected that a current flowing through their artificial-line coils j, j_1 in the direction of the arrows x, x_1, respectively, or through their line coils l, l_1 in the direction of the arrows y, y_1, respectively, will hold the polar relays open. That is, the polar-relay coils are so connected when the apparatus is first set up that this will be the case; hence, any current through either or both windings of the polar relay that will magnetize the relay in the same direction as the currents specified above, will hold the polar relay open, and any current that will reverse the direction of this magnetization will close the polar relay. This fact should be remembered.

In order to close the neutral relay, the intensity of the resultant magnetization produced by the current in the two coils must be equivalent to that produced by $\dfrac{300}{R}$ amperes through one coil only. It will be noticed that the arrow x_1 coincides in direction with the arrow y, and y_1 with x. The arrows x_1, y_1 are not absolutely necessary, but, by using them, the explanations are made clearer.

It may be well to add here that some of the current flowing over the line from the east to the west station may go to ground G_1 through the coils i, j and the artificial line, instead of through the pole changer and transmitter circuit to the ground G. This, however, is an advantage rather than a disadvantage, because the direction of this current through the artificial-line coils is always in the proper direction to assist the incoming current through the line coils.

4. **All Four Keys Open.**—When all four keys are open the negative poles of the short-end batteries, having an E. M. F. of 100 volts, are connected to the points h, h_1. Since these E. M. F.'s oppose each other, there will be no current in the line coils o, l, o_1, and l_1 of the four relays. There is current, however, of an intensity of $\dfrac{100}{R}$, in the direction of the arrows x, x_1 in the artificial-line coils of all relays. Since there is no current in the line coils of the four

relays, it is evident that the effective current has a strength of $\dfrac{100}{R}$ and flows in the direction of the arrow x in the artificial-line circuit AL at the west end, and in the direction of the arrow x_1 in the artificial-line circuit AL_1 at the east end. This current is not strong enough to close · the neutral relays and is not in the right direction to close the polar relays; hence, all the relays will be open.

5. **Key Pk Closed.**—Closing the key Pk, all other keys being open, reverses the short-end battery B_1, causing the potential at h to be $+100$. Consequently, B_1 and B_2 are in series, giving $\dfrac{200}{R}$ (y) in the line coils o, l, and $\dfrac{100}{R}$ (y) in the artificial-line coils i, j. The $\dfrac{200}{R}$ (y) in o and the $\dfrac{100}{R}$ (y) in i will give an effective current of $\dfrac{100}{R}$ (y) in o, because the current flows from h through the two coils o, i in opposite directions around the iron core; hence, the $\dfrac{200}{R}$ (y) in o neutralizes $\dfrac{100}{R}$ (y) in i and still has left $\dfrac{100}{R}$ (y) with which to magnetize the neutral relay NR. This current is too weak, however, to close this relay; hence, closing the pole-changer key Pk does not affect the neutral relay NR.

As in the case of the neutral relay NR, the current $\dfrac{200}{R}$ (y) in l and the current $\dfrac{100}{R}$ (y) in j flows in opposite directions around the cores of the polar relay, giving an effective current equivalent to $\dfrac{100}{R}$ (y) in l alone, which will, on account of the direction in which it flows, continue to hold the polar relay PR open. Or, we may consider that the current of $\dfrac{100}{R}$ (y) now flowing in the coil j tends to close the polar relay, but that the current $\dfrac{200}{R}$ (y) in the coil l tends to keep it open, and since $\frac{200}{R}$ is twice $\frac{100}{R}$, the resultant magnetism,

which is due to $\frac{100}{R}$ (y) in the coil l, will hold PR open; hence, closing the key Pk does not affect either of the home relays NR or PR.

There is a current of $\frac{200}{R}$ (x_1) in l_1 and $\frac{100}{R}$ (x_1) in j_1. The current in j_1 is the same in strength and direction as before and tends to hold the polar relay open, but $\frac{200}{R}$ (x_1) in l_1 tends to close the relay; hence, the resultant magnetism that is due to $\frac{100}{R}$ (x_1) in l_1 will close PR_1. The resultant of $\frac{200}{R}$ (x_1) in o_1 and $\frac{100}{R}$ (x_1) in i_1 is $\frac{100}{R}$ (x_1) in o_1. This is not sufficient current to close NR_1; hence, it remains open. Therefore, when Pk alone is closed, the only relay that responds is the polar relay PR_1 at the distant end.

6. Key Tk Closed.—Closing the key Tk, all other keys being open, connects the long-end battery, that is, both B and B_1 in series, to the point h at the west station; hence, we have -300 volts at h and -100 at h_1. The current in the line circuit will be $\frac{200}{R}$ (xy_1). The current in i and j will be $\frac{300}{R}$ (x); hence, the effective current that is due to $\frac{200}{R}$ (x) in l and $\frac{300}{R}$ (x) in j, will be $\frac{100}{R}$ (x) in j. The resultant magnetization is in such a direction as to hold the polar relay PR open.

The resultant magnetization of the neutral relay NR is also due to a current of $\frac{100}{R}$ (x) in the coil i. This current is not strong enough to close the neutral relay NR, hence it remains open.

Since the full battery, 300 volts, at the west station opposes the short-end battery of 100 volts at the east station, the effective E. M. F. in the line circuit, that is, the difference of potential between the points h, h_1, will be 200 volts in the direction of the arrows x, y_1; hence, the current in the line coils $l_1 o_1$ will be $\frac{200}{R}$ (y_1). The difference of potential between the point h_1 and the ground G_2 is 100 volts, due to the short-end battery B_2. This difference of potential tends to send a current of $\frac{100}{R}$ amperes through the artificial-line circuit $A L_1$ in the direction of the arrow x_1. Hence, the current in the artificial-line coils j_1, i_1 is $\frac{100}{R}$ (x_1). The currents in the line and artificial-line coils of the east relays circulate around the iron cores in such a direction that they help each other in magnetizing the relays; hence, the resultant magnetization due to a current of $\frac{200}{R}$ (y_1) in the line coils and a current of $\frac{100}{R}$ (x_1) in the artificial-line coils is equivalent to that produced by a current of $\frac{300}{R}$ (x_1) in the artificial-line coils j_1, i_1. The direction of this current in the coil j_1 is such that the polar relay $P R_1$ remains open, but $\frac{300}{R}$ in i_1 is strong enough to close the neutral relay $N R_1$; hence, when only the key $T k$ that controls the number of cells connected to the circuit at the west station is closed, the only relay closed is the neutral relay $N R_1$ at the distant east station.

7. Keys $T k$ and $P k$ Closed.—The positive pole of the whole battery at the west station is now connected to h, giving that point a potential of $+300$ volts, h_1 remaining at -100, as in all preceding cases. The current in the line coils o, l, l_1, and o_1 will be $\frac{400}{R}$ ($y x_1$), and the current in the artificial-line coils i, j will be $\frac{300}{R}$ (y); hence, the effective current, due to the difference between $\frac{400}{R}$ (y) in the line coils and $\frac{300}{R}$ (y) in the artificial-line coils, will be $\frac{100}{R}$ (y) in the line coils o, l.   A current of $\frac{100}{R}$ (y) in the coil l is equivalent in its magnetizing effect, both in direction and intensity, to a current of $\frac{100}{R}$ (x) in the artificial-line coil j, but a current in the artificial-line coil j in the direction of the arrow x will hold the polar relay open.   Therefore, the polar relay PR is held open by the effective current $\frac{100}{R}$ (yL).   Furthermore, this effective, current $\frac{100}{R}$ (yL) through the coil o is not strong enough to close the neutral relay NR.   At the eastern station the current in the line coils l_1, o_1 is $\frac{400}{R}$ (x_1), and in the artificial-line coils j_1, i_1 the current is $\frac{100}{R}$ (x_1); hence, the resultant current $\frac{300}{R}$ (x_1) is not only strong enough to close the neutral relay NR_1, but it is also in the right direction to close the polar relay PR_1; hence, the closing of the two western keys closes only the two eastern relays.

8.   Similarly, the currents in the line and artificial-line circuits and the relays affected by the other various positions of the four keys may be worked out.   But the space here can be used more advantageously.

The terminals of a local circuit containing a sounder and battery is connected to the front stop and armature of each polar relay, as shown in the polar duplex system.   The terminals of a local circuit containing a repeating sounder and battery are connected to the back stop and armature of the neutral relay and a second local circuit containing an ordinary sounder and battery is connected to the back stop and armature of the repeating sounder.   The latter arrangement, which has been fully explained in connection with the diplex system, eliminates false signals.

JONES QUADRUPLEX SYSTEM

9. The **Jones quadruplex system,** the circuits of which are shown in Fig. 2, is used by the Postal Telegraph Cable Company. The principle on which it operates is practically the same as that of the battery quadruplex, but the details are somewhat different. For operating the main-line circuits, four dynamos are used, adapted, respectively, to deliver current at about the following pressures: one at $+100$ volts, one at -100 volts, one at $+300$ volts, and one at -300 volts. On different lines, different voltages are used; the lower E. M. F. ranges from 75 to 135 volts, and the higher from 225 to 400 volts. Each machine has one terminal connected to a common ground G. The local circuits of the transmitting devices and sounders are supplied with current from one 40-volt dynamo. To avoid confusion two such dynamos LB, LB' are shown, but, in reality, only one machine is used. In each local circuit there is a noninductive resistance r_1, r_2, r_3, r_4, r_5 of such value as to give the desired current. The pole changer consists of two electromagnets PC, PC' connected in series and controlled by the one key Pk. When Pk is open, the -300-volt machine is connected to stop 5 and the -100-volt machine to stop 6. When Pk is closed, the polarity of 5 and 6 is reversed, for then the $+300$-volt machine is connected to 5 and the $+100$-volt machine to 6. The operation of the transmitter key Tk will cause the transmitter to shift the wire 7 from a 100- to a 300-volt dynamo, the polarity of the dynamo depending entirely on the position of the key Pk. Thus, the operator at Pk controls only the direction of the current that tends to flow in wire 7, while the operator at Tk controls only the strength of current that tends to flow in wire 7.

In order to thoroughly understand the quadruplex it is absolutely necessary to understand that the pole changer will reverse the direction of the current that tends to flow from the home generators toward the line relays, no matter in what position the transmitter may be, and similarly, that the transmitter will change the voltage at the home station

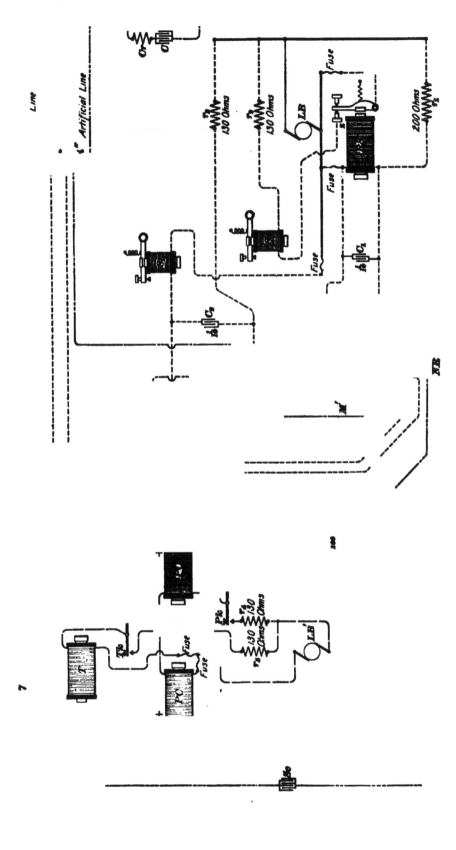

no matter in which position the pole changer may be. It must also be remembered that the operation of the home pole changer and transmitter will not affect the home neutral and polar relays, because they are differentially wound with respect to any currents originating at the home office; any change, either in direction or strength, of the E. M. F. applied at the home office will tend to energize both windings to the same extent, but in opposite directions, and, hence, cannot alone cause the operation of the home relays.

The transmitter and pole-changer magnets are very quick-acting, and are adjusted to have an extremely small gap between their front and back stops; hence, their tendency to interfere with signals due to the operation of the keys at the distant end is reduced to a minimum. The non-inductive resistances r are about 800 ohms each. They prevent too much change in strength of current when the system is used on lines of different resistances, and also protect the dynamos in case of accidental short circuits.

The so-called spark condenser Sc reduces the sparking at the contact points of the transmitting devices, because the discharge from the condenser opposes the spark that would otherwise be produced each time the circuit is broken. The switch Q is used for grounding the circuit through the coil Gc, which has a resistance about equal to that of the transmitting circuit between G and h, leaving all transmitting apparatus cut off. This is for the purpose of balancing the system and properly adjusting the relays. The resistances Rh, Cr and the condenser C constitute the artificial line.

10. Jones Neutral Relay.—NR is a special form of neutral relay; it is a triple-magnet relay wound differentially, but not polarized, and, therefore, responding to a current in either direction if it has sufficient strength. In order to attract its armature, the magnetism must overcome the pull of the spring s. On each core of the magnets M', M'' are wound two separate coils; one coil is connected in the main line and the other in the artificial line, as shown. The two coils on each magnet oppose each other when outgoing currents are flowing through them. As the two magnets pull in conjunction on the same armature lever, when an unbalanced current circulates through the line and artificial-line coils, their magnetic effects are added. *M* is a third magnet that acts on a third arm of the lever *l* and when energized assists the other two magnets. The lever of the relay is made of aluminum and is very carefully balanced. The retractile spring *s* and the magnets *M'*, *M''* have all the adjustments found in an ordinary relay.

R S is a pony relay, sometimes called a *bug-trap* relay, connected in a local circuit with the dynamo *L B*, and also to the neutral relay in such a manner that *R S* is short-circuited when the lever *l* touches the back stop *t*. Consequently, *R S* is energized only when this short circuit is open, that is, when the attraction of the neutral relay is sufficient to keep *l* away from *t*. An ordinary sounder *S* is controlled by the relay *R S* in the ordinary way. By this arrangement the tendency to produce false signals is reduced, for a reduction in the magnetizing force of the relay that will allow the armature to momentarily break away from the front stop will not produce a false signal by opening the sounder circuit, unless the time interval is sufficient for the relay armature to cross the gap between the front and rear stops, and thus make contact with the rear stop.

11. *I* is an induction coil having two primary coils *i'*, *i''* and one secondary coil *i*. The coil *i'* is connected in series with the line circuit, and *i''* in series with the artificial-line circuit. The induction coil is used to avoid the mutilation of the signals received on the neutral relay by preventing the armature from falling on the back stop *t* when it should not do so. Under the influence of equal currents flowing through *i'* and *i''* in opposite directions around the iron core, the secondary coil *i* has no current induced in it, because the two primary coils neutralize each other. But currents passing through only one primary coil, *i'* or *i''*, will induce a brief current in the secondary coil *i*, which passing through the magnet *M* will energize it for a moment. The effect of the current thus set up in the secondary coil and made to act on the neutral relay is adjusted so that the induced secondary current will not be sufficient to pull the armature from the back stop t. When, however, the armature lever is against the front stop w, and, therefore, in closer proximity to the core of the relay, the secondary current will be sufficient to momentarily hold up the armature, although the current through the coils on the magnet M', M'' consequent on a reversal of polarity, may momentarily cease.

Suppose that the armatures of the transmitter T and the pole changer PC, PC', and the armatures of the corresponding instruments at the distant end are all attracted, that is, all four instruments are closed; there is then a pressure of $+300$ volts at the point h at each end. Suppose that the distant pole changer is released due to opening the key corresponding to Pk at the distant end; then the E. M. F. at the distant end will be changed from $+300$ to -300 and there will, therefore, be a current through the line coils of the home relays due to $+600$ volts, but this current opposes the current, due to $+300$ volts, in the artificial-line coils of the same relays and hence the magnetism, due to the difference between that produced by $+600$ volts in the line and $+300$ volts in the artificial-line coils, which oppose each other, has been reduced to zero and then increased in the opposite direction, to the same strength as before. This reversal in the direction of the distant E. M. F., which produces a momentary reduction of the magnetism of the home relays to zero, will tend to suddenly release the armature of the neutral relay and reattract it, thus tending to break up the dash that the relay NR is receiving. The change in the magnetism of I due to the change in the current passing through the coil i' will induce a current in the coil i that will pass through the coil M and that will be at its maximum strength as the magnetism of I is reversed and therefore passes through zero. This current induced in i, magnetizes M and tends to prevent the armature l from fluttering or touching the rear stop when it should not do so. The current induced in the secondary coil i is always strongest when most needed; that is, when the current in the line is varying most rapidly, as it does when it passes through zero.

When the current in i' only changes in strength there is a tendency to induce a current in i'' as well as in i, but i'' is in a circuit of so much higher resistance that whatever current may be induced in i'' is too feeble to produce any apparent results.

12. PR is a differentially wound polarized relay, responding to positive currents coming from the distant station, or an equivalent current in the artificial-line circuit, but held against its back stop by negative currents from the distant station, or an equivalent current in the artificial-line circuit. Sounder S_1 is connected to the front contact stop of the polar relay PR in the usual manner. To reduce sparking and improve the operation of the sounders $\frac{1}{10}$-microfarad condensers C_1, C_2 are connected across the contact points of the relays.

Starting at the dividing point h, the line circuit may be traced through the coils o', o, f', f and the primary coil i' to the line; similarly, the artificial-line circuit may be traced from h through the coils e', e, j', j and the primary coil i'' to the ground G_1.

13. Both Keys Open.—When Tk and Pk are both open, the -100-volt dynamo is connected in the circuit and the path of the current may be traced from the -100-volt dynamo through r–4–6–7–u to h, where it divides between the line and artificial-line circuits. This produces in the coils of the distant relays a current that is neither strong enough to operate the distant neutral relay, nor, in the right direction, to operate the distant polar relay; hence, neither of them will be closed. The home relays will both remain open, provided the two distant keys are open, for the same reason.

14. Pole-Changer Key Closed.—When the key Pk is closed (the key Tk being open), current flows from the $+100$-volt dynamo over the path r–3–6–7–u to h, where it divides between the line and artificial-line circuits. This produces in the coils of the distant relays a current that will operate the distant polar relay, because it flows in the right direction, but it is not strong enough to operate the distant neutral relay.

15. Transmitter Key Closed.—When the transmitter is closed, due to closing the key Tk (the key Pk being open), a current flows from the −300-volt dynamo over the path r–1–5–7–u to h, where it divides between the line and artificial-line circuits. This produces in the coils of the distant relays a current that will operate the neutral relay at the distant station, because it is strong enough to overcome the spring; but it will not operate the distant polar relay, because it is in the wrong direction.

16. Both Keys Closed.—When both the pole changer and the transmitter are closed, due to the closing of both keys Pk and Tk, a current flows from the +300-volt dynamo over the path r–2–5–7–u to h, where it divides between the line and artificial-line circuits. This produces in the coils of the distant relays a current that is strong enough to operate the distant neutral relay and is in the right direction to also operate the distant polar relay, so that both relays at the distant station will be closed.

The strength and direction of the current in the line and artificial-line coils of all relays, two at each end, could be worked out if space would permit and it could be shown that any one key controls only one relay, but this has been sufficiently explained in connection with the battery quadruplex system.

SUBMARINE TELEGRAPHY

17. In **submarine-cable telegraphy,** the double-current system, that is, a system in which a signal is made by reversing the direction of the current, is invariably used. A current in one direction signifies a dot; a current in the opposite direction, a dash, no matter what may be the length of these signals; and no current signifies a space. A pole-changing key is used for transmitting the signals; the key may be operated by hand or by an automatic transmitter.

SIMPLEX CABLE SYSTEM

18. When cables are not duplexed, the principle of the arrangement is as shown in Fig. 3. K is a hand transmitting key that is extensively used. It consists of two long spring levers a, b, one operated by the index finger and the other by the second finger of the right hand. The two levers normally press against the top strip c. When the lever b is

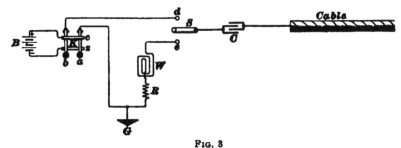

FIG. 3

pressed down it parts from the strip c and touches the strip s, thus connecting the negative terminal of B to d, the positive terminal remaining connected to ground G. Consequently, if S is closed on d, the battery side of the condenser C receives a negative charge; the cable side, a positive charge; and the ground at G, a positive charge. At the distant end the cable side of the condenser is negatively charged and the opposite side, including the ground, positively charged. When *a* is depressed the battery is reversed, that is, the positive terminal is connected to *d* and the negative to *G*. The polarity of all charges is now the reverse of that stated above and hence there has been a redistribution of the charges, or in other words, a flow of electricity through the various parts of the system. Thus, the signals are transmitted by sending charges of opposite polarity from one end of the cable to the other and are received, because these charges are sufficiently intense to affect a delicate receiving apparatus, represented by *W*, that is in series with a resistance *R* that can be adjusted to regulate the strength of the signals. At the end that is receiving, the switch *S* is placed on contact *e*. Condensers are very freely used in submarine telegraphy, because they make the signals sharper and reduce the disturbing effect of earth currents.

19. Cable Receiving Instruments.—In order to avoid the danger of injuring the insulating material of a long submarine cable, which is very expensive, an E. M. F. exceeding 40 or 50 volts is seldom, if ever, used. For this reason, and also on account of the large distributed electrostatic capacity of a long submarine cable, the current at the receiving end is usually too small to operate any kind of electromagnetic relay. Since it requires some time to charge a long submarine cable, the current at the receiving end increases in strength gradually and usually requires at least some fraction of a second after the closing of the transmitting key before it reaches an appreciable strength. Moreover, it is evident that the smaller the current that can be detected by the receiving instrument, the higher can be the speed of signaling.

At first *reflecting galvanometers* were used, the signals being read by the right-and-left deflections of a spot of light, a movement in one direction indicating a dot and a movement in the opposite direction a dash. The reflecting galvanometer has been replaced by the siphon recorder. On submarine cables of rubber used in the Philippine Islands and not exceeding 150 miles in length, polarized relays, requiring 12 milliamperes, are used and a speed of 25 words per minute is attained.

20. The *siphon recorder* consists of a coil of wire suspended by a fine fiber between the poles of a powerful permanent magnet, thus constituting a D'Arsonval galvanometer. The coil is attached to a siphon (a very fine and sort of inverted **U**-shaped glass tube) by a thread that moves the recording end of the siphon across the paper tape, which is kept moving uniformly past it in a direction at right angles to the motion of the siphon. The upper end of the siphon dips into a vessel containing ink and the lower end spurts ink on the paper. A system of very delicate springs tends to keep the siphon on one side, while the suspension of the coil tends to keep it in a central position. A charge passing through the coil in one direction increases the pull of the coil on the siphon and a charge in the opposite direction decreases the pull on the siphon and hence allows it to be pulled back beyond the central line that is made on the tape when no current is flowing through the coil. The siphon is kept vibrating to and from the paper tape by an electromagnet through which flows a current that is interrupted by a device resembling a common vibrating bell, the end of the siphon having attached to it a very small piece of iron that is attracted by the electromagnet each time it is magnetized.

21. Cable Alphabet.—The letters of the alphabet, figures, and other characters are formed by prearranged combinations of positive and negative currents that cause corresponding right-and-left movements of the recording end of the siphon. For instance, the letter "b" consists of one negative and three positive impulses (see Continental Telegraph Code), producing one movement of the siphon to the right and three to the left. On the tape these signals appear above or below the zero line. There is necessarily no return of the siphon exactly to the zero line every time between impulses. In the case of impulses of opposite polarity, the siphon will usually cross the zero line, but in the case of several impulses of the same polarity, the curve will merely fall back a little and move a little farther from the zero line each time.

The Continental code is used on all submarine cables. If this alphabet is deliberately sent over a very short cable, it will cause the siphon to trace a record about as shown in Fig. 4 (*b*). If the letters are sent continuously at ordinary speed, the actual record made by a siphon recorder connected to a long submarine cable is shown in Fig. 4 (*c*). Fig. 4 (*d*), which is an accurate reproduction of a portion of a message actually transmitted over a transatlantic cable with the accompanying translation, will more clearly convey the character of the recorder signals. The message is translated and written down by the operator as the tape glides in front of him.

CABLE DÚPLEX

22. In the **Muirhead cable duplex**, shown in Fig. 5, the condensers C, C_1 are used in place of resistances in the ordinary bridge duplex. A condenser C_1 is inserted in the bridge circuit ef. When connection is made at K with one pole or the other of the battery B, the condensers C, C_1, the cable, and the artificial cable are charged and the condenser at the distant station corresponding to C_1 is also charged. But if the charge of C is to the charge of C_1 as the charge of the cable is to the charge of the artificial cable, there will

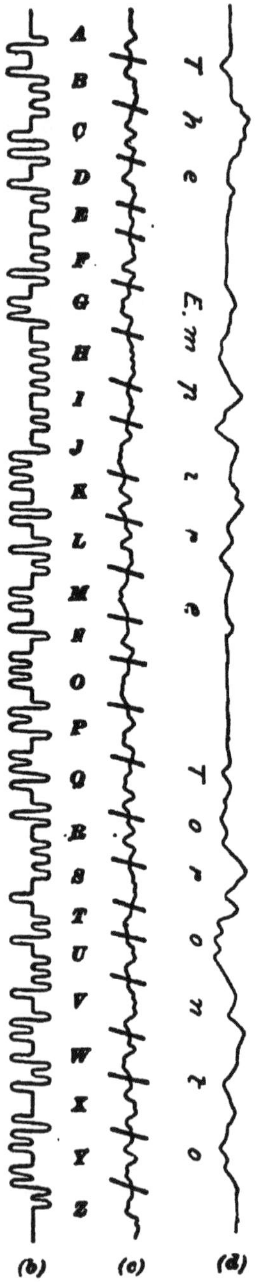

(*b*)   (*c*)   (*d*)

FIG. 4

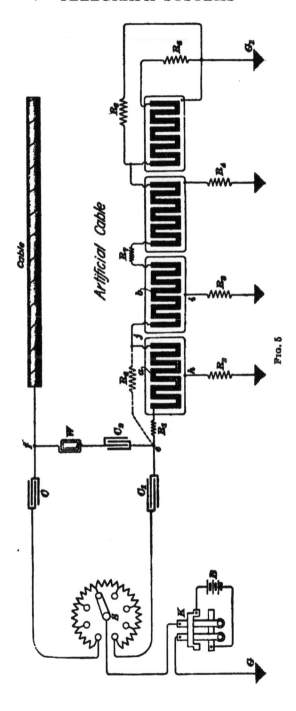

FIG. 5

be no charge given to C_2, because there is no difference of potential between the two points e and f to which the receiving circuit, containing W, C_2, is connected. Consequently, the receiving instrument W is not affected. It has been found by trial that the cable duplex is balanced properly when the capacity of C is to the capacity of C_1 as the square root of the resistance of the artificial cable, per microfarad, is to the square root of the resistance of the cable, per microfarad. By the resistance of the cable, or artificial cable, per microfarad is meant the total resistance divided by the total capacity.

S is an adjustable resistance that assists in obtaining a final balance. The arrangement of condensers and other apparatus, shown in this figure, is known as the *Muirhead double-block system.*

23. The **Muirhead artificial cable,** shown in connection with Fig. 5, consists of a very large number of sections, only four of which, however, are shown. A larger view of two sections is shown in Fig. 6. An artificial cable must have not only the same resistance and capacity as the cable, but the capacity must be distributed throughout its length so that it will not charge and discharge either quicker or slower than the cable, otherwise it will be practically impossible to balance the bridge cable duplex. Each section is made by placing on one side of paraffined paper rectangular sheets of tin-foil connected to earth, and on the other side zigzag strips of tin-foil. A great number of these sections, with paper between them, are piled together. The zigzag strips are connected so as to form one long conductor having a resistance and distributed capacity about equal to that of the cable.

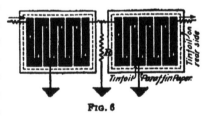

FIG. 6

Sometimes a resistance, as shown at R_1, Fig. 5, is connected in series with the zigzag strips to increase the resistance without increasing the capacity. The tin-foil being extremely thin and cut into quite narrow strips ($\frac{1}{8}$ to $\frac{3}{8}$ inch in width) has quite an appreciable resistance, and it is evident that sufficient resistance and capacity can be obtained by using enough sections. R_1 represents the resistance of the land line from the cable station to the cable itself; it is not usually very large. At the end of the artificial cable, the zigzag sheets are connected through the resistance R_2 to the ground G_1. In order to provide a small amount of leakage, equivalent to that in the real cable, a high resistance R_3 may be connected between one of the zigzag sheets and the ground. In some cases, a resistance R_4 is connected between the point e and some point in the artificial cable, as j. This is adjusted so that the magnitude of the charge and discharge due to the near portion of the cable may be regulated. Thus the artificial cable may be made to resemble the real cable: by having the same resistance, due to the zigzag sheets joined in series; the same capacity, due to the proximity of the zigzag sheets to the grounded rectangular sheets of tin-foil; the same rate of charge and discharge, due to the arrangement of resistance coils; and the same amount of leakage, due to the high resistances that are connected between the zigzag sheets of tin-foil and the ground. For a given cable the artificial cable seldom needs adjustment more than twice a year and the bridge arrangement, by means of the resistance S seldom oftener than once a day.

AMERICAN DISTRICT TELEGRAPH SERVICE

24. In all large cities there are companies with some title resembling the above that furnish their subscribers with messengers whenever one is desired. The subscriber, by turning the crank on a small *call box* placed in his office, causes a special signal or number to be sent to the central office. While the majority of call boxes are used to call a messenger, there are some that enable the subscriber to notify the central office whether a messenger, hack, policeman, fire department, or other service is desired. All types of call boxes may be fitted with a return signal, whereby the subscriber is notified by the ringing of a bell or by some other signal that his call has been properly received. Some of the district telegraph companies are now using specially connected telephones, which are installed in each subscriber's office and permit him to call up the central office and make his wants known. However, the central office cannot usually call up the subscriber, nor can one subscriber be. furnished with connection to any other subscriber. In other words, there is no provision or intention whatever to enter the telephone-exchange business.

Call boxes are made in an almost infinite number of ways and for various purposes. However, they are invariably connected in series in a circuit that does not normally use the ground as a return conductor. Morse ink or embossing registers are invariably used at the central office to record the calls sent in by the subscribers, and, in addition, a bell or gong is generally used to notify the central-office attendant that a call is coming in.

25. A diagram of connections used in the district telegraph service is shown in Fig. 7, in which the central office and four subscribers' call boxes *A*, *B*, *C*, and *D* are included.

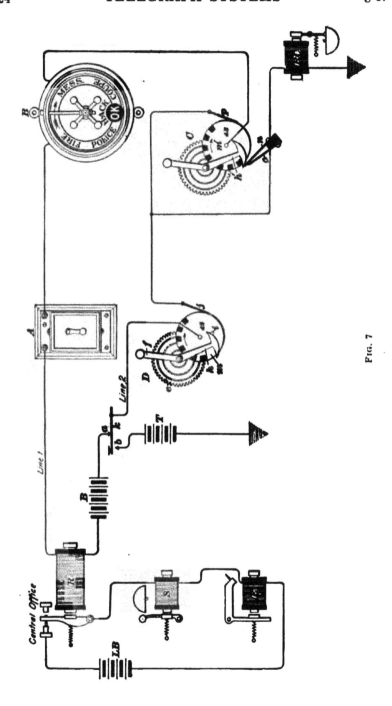

FIG. 7

At the central office, the line circuit normally includes the key k, battery B, and relay R. If call box *42* is properly operated, the line circuit, which is normally closed, will be opened four times, and after a proper interval two times, thus causing the armature of the relay to close, on its back stop, a circuit containing a local battery $L B$, a Morse ink register E, and a gong S four and two times. Thus, a record of the signal *42* is made on the register E and at the same time an audible signal is made by means of the single-stroke gong or bell S. If return-call boxes are used on this circuit, the attendant presses the key k, immediately after receiving the signal, thus sending a current from the return-call battery T over the line to the box that has been operated, where a momentary connection to earth affords a return circuit. The batteries $L B$ and T usually consist of Leclanché cells, and B of some form of closed-circuit cells, such as gravity, Gordon, Edison-Lalande, or storage cells, or a converter or motor dynamo may be used. A, D represent ordinary single-call boxes; B, C return-signal boxes, B, as indicated, having five distinct calls.

CALL BOXES

26. Single-Call Box.—The mechanism at D consists of a gear-wheel e having a spiral spring that is wound up whenever the crank f is turned in the act of calling for a messenger. When the crank is released, it returns to its normal position. When the handle f is turned, the stop h moves out of the path of the pin i and the spring propels the mechanism, causing the break wheel m to make one revolution in the direction of the arrow; the lever h coming in the path of the pin i stops the wheel at exactly the same place every time. As the wheel revolves, the circuit between the wheel and the brush j is broken every time an insulated segment on the periphery of the wheel comes under the brush. Thus, at station D, the circuit is broken four times, and then two times, causing the signal *42* to be sent into the central office.

27. Multiple-Call Box.—In a call box such as is shown at *B*, by means of which several different calls may be made, the circuit would be interrupted after the box number would have been sent in, once for a messenger, twice for a coupé, three times for a hack, etc. On the periphery of the break wheel there would be, besides those necessary for sending in the box number, such additional insulated segments as are required for the various calls.

28. Return-Call Box.—At *C* in the figure is shown the mechanism of a **return-call box.** It has two springs *o, n* that are normally insulated from each other, and also from the break wheel *m'*. When released, the break wheel makes two complete revolutions. On making the second revolution, the arm *h'* comes into such a position as to press the two springs *n, o* together, for a short time only, however. If, while these two springs are in contact, the attendant depresses the key *k*, a current from the return-call battery *T* will pass out over line *2*, through the springs *o, n* and the magnet *R C* of the return-call bell, and return through the ground. It is necessary with this type of box for the attendant to depress the return-call key as soon as the whole signal is received, otherwise he cannot give the return signal. Return-call boxes may be manual or automatic. In one form of the manual return-call box, the subscriber, after calling, must press his finger on a knob, or push button, in order that the office may signal back. A little ball tapping against a glass disk informs the subscriber that his call has been received.

29. Automatic Return-Call Boxes.—Box *B* in this figure is an **automatic return-call box.** When the subscriber calls, the O K disappears, and when the office signals back, the O K drops into view, signifying that the call has been received. The return-call magnet has a resistance of about 13 ohms and the normal current due to the battery *B* is not sufficient to operate it.

30. Many boxes have a provision for temporarily or permanently grounding the circuit in case of a break somewhere in the circuit. By also grounding both sides at the central office, all boxes are still in working condition, but with a ground return, instead of a complete metallic circuit. With metallic circuits, which are the ones generally used, one accidental ground on a circuit does not interfere with the operation of the system. In the case of two grounds on the same circuit only the call boxes between the two grounds are rendered useless until one or both grounds are removed. It is customary to make regular tests at the central office of every call box about once an hour, in order that grounds and breaks may be detected and removed. It is further customary to have extra relays at the central office that may be instantly cut in the circuit in place of the regular relays, in case any of the latter fail to work. All possible precautions are taken to keep all circuits always in working order.

Switchboards, resembling those in telegraph offices, are used for connecting the central-office relays, batteries, and test instruments with the various circuits. Although as many as 100 call boxes may be operated in one circuit, it is not customary to connect over 50 in the same circuit. With such a large number there is so much more danger of signals from more than one box being sent in at the same time. In district telegraph systems, usually no provision is made to avoid the interference of one signal with another. In case it happens and the attendant is unable to recognize one or both signals, there is no remedy, and one or both subscribers must repeat their calls. In case the subscriber has a return-call box he will know, from the absence of a return signal, that he should repeat his call.

Dynamos are now replacing batteries in district telegraph offices.

FIRE-ALARM TELEGRAPH

31. The plain **fire-alarm telegraph system** consists of a central alarm or steam whistle, operated by an automatic electric striker or whistle-blowing machine, and a number of boxes located on the streets at points desired. There are also alarm gongs and indicators in the engine houses; the indicators show the number of the box from which the alarm is sent, and the gongs ring the number, repeating it from three to five times. The boxes are connected, in series, by an iron or copper wire and with a central office, where the battery and testing instruments for operating the system are located. This is called a plain automatic one-circuit system. From this has grown the larger and more elaborate systems now in use in the larger cities. As more stations are required it becomes necessary, for safe and satisfactory operation, to divide the long circuits into two or more independent circuits, each circuit having a separate battery. This necessitates a central-office repeater that automatically repeats an alarm from any one street-box circuit over each engine-house circuit and in such a manner that each circuit gives the alarm at the same time. If one circuit is damaged it does not disable the rest of the system. The central office for this system must be supplied with the necessary protecting devices for safe-guarding the apparatus from lightning or excessive currents due to crosses with other circuits and also instruments for testing, etc., but no attendant is required, since the work is done automatically. When the system becomes too large to be worked automatically a central-office attendant is required. In this case, alarms are received on a telegraph register and then transmitted manually on a transmitting machine, which sends the alarm direct to all engine houses. In the larger and more recently installed fire-alarm systems, storage batteries are being used in place of primary batteries.

32. Street signal boxes are usually of cast iron, cottage-shaped, and contain clockwork, with spring or weight motors, so arranged as to operate a circuit-breaking wheel to open and close the electric circuit, with which the box is associated, a definite number of times at certain intervals, indicating its number on the alarm bells, gongs, and whistles. The make-and-break mechanism is about the same as that in a district messenger box, but it is more substantial, better protected, and contains additional testing apparatus. All the clockwork and finer mechanism is carefully secured and protected. The opening of the door of the outer case usually exposes nothing but a square iron box with a lever or hook projecting from its door, the turning of the lever, or the pulling of the hook, once only, is all that is needed to transmit the alarm. All the boxes are arranged to repeat the box number as often as required, but three rounds are usually regarded as sufficient. Some fire-alarm boxes are now made *non-interfering;* that is, if a signal is being sent from one box, a signal from another box on the same circuit cannot be sent until the preceding signal has been completed. So-called *successive boxes* are not only non-interfering, but will send in the signal correctly as soon as a box that may have been pulled just a moment before has completed its signal. In some boxes the case contains, in addition to the signal box, a lightning arrester, a telegraph key, and a call bell for engineers, or police signals; thus, every signal station may be used as a telegraph office for any city purpose.

33. Automatic fire-alarm transmitters, used in central fire-alarm offices, will transmit a signal over one or more lines, up to eight, simultaneously. The operator at a central office throws a controlling switch that connects the wire on which a signal is being received to the repeater; it is thereupon automatically set in operation and transmits the remaining rounds of the signal over all the connected circuits, coming to rest at the completion of the signals. The mechanism is spring-driven and requires winding only at long intervals.

Manual transmitters are used for transmitting special signals, for retransmitting one that has been imperfectly transmitted or interfered with by another signal.

34. The auxiliary fire-alarm system consists of auxiliary signal boxes that may be placed wherever desirable in factories and institutions without running the main street circuit into the buildings. The auxiliary boxes are connected in independent circuits in such a manner that the nearest street signal box may be operated from any one of the auxiliary boxes associated with it; hence, faults that may develop in the auxiliary circuit do not interfere with the proper operation of the street signal circuit. In case of fire, it is only necessary to break a small glass in the front of an auxiliary box, pull down the ring, and the alarm is given; a loud buzzing, called a return signal, indicates that the alarm has been properly received at the central office.

WIRELESS TELEGRAPHY

35. The most successful method for telegraphing through space without a connecting wire has been perfected by Guglielmo Marconi, whose method depends on the coherer devised by Branly (about 1890) and on discoveries made by Lodge and Righi. To Maxwell and Hertz is due the credit of predicting and demonstrating, respectively, the fact that electromagnetic waves are propagated through space with the velocity of light. These waves, which are frequently called *Hertzian waves*, are identical in some respects to light waves, but have different frequencies and wave lengths and other different properties. For instance, while wood and many other substances are not transparent to light waves, they are transparent to electromagnetic waves. Most metals, however, are opaque to almost all waves. Our optical nerves are capable of detecting light waves but cannot detect electromagnetic waves. On the other hand, a coherer is insensible to light waves, but will readily detect some electromagnetic waves.

SIMPLE TRANSMITTING AND RECEIVING
APPARATUS

36. In Fig. 8, *p* and *s* represent the primary and secondary winding of an induction coil, having an ordinary interrupting device *c d* consisting of a stop *d* and a small piece of iron *c* fastened to a flat spring. With the key *K* open, *c* rests against *d* and so tends to keep the circuit closed between them. The ends of the secondary coil are connected to two brass balls or spheres *a, b,* between which there is a small air gap of about 1 inch.

When the key *K* is closed, the iron core *t* is magnetized and attracts *c*, thereby opening the circuit between *c* and *d*, which allows the core to immediately demagnetize and

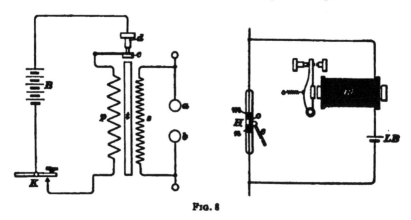

Fig. 8

release *c*, allowing the latter to spring back against *d* and again close the circuit, when the core will again attract *c*. The constant repetition of this action causes the circuit to be broken many times a second, and a torrent of sparks will pass across the gap between *a* and *b* as long as the key *K* remains closed. Although the primary circuit is broken only an appreciable number of times per second, it has been shown by mathematical and experimental demonstrations that the current between the balls *a, b* surges back and forth many millions of times per second and that electromagnetic, or Hertzian, waves are sent out in all directions into space on account of the disturbance or spark that is produced between the spheres. The waves ordinarily produced have a frequency of many millions of vibrations per second.

37. Coherer.—H is a small glass tube, called a **coherer,** that contains two small metal plugs m, n, the small space o between which is partially filled with loose, coarse, metal filings. The resistance of these filings is ordinarily very high, thousands of ohms, and, consequently, enough current cannot flow from the single cell LB to energize the relay R. However, if Hertzian waves of sufficient intensity, or amplitude, are emitted from ab and reach the coherer H, oscillatory currents surge through the coherer and the resistance of the filings may be reduced from several thousand ohms to from 500 to 100 ohms. Therefore, enough current can now flow from the local battery LB through the coherer H and the relay R to energize the latter, and thus close a local circuit, not shown, however, in this figure. The current through the coherer will continue to flow, even if the waves from the transmitter ab cease, but the least mechanical jarring of the coherer after the waves from ab have ceased will restore the filings in the coherer to their normal high resistance and hence cause the relay armature to open the local circuit. A convenient way of jarring the coherer is to arrange an ordinary vibrating bell, the hammer of which is here represented by e, so that it will continually tap the tube lightly as long as the relay holds the local circuit closed. This tapping of the tube does not restore it to its normal high resistance as long as sufficiently intense waves from ab strike it, but the first tap after the waves cease restores it to its normal high resistance and hence the relay opens and the tapping ceases. R is rather a high resistance and sensitive relay, and LB a local battery of one cell.

38. Explanation of the Action of a Coherer.—The following explanation of the action between the filings in a coherer, which is given by Professor Lodge in his book "Signaling Through Space Without Wires," is the one generally accepted:

"Suppose that there are two fairly clean pieces of metal in light contact connected in series with a single voltaic cell; a film of what may be called oxide intervenes between the surfaces so that only an insignificant current is allowed to pass, because a volt or two is insufficient to break down the insulating film. Now let the slightest surging occur, say by reason of a sphere being charged and discharged at a distance from the coherer; the film at once breaks down; perhaps not completely—that is a question of intensity—but permanently. Apparently more molecules get within range of each other and a momentary wave seems to weld them together. It is a singular variety of electrical welding. A stronger wave enables more molecules to hold on and the change in resistance seems to be proportional to the energy of the electric radiation from a source of given frequency. It is to be especially noted that the battery current is not intended to effect the cohesion, only to show that it has taken place. The battery can be applied after the spark has taken place and the resistance will be found to have changed as much as if the battery had been on all the time. The cohesion electrically caused, can be mechanically destroyed. Ground vibrations or any other feeble mechanical disturbances, such as scratches or taps, are well adapted to restore the contact to its original high resistance and sensitive condition. The more feeble the electrical disturbance, the slighter is the corresponding mechanical jar needed for restoration."

For the filings, Marconi has used a mixture of nickel and silver, or carbon dust and cobalt, and also carbon alone. When carbon is used, the coherer is self-decohering; that is, no mechanical vibration is required to restore the coherer to its normal condition when the waves cease. Marconi says that self-decohering coherers are not as sure to act nor as reliable as those requiring mechanical vibrations to decohere them; moreover, he says, relays and recording instruments cannot be so satisfactorily used with self-decohering coherers, telephone receivers generally being required. In most cases, reliability is more essential than extreme sensitiveness.

39. Metals Opaque to Hertzian Waves.—Waves cannot get at a coherer that is completely shut up in a metallic box, but if wires are led to it from outside, the waves seem to run along the wires into the box and the coherer is nearly but not quite as sensitive to the external waves as if no enclosing box had been used. To screen it perfectly, according to Professor Lodge, it is necessary to have no opening of any kind in the box. Even the joints should be soldered. A lid, if securely clamped, using pads of tin-foil to secure perfect joints, may suffice. The inside of the box is then said to be electrically dark.

THEORY OF HERTZIAN-WAVE TELEGRAPHY

40. The phenomena of wireless telegraphy are very complex and appear to be a combination of several effects, one or the other of which predominates, according to the conditions. Electric oscillations are produced along the wire and in the space between the vertical wire and the earth. From the seat of this disturbance originate waves that are propagated through the surrounding space. The Hertzian, or electromagnetic, waves form surfaces of revolution around the vertical wire. The lines of electric force are in meridional planes and are perpendicular to the earth; the electromagnetic lines of force are in horizontal circles having the vertical wire as a common axis. As a result of the effect of concentration, well known in the propagation of waves along conducting surfaces, the electric density is much greater at the surface of the earth, which is a good conductor compared with the atmosphere, directly connected with the oscillator, than in the atmosphere, and in large part the magnetic lines appear to slip along the earth. In the case of a hill intervening, it is supposed that the waves slide up and over it. This concentration, moreover, is the greater the more perfect the conductivity of the surface over which the waves proceed, and the loss of energy in this transmission is thereby lessened over a comparatively smooth and good conducting surface, like the ocean. Yet this concentration does not prevent the diffusion of an important part of the energy into all space, under the form of hemispherical waves, the effects of which are less intense than those near the earth, but, nevertheless, noticeable.

One of Marconi's assistants stated that the length of the waves generally used proved to be about four times the length of the vertical wire. This would make the length 600 feet for the waves that leave a vertical wire 150 feet high. Professor Fleming states that Marconi's transatlantic waves were about 1,000 feet long, which was not small compared to the obstacle they had to encounter; that is, the hill of water formed by the curvature of the earth, which he calculated was about 110 miles above a straight line joining the Lizard and Newfoundland.

The receiving wire, cut at all points by the lines of magnetic force, is the seat of a resultant E. M. F. proportional to the intensity of the field and to the rapidity of the oscillations. The higher the vertical wire the more lines of force are cut. With a given length, fewer lines are cut as we ascend farther from the earth. It is not necessary that the receiving vertical wire be connected to the earth, but the range appears to be greater, due, perhaps, to conduction over the surface of the earth or to the greater electrostatic capacity of the system that connection with the earth produces.

41. The energy represented by the waves that leave the transmitting station and are received at the receiving station is said to vary as the square of twice the distance between the oscillator and coherer; hence, the energy required to transmit signals increases enormously as the distance becomes greater.

The quality of communication is about the same whatever may be the conditions of fog, rain, wind, etc. The heights of vertical conductors required over land are always considerably greater than those sufficing for sea communication over the same distance. Messages may be sent with the same apparatus about three times as far on the ocean as on land.

It seems as though the curvature of the earth can have but little influence, if any, on the height required for the vertical wire. All other conditions being equal or favorable, the distance over which it is possible to signal varies with the square of the height of the vertical wire when the vertical wires at the transmitting and receiving stations are equal, or to the product of their heights provided they are approximately equal.

42. Results Attained.—As the field for wireless telegraphy seems to be between the shore and ships, stations have been placed along the coast, and transatlantic liners, when properly equipped, can keep in communication with America or England all the way across the ocean.

In December, 1901, the first signals, which consisted only of the letter *s* in the Continental code, were sent by the Marconi system across the Atlantic Ocean from Poldhu, England, to St. Johns, Newfoundland. In February, 1902, messages were sent from Poldhu, England, to the ship "Philadelphia" as the latter proceeded to the United States. At Poldhu was a very powerful transmitting station having fifty almost vertical wires about 48 meters high. The vertical conductor on the ship extended about 60 meters above the sea level. The signals were recorded on a Morse register. Readable messages were received up to a distance of 1,551 miles from Poldhu and simple signals as far as 2,099 miles. In December, 1902, Marconi sent messages both ways between Poldhu and Glace Bay, Nova Scotia.

SYNTONIC SYSTEMS

. **43.** A serious fault charged against wireless telegraphy, and one that has prevented its competing with systems using connecting conductors, is the fact that, with the ordinary arrangement of apparatus it is impossible to obtain, at the same station, two independent communications; every receiver placed in the radius of action of a transmitter is acted on by the waves sent out by the one transmitter. Various systems, founded on different principles, have been proposed with a view to avoid this fault.

If a circuit having certain resistance, inductance, and capacity be placed in a region in which waves are passing in such a position that the successive waves can induce currents in it, then each wave will tend to slightly increase the intensity of the preceding wave, provided the waves have a certain particular frequency. The oscillations will increase in intensity, just as small pushes given to a pendulum at the proper times will make it swing violently. Such a system is said to be in tune, resonance, or syntony with the waves or the generator that emits the waves. The generator and receiver are also said to be in tune, resonance, or syntony with each other.

44. Resonance.—Research has shown that in· some circuits the electric oscillations die away very rapidly. In all so-called good radiators of electric waves, electric oscillations set up by an ordinary spark discharge are very apt to cease, or be damped out very rapidly, not necessarily by the resistance, but by electrical radiation removing the energy in the form of electric waves. In order to set a heavy pendulum in motion by means of small thrusts or impulses, the latter must be timed to agree with the period of oscillation of the pendulum, since otherwise its oscillations will not acquire any perceptible magnitude, or *amplitude*, as it is called. The same effect is obtained in a very small fraction of a second when an attempt is made to induce electrical impulses in a good resonator. Electrical resonance, like mechanical resonance, essentially depends on the accumulated effect of a large number of small impulses properly timed.

That under certain conditions various receivers will respond to one transmitter, even if the periods of the receivers be different from that of the transmitter, may be due to the fact that all the energy of the transmitter is radiated in only one or two impulses, with the result that oscillations may be induced in receivers of different periods, whereas, if the same amount of energy be distributed in a great number of individual and more feeble impulses, their combined effect can only be utilized or detected by a receiver tuned so as to respond to their particular frequency. The tuned receiver will not then respond with sufficient intensity to the first two or three oscillations, but only to a longer succession of properly timed impulses, so that only after an accumulation of a number of impulses will the E. M. F. developed become sufficient to break down the coherer and cause a signal to be recorded.

MARCONI TUNED SYSTEM

45. Marconi's transmitting circuits are shown in Fig. 9 (*a*). *I* is an ordinary induction coil, capable of producing very heavy, or thick, sparks across the air gap *a*. The condenser *C*, coil *c*, and gap *a* constitute the oscillator, whose inductance, resistance, and capacity are adjusted for a desired frequency. The induction coil *c b* contains no iron. *L* is a variable inductance and *N* a surface of metal wire or gauze, having, therefore, considerable capacity, and is elevated in the air. *N–L–b–G* is in tune with *C–c–a*. Sometimes cylinders are used in place of *N*, the inside cylinder being grounded and the outside cylinder connected to *L*. This gives considerable capacity. The cylinders may be 20 feet in length, the inside one 1½ feet and the outside one 3 feet in diameter, respectively. A ·similar cylinder, similarily connected, would also be used at the receiving station.

46. The **receiving circuits** are shown in Fig. 9 (*b*). The secondary of an ironless induction coil consists of two equal coils *f*, *g*, while *d*, *e* are adjustable inductance coils. No coils contain iron unless shown and so stated. The circuit for the oscillating current may be traced as follows: *f–d–C₁–e–g–C₁*. When the oscillations in this circuit have attained sufficient intensity, the potential difference across the condenser *C₁* becomes sufficient to break down the coherer *H*; then sufficient direct current flows from *B* through *i–f–d–H* *–e–g–j–R–B* to operate *R*. *i*, *j* are very small coils wound on very small iron-wire cores, but they have so much inductance as to exclude the high-frequency oscillatory currents, which are therefore compelled to take the path traced out for them. n is a non-inductive resistance shunting the relay to absorb the kick due to the inductance of the relay coil.

The best results are obtained when the length of wire constituting the secondaries b, m of the ironless induction coils is equal to the length of the vertical wire used at the transmitting station. Marconi says that experiments demonstrate that the receiving induction coils having the secondary wound in one layer and the turns a certain distance apart, say 2 millimeters (to cause the capacity to be so small as to be negligible), have a time period approximately equal to that of a vertical wire of equal length. Therefore, an induction coil having a secondary wound with a wire 40 meters long at the receiving station should be used with a vertical wire 40 meters long at both transmitting and receiving stations.

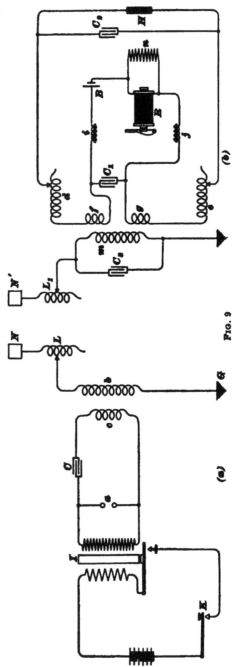

Fig. 9

Then it is only necessary to adjust the capacity of the condenser C, Fig. 9 (a), at the transmitting station until the signals are recorded at the receiving station.   By following the above, using cylinders in the air or enough vertical wires, and having previously proportioned the capacity and inductance of all the transmitting and receiving circuits shown in Fig. 9 (a) and (b), Marconi claims to have solved the problem of selective signaling.   For moderate distances this is doubtless true.   It is doubtful, however, if a near-by and powerful transmitting station, especially if not controlled by the same company, would not affect a sensitive receiver tuned for a distant transmitting station.   Up to 1904, selective signaling does not seem to have been commercially very successful. In order for two systems to be in tune, it is necessary, if the resistance is assumed to be negligible, as is generally the case, that the product of capacity and inductance of all circuits at both transmitting and receiving stations should be equal.

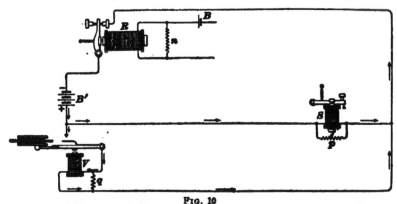

Fig. 10

47.   Local Circuits.—It is necessary to suppress all sparks and kicks due to inductance in the local circuits. This is accomplished by connecting non-inductive high resistances across all contact points and all magnet coils.   Furthermore, the coherer must be protected from the transmitter at the same station, while transmitting.   This is accomplished by enclosing all the receiving circuits, except the Morse register that is used to record signals, in a specially constructed metal box that is practically wave-proof when the vertical wire is removed and the hole through which it made connection with the receiving apparatus is closed with a metal door.

The arrangement of local circuits, said to be used by Marconi, is shown in Fig. 10. The relay controls two circuits in parallel with each other, both operated by the same battery B', however. One circuit contains a vibrating bell V, the hammer of which is arranged to tap the coherer. The bell V and register S should have high resistances, about 1,000 ohms, and should be shunted by non-inductive resistances q, p of about 4,000 ohms each. It is preferable to connect across the contact points of the relay and vibrating bell non-inductive resistances of somewhat less resistance than that across the bell or register. The relay coils should have a resistance of about 1,200 ohms and should be shunted by a non-inductive resistance n of about 5,000 ohms. The battery B in this figure is the same as B, Fig. 9 (b), and consists of only one cell. The battery B' may consist of any desirable number of cells, usually twelve or more.

48. Results Attained in Selective Signaling. While it is possible to make a receiver respond to only one transmitter, up to the present time it has not given very good results. In the fall of 1900, Marconi succeeded in simultaneously sending two different messages between two stations in England, 30 miles apart, and they were recorded on Morse registers without mistake. Each receiver in this case was connected to the same aerial wire about 40 feet high. Other trials showed that messages could be sent from one station to another, while between two other stations, messages were also being sent, the line between the first two intersecting the line between the second two stations. In March, 1901, Marconi stated that by his system, messages from five different places could be received simultaneously at one point. There seems to be no system whose signals cannot be broken up by a rival company when it so desires; in fact, it seems to be difficult for more than one company to operate in the same territory without interfering with one another. Marconi also states that his successful signaling across the Atlantic Ocean during 1902 was due to the use of his tuned system.

There are a number of wireless telegraph systems for which much has been claimed and some of which have undoubtedly proved successful. Many details that are necessary for the successful operation of any system are kept as secrets by the various inventors and companies.

TELEGRAPH LINES

BARE OVERHEAD LINES

WIRE

49. For telegraph lines galvanized iron, steel, and hard-drawn copper are used.

Galvanized iron wire is used more than any other, although the use of hard-drawn copper for this purpose is rapidly increasing. Aluminum wire is used very little for either telegraph or telephone lines, but as it becomes lower in price it may come more into favor.

50. Grades of Iron Wire.—The various grades of iron wire on the market are termed "Extra Best Best," "Best Best," and "Best." A steel wire, which is cheaper and of higher resistance than iron, is also used. The steel wire has the advantage of possessing greater tensile strength. *Extra Best Best* (E. B. B.) is the highest grade iron wire obtainable. It has the highest conductivity and is very uniform in quality, being both tough and pliable. *Best Best* (B. B.) is less uniform and tough than Extra Best Best, but is a fairly good grade of wire. *Best* (B.) is the poorest grade of wire. It is harder, more brittle, and has a lower conductivity than the better grades. *Steel wire* is lower in conductivity but has a greater tensile strength than any of the grades of iron wire. For a short telegraph or telephone line, or for an especially long span, this wire is sometimes used in preference to iron wire, as its lower conductivity may not be a great objection for a short distance, while its tensile strength may be a decided advantage.

51. Galvanizing.—Iron is very susceptible to corrosion, due to moisture and other elements in the atmosphere, and in order to protect the iron wires used in outdoor work, they are covered with a thin film of zinc. The process by which this is done is called *galvanizing*. The thin zinc coating, on being exposed to the atmosphere, becomes oxidized, and as oxide of zinc is not soluble in water, it forms a protection against moisture. However, when the zinc is exposed to the action of sulphur or chlorine in smoke or salt spray, it is converted into zinc sulphate or chloride, which readily dissolves in water. Under especially adverse conditions iron wire will last only a few years, and in some cases only a few months; it is therefore desirable to use copper wire in such cases, as the latter is practically indestructible.

52. Size of Line Wire.—No definite rules can be given for the size of wire to use on overhead lines. The following sizes are those in use: No. 10 B. & S. hard-drawn copper and No. 4 B. W. G. galvanized-iron wires are now used on important quadruplex circuits. Formerly, No. 6 B. W. G. galvanized-iron wire was used for this purpose. No. 6 B. W. G. galvanized-iron wire is used for important circuits between cities. No. 8 B. W. G. galvanized-iron wire, or No. 12 B. & S. hard-drawn copper wire, is much used for circuits of 400 miles, or less, in length. No. 9 B. W. G. galvanized-iron wire was formerly used for this purpose. No. 9 B. W. G. galvanized-iron wire was, until recently, the size most generally used in the United States. It is now used on short circuits where No. 8 is not considered necessary. Nos. 10 and 11 B. W. G. galvanized-iron wires are used for still shorter circuits and for railway telegraph, police, fire-alarm, and private lines. No. 12 B. W. G. galvanized-iron wire is also used for these purposes and for telephone lines. Nos. 13 and 14 B. W. G. steel wires are used for short private lines, for telephone lines, and where strength is especially necessary. No. 8 B. & S. copper wire should be used for permanent ground wires in terminal telegraph offices.

53. The increasing use of copper for telegraph-line work is largely due to the fact that copper can be drawn into wire, called hard-drawn copper wire, having very nearly the same tensile strength as iron wire, and having only about one-seventh the resistance. The fact that copper wire is practically non-corrosive when exposed to ordinary atmospheric conditions, while even the best grades of galvanized-iron wire are always, sooner or later, rendered useless by corrosion, is another very strong argument in favor of copper. The size of copper wire used for telegraph-line work varies from No. 9 B. & S. gauge to No. 14 B. & S. A No. 9 B. & S. hard-drawn copper wire was strung from New York to San Francisco by the Postal Telegraph Cable Company in 1903.

54. Joints.—Iron wires are usually connected in this country by the American telegraph joint, as shown in Fig. 11. The wires are first placed side by side, and then each end is wound around the other. The joint should then be soldered, to insure the maintenance of a perfect electrical

FIG. 11

contact. The joint may be cleaned for soldering by a wash of muriatic acid in which zinc has been dissolved. If the joint is to be covered with insulation, it should not be soldered with acid, but with resin or some soldering compound that contains no acid.

Another joint that is in general use especially for hard-

FIG. 12

drawn copper and aluminum wires, and even for iron wire, is the McIntire sleeve joint. The ends of the wire are slipped into a double sleeve of the same metal as the wire, as shown in Fig. 12, and the two are then twisted through several turns, making a joint like that shown in Fig. 13. These joints give excellent service, always keeping good

Fig. 13

electrical contact without the use of solder. The heat required to solder an American telegraph joint generally softens and weakens hard-drawn copper wire.

POLES

55. Overhead telegraph and telephone wires are supported on glass insulators generally on wooden poles, as shown in Fig 14, unless appearance or mechanical strength compels the use of an iron pole. Wood is preferable on account of its cheapness and its insulating qualities. The woods used are pine, cypress, red and white cedar, redwood,

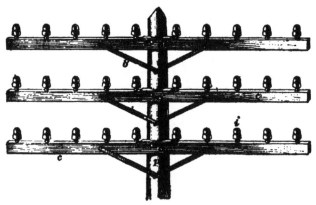

Fig. 14

and chestnut. Poles should be of the best quality, as straight as possible, and well proportioned. Their lengths depend on the nature of the locality in which they are erected and the ordinances governing the district. In cities, they should be so tall that the wires will not interfere with the possible work of the fire-department, and so that the wires are not dangerously near roofs, house fronts, trees, or trolley wires, which last ought always to be provided with guard wires when passing beneath overhead conductors.

56. It is usually required that poles on which cross-arms are to be carried shall have a diameter of not less than 7 inches at the top. This dimension is, as a rule, sufficient to specify for any length of pole, for the natural taper of the tree will insure a proportionately larger diameter at the butt. Larger poles should be used for corners and curves, and, when possible, should be raked from the inside of the bend. Poles are becoming more scarce and higher in price each year, and consequently those that formerly would be rejected are now being accepted. Poles 5 inches in diameter at the top are now being used for telephone poles in some places. Steel poles are also coming into use.

Before erection, the poles should be prepared by having the bark peeled and the knots trimmed close; then they should be cut to the proper lengths, the butts squared, and the tops chamfered or coned. In cities, or in conspicuous places, the specifications may call for octagonal or sawn poles. It is in some cases desirable to treat the poles by creosoting, or by injecting some preservative into the pores of the wood, after which they may be painted. It is always good practice to coat the butt of the pole with pitch for a distance of 6 feet.

CROSS-ARMS

57. The **cross-arms** c, c, Fig. 14, should be made of sound, well-seasoned, straight-grained timber. Some prefer red or black cypress for this purpose, though yellow pine, especially the long-leaf variety, creosoted white pine, Oregon fir, and yellow poplar make excellent cross-arms. All cross-arms should be given two coats of good oil paint before leaving the factory.

The size and length of cross-arms depend on the load they are to carry. Two regular sizes, however, are made, one termed the *standard cross-arm*, and the other the *telephone cross-arm*. The standard cross-arm is used for all heavy work and in constructing a line that is expected to last.

The standard cross-arm is $3\frac{1}{4}$ by $4\frac{1}{4}$ inches and varies in length from 3 to 10 feet. It is usually bored for $1\frac{1}{4}$-inch wood pins or for $\frac{1}{2}$-inch steel pins and provided with holes for two $\frac{1}{2}$-inch bolts.

The so-called telephone cross-arms are lighter, being made from $2\frac{1}{4}$- by $3\frac{1}{4}$-inch stuff, sometimes 3 by $4\frac{1}{4}$ inches, and bored for $1\frac{1}{4}$-inch pins, and provided with two $\frac{1}{2}$-inch bolt holes. For light lines, these arms give excellent satisfaction, but are not, of course, as durable as the heavier arms, and are seldom used for telegraph lines.

58. Gains, or slots for the cross-arms, are cut in the poles at intervals usually of 24 inches for telegraph poles and 20 inches for telephone poles. These should be $1\frac{1}{4}$ inches deep and wide enough to fit the cross-arm snugly. They should be carefully cut, and afterwards painted with white lead or oil paint, to prevent rotting due to the collection of moisture. In each cross-arm are bored two holes for the lagscrews or carriage bolts s, s, that secure the arm to the pole. The carriage bolts are the better, because they injure the pole less and new cross-arms can be more securely put on if necessary. The screws with washers are then inserted and screwed up. Lagscrews should never be hammered more than a short distance into the pole, as hammering destroys the holding quality of the fiber of the wood. The cross-arms may also be braced by iron braces b that extend from the pole 18 inches below the arm to 19 inches from the pole on each side. The braces are $1\frac{1}{4}$ inches wide, $\frac{1}{4}$ inch thick, and 28 inches long. When the poles are up, the cross-arms of adjacent poles should be alternately on the same side and on the opposite side from each other, so that if one arm gives way, either the arm next to it or the next but one will successfully withstand the shock.

BRACKETS AND PINS

59. Wooden pins for the insulators (see d, k, Fig. 15) should be made preferably of oak or locust. They are turned up, with a coarse thread on one end, on which the insulator is to be screwed. The shank *k* is turned to a diameter of 1¼ inches for telephone cross-arms and 1⅜ inches for standard cross-arms, so as to fit the hole in the arm. If the pins are loose in the hole, they should be fastened

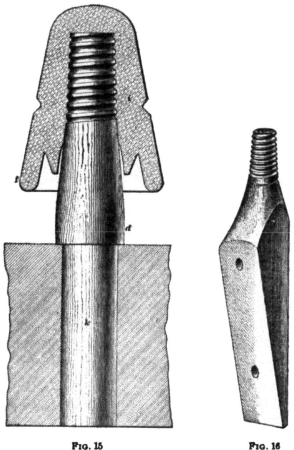

FIG. 15   FIG. 16

by driving a nail through the arm and through the shank of the pin.

If only one wire is to be placed on a pole, a **bracket pin,** Fig. 16, is used. This is so shaped at the lower portion as to lay against the pole at an angle, and is secured to the pole by two heavy spikes.

60. **Insulators** in the United States are usually made of glass. For very wet climates, porcelain may be a better insulator, but is more costly, and under the action of cold the glazed surface is apt to become cracked. When this happens the moisture soaks into the interior structure, and its insulating quality is greatly impaired, if not rendered useless. The form of insulator shown in Fig. 15 is much used in telegraph and telephone work. It is bell-shaped, with an interior thread, and a single "petticoat" *t*. The resistance offered by insulators follows the same law as the resistance of conductors. The larger the path afforded and the less its cross-section, the greater resistance it will give to leakage from the line. When moisture forms on the glass, it must first cover the outside of the cup, and then proceed up the inner curve before it causes leakage.

In cases where greater insulation resistance is required, double petticoats may be used. The moisture on such an insulator must extend beyond two inner curves before causing leakage.

61. Tying Wires to Insulators.—Fig. 17 shows the most common method of attaching iron line wires to insulators. The line wire is not passed around the insulator, but is simply laid in the groove and tied there by a piece of the same size of wire used for the line wire. In cases where a very heavy line wire is being used, however, the tie-wire may be somewhat smaller. Some advocate to start wrapping one end of the tie-wire over and the other under the line wire.

FIG. 17

A tie that is now being largely used in telephone work, and which is probably the best for hard-drawn copper wire, is shown in Fig. 18. In this, the line wire is laid in the groove of the insulator, and the tie-wire is laid in the groove and passed once entirely around the insulator. One end of the tie-wire is then brought down over the line wire, while the other end is brought up under it in an opposite direction, the two ends being wound around the line wire, as shown in the figure.

62. Lightning Conductors.—It is customary to run a grounded wire up certain poles, allowing 3 inches to project above the top of the pole, for a protection against lightning. A No. 8 B. W. G. galvanized-iron wire is recommended for this purpose, and at least 6 feet of it should be laid in zigzag shape (not in a coil) in the bottom of the pole hole. It is generally considered advisable to put a lightning conductor on at least every fifth telegraph pole. Telephone engineers generally specify a lightning conductor of No. 6 B. & S. copper wire on every tenth telephone pole.

Fig. 18

TELEGRAPH AND TELEPHONE CABLES

63. Where it is necessary to run a greater number of wires than can be accommodated by the bare-wire construction, cables become necessary. For both indoor and outdoor work, the use of a cable makes it possible to easily run a large number of wires where the same number by the ordinary construction would be out of the question. Moreover, for the problem of underground and underwater work, where it is impossible to use bare-wire construction, the cable forms the only solution.

RUBBER-COVERED CABLES

64. Cables composed of rubber-covered wire, without a lead cover, are occasionally used for overhead telegraph circuits, although the paper-insulated cable is much more desirable. A good **rubber-covered cable** manufactured in the United States is made as follows: The wires are of copper Nos. 14, 16, or 18 B. & S. gauge, having a conductivity of 98 per cent. that of pure copper. The wire, having been thoroughly tinned, is given a double coating of rubber insulation, and then taped. After the requisite number of conductors are bunched, the cable is double-taped and covered with tarred jute, over which is placed a heavy braid of cotton saturated with weather-proof compound, which serves not only to protect the rubber from the action of the air, but to protect the entire cable from mechanical injury.

The insulation resistance of wires in this cable will, if the cable is in good condition and the proper materials are used, vary from 300 to 500 megohms per mile, at a temperature of 60° F., after the cable has been immersed in water for 24 hours. For use in mines or other wet places, or where it is necessary to pass a large number of wires on poles through the foliage of trees, this cable should give good results. One objection to cables of this kind is the high electrostatic capacity of the conductors. Rubber is an excellent insulator, but has a very high inductivity, thus greatly increasing the electrostatic capacity of the wires that it serves to insulate.

While high electrostatic capacity is undesirable in a telegraph cable, it is not so serious an objection as in a telephone cable. On the other hand, a low resistance is more desirable in telegraph than in telephone cables, consequently telegraph-cable conductors are seldom smaller than Nos. 14 or 16 B. & S.

PAPER CABLES

65. Methods of Reducing Capacity.—To reduce the electrostatic capacity of the conductors of a cable, three methods are available: First, the wires may be placed farther apart; second, the wires may be made smaller; and third, an insulating medium having a low inductivity may be used.

To place the wires farther apart would be to defeat the principal object for which the cable is employed; that is, compactness. The sizes of the wire may be reduced to a certain limit, but beyond that limit the mechanical strength of the conductor and its ohmic resistance forbids us to go. As a result of this reduction in· the size of wires, Nos. 19 to 24 B. & S. gauge are commonly used in dry-core telephone cables. In following the third method of reducing the electrostatic capacity, various materials having a lower inductivity than rubber have been tried, and have been found to give far better results so far as the electrostatic capacity is concerned, and, in fact, in all other respects, when proper care is exercised in their manufacture and maintenance.

66. Saturated-Core Cables.—Underground or overhead cables for telegraph purposes are now usually insulated with paper or cotton fiber, and saturated with an insulating compound. Paper is preferable to cotton. The composition of the insulating compound is a trade secret, known only by the companies that manufacture the cables. These cables are commonly termed **saturated-core cables,** in order to distinguish them from the dry-core paper cables used for telephone systems.

The advantage of a saturated-core cable over a dry-core is that if only the lead is injured, a small portion of the cable (sometimes only a few inches) will be lost, and the injury may be located and repaired before much damage is done. Sometimes the conductors in a telegraph cable are made of Nos. 16 or 18 B. & S. gauge, but generally of No. 14 B. & S. gauge, and each wire is insulated to a thickness of $\frac{1}{32}$ inch. The conductors are laid up in layers, each layer being wound in a contrary direction to the preceding one. The bunch of insulated conductors should be covered with the same thickness of insulation as is used on each wire, and the whole encased in a lead sheath, usually $\frac{1}{8}$ inch thick.

Fig. 19 shows a saturated paper-core telegraph cable made by the National Conduit and Cable Company, which makes a large variety of paper-insulated cables—dry-core cables for telephone purposes, and saturated-core cables for telegraph, electric-light, and power purposes. Paper-insulated cables have proved very satisfactory and successful.

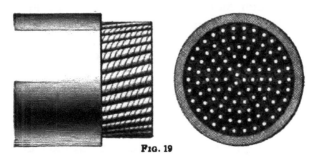

Fig. 19

The cable illustrated in this figure contains 100 conductors of No. 14 B. & S. copper wire, each conductor being covered with a paper insulation that has been treated with an insulating compound. The outside diameter of the cable is 1⅝ inches.

67. Dry-Core Cables.—The telephone cables now most commonly used are known as **dry-core cables**. They are made by insulating the various wires with a loose wrapping of very porous dry paper, after which the wires are twisted in pairs and bunched into a cable. A sheath of lead is then placed over the cable in order to exclude all moisture and also to prevent mechanical injury. The loose wrapping of the paper and its porous nature insure the inclusion of a great amount of dry air in the cable, which, as we have seen, possesses the lowest inductivity of any well-known insulating material. Two or three feet at each end of the dry-core cable is always saturated or sealed up tight, to exclude moisture, with paraffin; or, better, with some of the special compounds that are made and used by the cable manufacturers. Immediately after testing in the factory, the lead sheath at each end is hermetically sealed by a plumber's joint.

The electrostatic capacity of the wires in a cable built in

this manner is often as low as .06 microfarad per mile, and it is customary, in making specifications for telephone cables using No. 19 B. & S. gauge wire, to specify that the electrostatic capacity of each wire shall not exceed .08 microfarad per mile. All cables of this description are made in twisted pairs, the conductors being twisted together so as to give one turn in about 6 inches.

The dry-core cable represents the highest development in the line of telephone cables. The high insulation obtained by the dry air and paper, the low electrostatic capacity, and the compactness of the cable as a whole render it admirably adapted for both underground and aerial work.

A puncture in the sheath of a dry-core cable allows the entrance of moisture, which, due to capillary attraction, will soon penetrate the entire length of the cable, thus totally ruining the insulation. When moisture first enters, immediate steps should be taken to expel it and to repair the fault, or the cable will soon be worthless. This point shows the necessity for making frequent insulation tests on cables of the dry-core type, so that if moisture enters its presence may be detected before it has time to do serious damage.

68. Telephone and telegraph cables are usually drawn into vitrified clay, cement-lined iron, or wooden conduits underground in the central part of large cities and overhead on poles, being suspended from a steel cable fastened to the cross-arms or poles, in the outlying districts of large cities and in the central parts of smaller cities. The overhead cable construction is cheaper than underground; some claim that it is even cheaper than the use of bare overhead wires where there are many wires to be run on one pole line.

The only objection, outside of the cost, to the use of a cable is the greater electrostatic capacity it gives to the line circuit. This is an especially serious objection for long telephone circuits, for which the use of cables is to be avoided as much as possible. The articulation becomes poorer the greater the electrostatic capacity of the line circuit.

69. It is frequently necessary to extend telegraph lines under water, either in crossing rivers, bays, or lakes, or in extending lines from the mainland to neighboring islands. For short lengths of cable across rivers or bays having smooth bottoms and slow currents, cables of the ordinary lead-covered type, having rubber or gutta-percha insulation, are sometimes used, no special armor for the mechanical protection of the cable being necessary. It is well in such cases to have an extra heavy lead sheath, and also to have

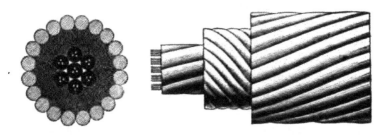

FIG. 20

the lead sheath covered with a heavy braiding of fibrous material saturated with waterproof compound.

In order to meet more severe conditions, special armored cables of the best rubber or gutta-percha covered wire are required, the whole bunch being embedded in rubber insulation or a heavy wrapping of jute, which is afterwards served with an armor composed of iron wire of about No. 10 B. & S. gauge, affording a continuous mechanical protection for the wires and insulation within. This construction is shown in Fig. 20.

70. A submarine cable consists of a core that comprises the conductor, made of copper wire, and its insulating covering of gutta percha, over which is placed a tanned jute yarn covering, to protect the gutta percha from the steel-wire sheathing. As a protection against the Teredo bug, some cables have the gutta percha covered with a layer of white canvas tape and then with brass tape. This method

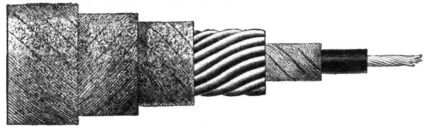

FIG. 21

has successfully protected a cable that was laid in 1879 in the Straits of Malacca and in Java. Over the gutta percha

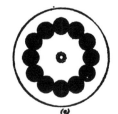

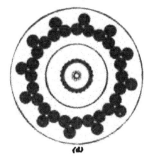

FIG. 22

and jute yarn is wrapped the steel-wire sheathing, and this, in turn, is enclosed in jute yarn and a bituminous compound.

The different coverings of an intermediate cable are shown in Fig. 21, the specimen being cut at intervals to show each covering in succession.

There are as many as seven types of sheathing, increasing in strength and protective power as the shallow water is reached. Four types are shown in Fig. 22, in which (a) is the deep-sea type, with a sheathing of many small steel wires. This type weighs about 1½ to 2 tons per knot. In the intermediate types (b) and (c), the sheathing wires become gradually larger, and finally in the shore-end type (d), the deep-sea sheathed cable (a) is again sheathed with strands, each made of three steel wires. It will be noticed, however, that the core of copper wires and the gutta percha are the same size throughout.

An intermediate cable, for use in a depth of from 500 to 1,000 fathoms,* would weigh about 3 tons per knot, a heavier intermediate cable, for use in from 50 to 100 fathoms, about 6¼ tons, and a shore-end cable about 10¼ tons per knot. The transatlantic cable laid in 1894 is 2,161 knots long and weighs 5,460 tons. Another transatlantic cable (from Canso to Waterville) has a total resistance of 6,997 ohms, an electrostatic capacity of 876 microfarads, and a length of 2,346 knots.

*A fathom is equal to 6 feet and a knot to 6,080 feet.

TELEPHONE SYSTEMS

(PART 1)

TELEPHONY

1. Telephony is the art of transmitting articulate speech and other sounds between distant points by means of fluctuations of an electric current flowing between those points.

SOUND

2. Sound is produced by vibratory movements in elastic bodies. It is usually produced by a vibrating solid, and is transmitted through the atmosphere to the drum of the ear; it may, however, be produced by a vibrating liquid or gas. As the transmitting medium must possess elasticity, and as all solids, gases, and liquids have this property to a certain extent, sound is transmitted through them with more or less facility. Its transmission through air may be explained as follows:

Imagine the thin elastic plate, or *diaphragm P*, Fig. 1, that is held between supports *M*, *N*, to be in rapid vibration. As it moves to the right there is a condensation of the atmosphere at *C* and a rarefaction at *R*. The condensation at *C*

NOTE.—While it is desirable for an electrical engineer to understand the principles involved in telephone systems and apparatus, it is not necessary for him to know about all the systems and their operation; this knowledge is only essential to the telephone engineer. Therefore, only such systems and apparatus as will probably be of interest to the average electrical engineer will be considered.

is communicated to the particles of air at *A* at the right of *C*, and by them to the particles of air at their right, and so on. A wave of condensation, therefore, travels through the air, gradually diminishing in intensity, until it is finally lost. In its movement to the left, the plate causes a rarefaction at *C*, and the particles of air from *A* rush into it, thus causing a rarefaction at *A*, which, in turn, is filled by particles in the space at its right. Thus, a wave of rarefaction follows the wave of condensation, and this, in turn, is followed by another wave of condensation, and so on as long as the plate continues to vibrate. A similar set of

FIG. 1

waves is also sent in the opposite direction by the other side of the plate. Waves of this description that are capable of acting on the ear to cause the sensation of sound are called **sound waves.**

CHARACTERISTICS OF SOUND

3. All sounds have three characteristics, variations in which enable us to distinguish between different sounds. They are *loudness*, *pitch*, and *timbre*.

Loudness is that characteristic of sound that depends on the amplitude of the sound wave. Thus, a certain note on the piano may be loud or soft according to whether we make the string vibrate violently or not. Loudness is the intensity of sound.

Pitch depends entirely on the rate of vibration of the body producing the sound. A low rate of vibration produces low tones, and a high rate shrill or high tones. The difference between the sounds emitted by long and short strings of the same material and of equal size and tension is one of pitch.

Timbre is the quality of sound, and depends on the form of the sound wave. The difference between a certain tone on the flute and the same tone on the violin is one of timbre.

ARTICULATE SPEECH

4. The successive vibrations of the human voice that form distinguishable and intelligible sounds are called **articulate speech.** The vibrations serving to make up spoken words are probably the most complex in the whole realm of sound. The difference between a simple and a

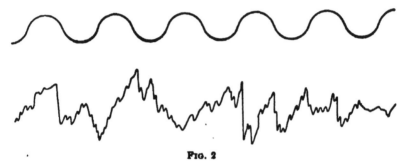

FIG. 2

complex series of sound waves is indicated in Fig. 2, in which the upper line represents the waves of a simple musical sound, as, for instance, that of a tuning fork, and the lower line represents the vastly more complex waves set up by the human voice.

ACTION OF THE TELEPHONE

5. In electric telephony the transmission of sound is accomplished by making one plate, or diaphragm, at the transmitting end of the line take up, or respond to, the waves of the sound to be transmitted, and causing, by electric means, another diaphragm at the receiving end of the line to vibrate as nearly as possible in exact accordance with the first.

THE MAGNETO-TELEPHONE

6. Action as a Receiver.—The production of the first successful speaking telephone is usually accredited to Prof. Alexander Graham Bell. The action of Bell's instrument, which is the simplest form of electric telephone, will be understood by reference to Fig. 3 in connection with the following description:

47—27

A thin, soft-iron diaphragm P is mounted close to, but not touching, one pole of the permanent magnet NS, about one end of which is wound a coil of fine insulated wire C, the terminals of which are connected directly in the circuit in which the instrument is to be used. It is evident that the diaphragm will normally be strained slightly toward the magnet by the attraction of the latter. If a current is sent through the coil in such a direction that the lines of force set up by it coincide with those of the permanent magnet, the strength of the magnet will be increased, and the diaphragm will be pulled still closer to the pole. If, however, a current is sent through the coil in such a direction as to set up lines of force opposing those of the magnet, the

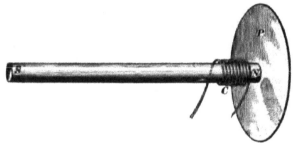

Fig. 3

strength of the magnet will be diminished and the diaphragm allowed to spring farther from the pole.

If an undulating current that always flows in the same direction is sent through the coil, the lines of force induced by it in the magnet will increase while the current is increasing, and decrease while the current is decreasing. Thus, whether the lines induced by the coil are in the same direction as those of the magnet or not, a varying pull on the diaphragm will cause in the latter vibrations that will be in harmony with the changes in current.

If the current is an alternating one, that is, one that flows first in one direction and then in the other, the lines set up by it in the magnet will change their direction every time the current changes its direction. They will thus, while flowing in one direction, add to the strength of the magnet, and while flowing in the other diminish it, thus producing the same kind of an effect on the diaphragm as in the case of the undulatory current.

7. Action as a Transmitter.—From the foregoing, it is seen that an alternating current sent through the coil will cause the diaphragm to vibrate. The converse of this statement is true; for if the diaphragm of the instrument is caused to vibrate by some external means, as by talking loudly into the receiver, corresponding alternating currents will be caused to flow in the coil, provided, of course, its circuit is closed. This is true because the movement of the diaphragm toward the pole of the magnet reduces the length of the air gap; hence, it reduces the reluctance of that part of the magnetic circuit, and, consequently, increases the number of lines of force, or at least causes a redistribution of the lines of force passing through the coil. It is a well-known fact that the changing of the number of lines of force through a closed circuit will cause currents to flow in that circuit. Similarly, when the diaphragm moves from the pole, the number of lines passing through the coil will diminish, or be redistributed, and thus cause a current to flow in the opposite direction. Thus, for an outward movement of the diaphragm the current in the coil is in one direction, while for an inward movement the current is in the other direction.

8. Two instruments like that shown in Fig. 3, if connected in one circuit, as shown in Fig. 4, represent the apparatus by which Bell brought about successful telephone transmission.

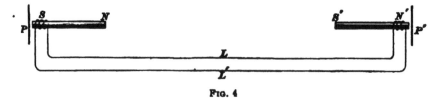

Fig. 4

When the soft-iron diaphragm P is spoken to, it takes up the vibrations of the sound waves, and thus causes changes in the strength of the magnetic field in which the coil lies. These changes, as just shown, cause currents, first in one direction and then in the other, that is, an alternating current, to flow in the circuit. These currents vary in direction, strength, and frequency of alternation, that is, in every way in unison with the movements of the diaphragm, and therefore have all the characteristics corresponding to the relative loudness, pitch, and timbre of the sound causing the diaphragm to vibrate. Passing along the line wire, these feeble currents alternately strengthen and weaken the permanent magnet at the other instrument, and cause it to exert a varying pull on the diaphragm, which thus vibrates in unison with the changes in current, and, therefore, reproduces the first sound.

The action described as taking place in the instrument spoken to is that of a *transmitter*, and that in the instrument that repeated the sound is that of a *receiver*.

9. A **telephone transmitter** is an instrument that takes up the vibrations of the sound to be transmitted, and causes corresponding fluctuations of electric current to flow in the circuit in which it is connected.

A **telephone receiver** is an instrument that receives fluctuating currents corresponding to sound waves, and translates them into distinguishable sounds.

10. The single instrument so far described, and consisting of a permanent magnet, a coil, and a vibrating iron diaphragm, is called a **magneto-telephone.** Many forms of it have been devised, but all contain the same essential parts, though frequently these parts are duplicated. As a receiver, the magneto-telephone is marvelously efficient and one of the most sensitive instruments known in science. Its sensitiveness depends, however, on its construction, adjustment, and the frequency of the current.

As a transmitter, however, it has not proved generally successful, as the amount of energy derived from the sound waves is so extremely small that the electromotive force (E. M. F.) generated by the movement of the diaphragm is too feeble to produce a sufficient current in lines of considerable length.

THE BATTERY TRANSMITTER

11. To overcome the difficulty due to the weak transmission of the magneto-instrument, a class of instruments depending on an entirely different mode of operation was devised. These instruments, instead of causing the transmitter to act as a generator of electricity, serve to produce variations in the strength of a current already flowing from some other source. These instruments, known as **battery transmitters**, or **microphones**, depend for their action on the fact, first pointed out by Professor Hughes, that the electrical resistance between two bodies in light or loose contact was made to vary greatly by slight changes in the pressure

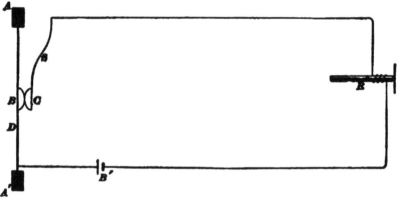

FIG. 5

between them. Though all conductors have this property to a certain extent, it is found to be greatest in carbon.

In Fig. 5, D is a diaphragm carrying a carbon button B, while C is a similar button carried on a light spring S in such a manner as to bear lightly against the button B. The diaphragm and spring form the terminals of wires leading to a receiver R, one of which wires includes a battery B'. While the instrument is at rest, a steady current flows through the circuit, of which the two carbon buttons and the coil of the receiver form a part. If, however, the slightest jar is given the diaphragm, a variation in pressure at the point of contact between the buttons will take place, which will cause a corresponding variation in the resistance of the contact. This, in turn, produces a fluctuation in the strength of the current, which, as before, travels along the line wire and acts on the diaphragm of the receiver. Sounds, such as spoken words, uttered before the diaphragm D, will cause it to vary the pressure between the buttons in such a manner as to produce fluctuations in the current flowing in the circuit from the battery B', which fluctuations will be in accordance with the sound waves, and will, therefore, cause the diaphragm of the receiver to vibrate in unison with that of the transmitter. We thus see that the battery transmitter serves, not as a generator of electricity, but as a valve to control the flow of current from a separate generator in the same circuit.

12. Theory of the Battery Transmitter.—Transmitters that depend for their action on the variation in pressure between two or more electrodes are called microphones. Their action has been the subject of much discussion. Undoubtedly the variation in resistance is due directly to changes in the *area of contact*, brought about by changes in pressure between the electrodes. If two rubber balls be touched lightly together, the area of their contact will be but a mere dot. If the pressure between them be increased, more and more of the surfaces of the balls will be in contact, and the amount of area so in contact will depend on the pressure and nature of the material. It is easy to see that in the case of an elastic conducting material like carbon, the surface of contact would vary in the same way with the pressure, though not to such an extent as between the two rubber balls, and that the electrical resistance between them would decrease as the area of contact increased. This is a perfectly satisfactory theory for the phenomena of the microphone, and experiments recently performed tend to show that the variable resistance is due to variation in surface contact entirely, to the exclusion of all the other more complicated theories that have been advanced.

THE INDUCTION COIL IN TELEPHONY

13. One more step will explain the principles involved wherever an induction, or repeating, coil is used in telephone systems. Assume that in Fig. 5 the resistance of the entire circuit is normally 1,000 ohms, and that the transmitter is capable of producing a change of 1 ohm in the resistance of the circuit. It will be evident that the transmitter will be able to vary the resistance of the entire circuit by only $\frac{1}{1000}$ of its total value, and that the variation of the current flowing will be but $\frac{1}{1000}$ of its value.

Consider, now, what will be the effect if the transmitter T is put in a local circuit containing a battery B and the primary coil P of an induction coil, as shown in Fig. 6. The resistance of the local circuit may be made very low, say, 10 ohms. The transmitter will still be able to produce a variation in the resistance of the circuit of 1 ohm, from which

FIG. 6

it follows that the total change in the strength of the current will be one-tenth of its normal amount, or one hundred times as great an amount as in the case of Fig. 5. When it is considered, in addition to this, that the normal value of the current in the local circuit is far greater with the same battery power, owing to its low resistance, than when the battery is directly in the line, it will be seen that the difference between the two systems is still further augmented. The fluctuations in the current in the local circuit in Fig. 6 induce corresponding currents in the secondary winding S of the induction coil, which currents flow over the line circuit and cause the diaphragm of the receiver at the distant station to vibrate.

14. The primary winding of the induction coil is usually composed of a small number of turns of comparatively coarse wire while the secondary winding is composed of a much greater number of turns of fine wire. The E. M. F. in the secondary is therefore many times that in the primary, thus enabling the secondary currents to overcome great line resistances to better advantage than the low-voltage primary currents.

The current in the primary, or local, circuit in Fig. 6 is *undulating, but not alternating;* it never changes its direction. The current in the secondary, however, is *alternating* in character. This fact is usually a little puzzling but its explanation is very simple. When the current in the primary winding increases, due to a decrease in the resistance of the transmitter, the number of lines of force through the core of the induction coil increases. When the current in the primary decreases, the number of lines of force in the core also decreases. The direction of an induced current in a coil depends on whether the lines of force through that coil are increasing or decreasing in number, provided, of course, that the lines continue in the same direction. In the induction coil, as shown in Fig. 6, the lines of force remain in the same direction, because the current producing them does not change its direction; but they are alternately increasing and decreasing in number, therefore the induced current in the secondary must flow in one direction as long as the lines of force are increasing, and then in the other as long as they are decreasing in number, and, therefore, the current in the secondary must be alternating. The action of induction coils used in central-energy systems will be considered in connection with such systems.

DIRECT AND ALTERNATING CURRENTS IN PORTIONS OF THE SAME CIRCUIT

15. As condensers, inductance coils, and non-inductive resistances are used extensively in telephone systems, their action on direct and alternating currents is briefly explained. It might be well to say that a coil wound, in one direction only, around an iron core possesses inductance and hence acts as an inductive resistance, whether the iron core is part of a relay or any other electromagnetic mechanism, or is simply used to increase the inductance of the coil. Furthermore, the total opposition, called *impedance*, that an inductance coil offers to an alternating, or rapidly fluctuating, current increases with the resistance, number of turns, and frequency of alternation. Where the frequency is very high and the resistance very low, the resistance may be an entirely negligible part of its impedance. On the other hand, the steady direct current that will flow through the inductance coil will depend on its resistance and not on its inductance.

A condenser, on account of the very high resistance of its insulating sheets, will not allow a direct current of appreciable strength to flow through it, but it is said to allow an alternating, or rapidly fluctuating, current, such as is produced when telephoning, to flow through it. This does not mean that any current actually flows through the insulating sheets of the condenser, but rather that the plates of the condenser are charged and discharged; that is, the charges on the plates are reversed in polarity every time the potential is reversed at the condenser's terminals and the quantity of electricity on the plates changes every time the potential changes in value. Whenever there is a change in the amount or polarity of the charge on the condenser plates, some electricity must flow into one side and out of the other side of the condenser and through the circuit, and just as much electricity must flow into one side as flows out of the other, that is, the positive charge on one side is always equal to the negative charge on the other side of the condenser. Moreover, the opposition that a condenser offers to the flow of an alternating, or rapidly fluctuating, current, decreases as its capacity and the frequency increase. In telephone circuits it will not generally do to use too large a capacity, because the voice currents become so much distorted in form as to become indistinct, even though their strength may not be appreciably affected.

16. A non-inductive resistance offers exactly the same opposition to an alternating, or rapidly fluctuating, current as to a direct current. This opposition is equal to its resistance only; that is, the frequency has no influence on the strength of current that a non-inductive resistance will allow to flow through it. The usual manner of representing *inductance*, *choke*, *retardation*, or *impedance* coils, as they are variously called, is shown at R or R'', Fig. 7. The addition of an armature, as shown at R', makes a relay or other electromagnetic device out of it. Non-inductive resistances are represented as at r, r'.

The direct current from the battery B will flow through R, r, which are in parallel, then through R', r', R'' back to the battery. The alternating current generated by the alternating-current dynamo D, or any other source of alternating currents, as the secondary winding of an ordinary telephone

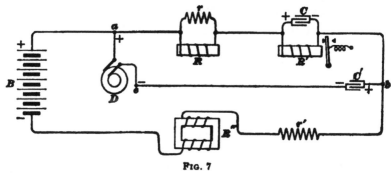

FIG. 7

induction coil, will flow most readily through $r-C-b-C'-D$. Some may flow through $R-R'$ and also through $a-B-R''-r'$. But if the frequency of the current and the inductances of R, R', R'' be made sufficiently great most of the alternating current will flow through r and most of the direct current through R. By making the capacity of C large enough practically all the alternating current, if not too low in frequency, will pass through it in preference to passing through R', and all the direct current must pass through $R'-b-r'-R''-B$. By making the inductance of R'' large enough, all the alternating current may be practically excluded from the circuit $a-B-R''-r'-b$.

When D produces a positive potential at a and a negative potential at e, a positive charge flows to the plates of the condenser C connected toward a and a negative charge to the side of C' connected toward e; that is, a current then flows momentarily from C' through e–D–a–r to C. The positive charge on one side of C attracts a negative charge to the other side, and the negative charge on one side of C' attracts a positive charge on the other side of C'; hence, a current flows momentarily from C through b to C'. When D reverses the polarity of a and e, the polarity of the charges and the direction of the flow of the currents are also reversed. When only a direct current is flowing, C will merely remain charged to the difference of potential across the coil R'; there will be no flow of charges in and out of it as long as there is no change in the strength of the current flowing in R'. The difference of potential across the coil R' for a given current will increase as the frequency increases.

A very rapidly fluctuating, or undulating, direct current may be considered as an alternating current superimposed on a direct current. As a matter of fact, the direction of such a current in a circuit does not change. But the rapid variations in its strength are readily transmitted through a non-inductive resistance or condenser, while only the steady portion of the current can readily pass through an inductive resistance. Hence, a condenser or non-inductive resistance in parallel with an inductive resistance will allow the transmission of a very rapidly fluctuating direct current through the combination, because the fluctuating portion may be considered as being transmitted through the condenser or non-inductive resistance, and the steady portion through the inductive resistance. For such purposes the non-inductive resistance will usually be relatively high and the inductive resistance relatively low.

TELEPHONE APPARATUS

TELEPHONE RECEIVERS

.17. While the most important consideration in the construction of a **telephone receiver** is its electrical efficiency, nevertheless many considerations of a mechanical nature enter into the design of a good receiver. It must be adapted to withstand rough usage of careless people, and, at the same time, maintain a permanent adjustment. Two points in its construction that have given much trouble are the adjustment of the diaphragm with respect to the pole pieces and the attachment and construction of the binding posts. Hard rubber and steel expand and contract, when the temperature varies, at very different rates, thus causing the distance between the pole pieces and the diaphragm to vary with changes in temperature, sometimes to such an extent as to noticeably affect the talking quality of the instrument. This may be avoided sufficiently by supporting the entire magnetic system in the shell near the diaphragm, for instance, at the base of the coils. The chamber containing the diaphragm should be of such shape as not to muffle the sound, and the shell should preferably be made of the best grade of hard rubber. However, good composition material can now be obtained and comparatively little hard rubber is used on account of its high price and the severe competition between the manufacturers of receivers.

It is now considered advisable to have no binding posts or other metal exposed on the outside of a receiver where it can come in contact with the hand in the ordinary use of the receiver. This is to avoid all possibility of the user receiving disagreeable or dangerous electric shocks due to lightning, or crosses between the telephone line wires and lighting or power circuits. The absence of exposed binding posts also reduces meddling with the receiver and the working loose of such connections. The terminals of the receiver coil should be connected, inside the receiver shell, to flexible wires, or *receiver cords*, as they are called, that are long enough to connect to binding posts on the telephone box or case. A strong cord is bound in with the flexible connecting wires and fastened to the receiver magnets and the telephone box, so that it receives all the strain should the receiver be dropped or left hanging by the cord.

Many claim that subsequent adjustment of a receiver is a detriment rather than an advantage; hence, many receivers are now constructed so that the original adjustment will remain permanent throughout the life of the receiver.

SINGLE-POLE RECEIVERS

18. By far the greatest number of receivers in use in the United States up to about 1898 were of the **single-pole** type, such as is shown in Fig. 8. The single-pole receiver usually has a compound magnet *M* composed of two pairs of permanently magnetized steel bars with like poles

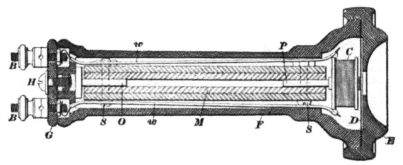

FIG. 8

together. These pairs of bars have clamped between them at one end a soft-iron pole piece *P*, and at the other end an iron block *O*. The parts of the pole piece *P* and of the block *O* that lie between the pairs of magnets are flattened in such a manner as to present a large contact surface to the magnets that rest against them. The projecting end of the pole piece P is cylindrical and has wound on it a coil C of fine, insulated, copper wire, usually No. 38 B. & S. gauge. The magnet and its coil are encased in a hard-rubber shell composed of three parts, E, F, and G. Between E and F, which screw together in the manner shown, is clamped a soft-iron diaphragm D, the space between the diaphragm and the pole piece being about $\frac{1}{74}$ inch. This diaphragm is stamped from ferrotype metal, such as was once used by photographers in making tintypes. This metal is rolled from a very fine quality of iron, the thickness being from .009 to .011 inch. Two binding posts B, B are secured to the tail-piece G of the shell, and are connected by heavy wires w, w to the terminals of the coil C. All connections inside the instrument are soldered. The resistance of receivers varies from 10 to 125 ohms, depending on the system in which they are used; the most common resistance is about 75 ohms.

BIPOLAR RECEIVERS

19. A form of receiver presenting two poles to the diaphragm has now come into extended use, and has practically superseded the single-pole instrument. In these the magnet is made in substantially horseshoe form, and each pole carries a flat coil, the two coils being brought quite near together and close to the diaphragm. The two coils are connected in series, and usually have a joint resistance of 60 to 100 ohms, although for certain purposes it is sometimes made as low as 10 ohms.

20. The **Kellogg receiver**, which is a good representative of a modern double-pole receiver, is shown in Fig. 9. The permanent magnets m, m', which are two straight bars, have a cylindrical piece of iron o held tightly between them at one end by a bolt and nut. The pole pieces p, p', said to be made of soft Norway iron, fit into a slight recess on each side of a casting a of non-magnetic metal that fills the space between the pole pieces so that the latter bear tightly against the permanent magnets when the nut is tightly screwed on the bolt b that passes through the two pole

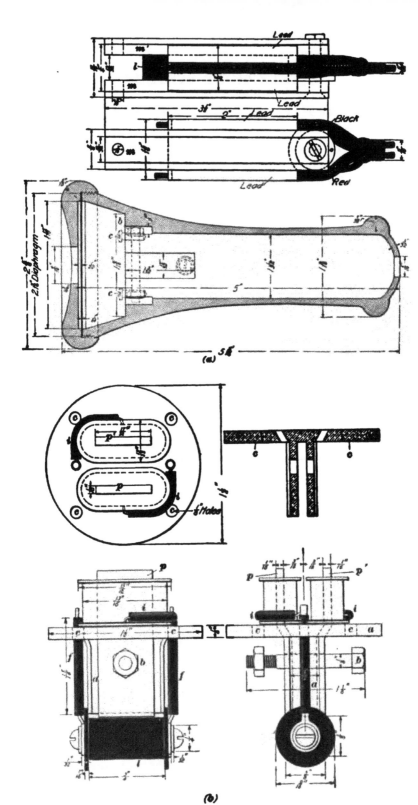

FIG. 9

pieces, permanent magnets, and the casting *a*. The casting is securely fastened by four brass screws, at *c*, to the hard-rubber shell at a point near the diaphragm, so that there is little chance for any change in adjustment due to the rough handling of the receiver or to the unequal amount of con-traction and expansion of hard rubber and metal when the temperature changes. The metal end of each coil spool is fastened by a bit of solder to its pole piece. There are no binding posts on the outside of the shell, instead the receiver cord passes through a hole in the shell. A cord bound in with the conductors is tied around the iron cylinder *o* so as to relieve the conductors and shell of all strain. Between the two permanent magnets is a piece of hard fiber *l*, on each side of which the receiver conductors are electrically connected by brass screws and washers to short pieces of stout wire *f, f* that are soldered at their upper ends to the terminal wires *i, i* of the coils. Any standard receiver cord may be used with this receiver. All connections are enclosed within the shell so that no metal whatever is exposed on the outside. To get at the connections, it is necessary to remove the cap at the diaphragm end and the four screws *c* that hold the casting, magnets, pole pieces, and coils to the rubber shell. To make the receiver heavier, the space between the bar magnets is about filled with lead, as indicated in the figure, thus insuring good wiping con-tacts in the hook switch, because the additional weight allows the use of a stronger spring to raise the hook.

21. The **watch-case receiver** is a form invariably provided for the use of operators at all telephone exchanges. It is a small, compact light instrument and usually has a steel spring band that passes over the head to hold the receiver to the operator's ear all the time she is tending to the switchboard. It is usually double-pole, the magnets being small, flat, and circular in shape so as to go into the flat, circular case.

TRANSMITTERS

THE BLAKE TRANSMITTER

22. The **Blake transmitter** is a form of microphone that was more extensively used in the United States for 20 years than any other one transmitter; not because it was better, or even as good as others, but because it was fairly efficient, and required little battery power. It is used so very little at the present time that it is not of sufficient importance to warrant giving a description of it here, especially as the space can be devoted to much better advantage to a description of more modern apparatus. The Blake transmitter is constructed very much on the same principle as the transmitter illustrated in Fig. 5.

THE HUNNING TRANSMITTER

23. Hunning introduced the idea of using granulated material, preferably carbon in a loose state, as the variable resistance medium of transmitters, and nearly all transmitters are now constructed on this principle. Hunning's original device was very similar to that shown in Fig. 10. Clamped between the wooden block *B* and the mouthpiece *A*, which may also be of wood or of hard rubber, is a thin diaphragm *D* of some elastic non-corrosive conducting material. Hunning used a platinum diaphragm, but this, on account of its high cost, has generally been superseded by a diaphragm of thin

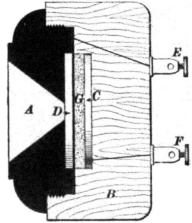

Fig. 10

carbon, iron, or aluminum. Within the chamber behind the diaphragm is mounted a block of carbon *C*, and the space between this and the diaphragm is filled, or partially filled, with granulated carbon, resembling in appearance ordinary gunpowder. Finely pulverized carbon was first used, but this packed so easily that it was soon superseded by granulated carbon. The two binding posts E, F are connected, respectively, with the diaphragm and with the carbon block, the circuit between them being completed by the carbon granules.

In this we have a large number of loose contacts between the granules and the diaphragm, between the granules and the back block, and also between the granules themselves. The sound waves entering the mouthpiece produce vibrations of the diaphragm, which vary the pressure at the multitude of loose contacts, and thus vary the resistance of the circuit. In these transmitters it is important that the granules lie loosely in their chamber, and to this end the chambers are only from one to two-thirds filled.

24. Packing of Transmitters.—The principal objection to transmitters of the Hunning type, but one that in good instruments is now practically overcome, is known as *packing*. An instrument is packed when the granules become wedged between the back block and the diaphragm in such a way as to prevent the free vibration of the latter. Sometimes granules may become packed by reason of some sudden mechanical shock, sometimes by reason of the slow settling down of the small particles between the large ones, thus forming a compact mass, and sometimes by reason of moisture from the breath or some other source. When due to the first or second causes, a shaking of the transmitter sidewise will usually remedy it; but when due to the third, the instrument should be taken apart and refilled.

SOLID-BACK TRANSMITTERS

25. Many of the granular carbon transmitters now made are called **solid-back transmitters** because the back electrode is rigidly supported by the frame of the instrument. The White transmitter, used by the licensees of the American Bell Telephone Company, and the transmitter made by the Kellogg Switchboard and Supply Company, are solid-back instruments of similar construction and probably as good as any transmitter now made.

26. The **Kellogg solid-back transmitter** is shown in Fig. 11. On the frame p of the instrument is fastened a bridge piece h of rather heavy sheet brass, used to support

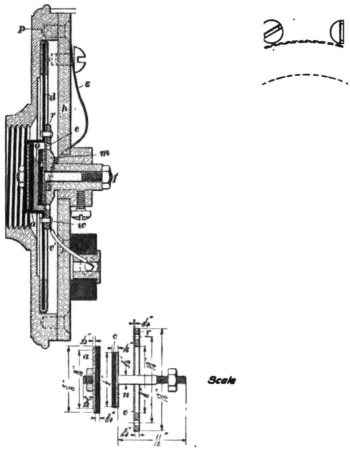

FIG. 11

the rear carbon electrode c, which is cemented to a brass piece n. The center portion of a mica diaphragm e is held tightly between the two brass pieces m, n, forming part of the rear electrode; its outer edge is held between the hard-drawn aluminum diaphragm d and a small aluminum ring r, which are held together by rivets w. The diaphragm, which is about .025 inch in thickness, has a cup-shaped recess at the center in which is placed the front carbon electrode a; this carbon is also cemented to a brass piece that is fastened tightly to the diaphragm by a nut on a projecting screw. The diaphragm has a rubber band around its edge and, including the front carbon electrode, is thus insulated from the frame of the instrument. It is held in its seat solely by two damping springs t, s, secured by screws to the frame. The end of each spring is covered with rubber, to the diaphragm side of which is glued a small felt cushion. The spring s bears directly against the diaphragm and the spring t against the rubber band around the edge of the diaphragm. The damping springs prevent an undue vibration of the diaphragm, corresponding to some particular rate at which it might be especially adapted to vibrate.

Both carbon electrodes are very highly polished, which is one of the most essential features of the instrument, and the rear one is somewhat smaller in diameter than the front one. The space o between the two electrodes is partly filled with a special grade of hard granular carbon very uniform in size. The mica diaphragm insulates the front and rear electrodes from each other and the current must pass through the frame of the instrument, rear electrode, granular carbon, front electrode, diaphragm, and connecting wire v', to the insulated terminal v. The mica diaphragm serves not only to retain the carbon granules but also is an auxiliary diaphragm, because as the aluminum diaphragm vibrates the mica diaphragm must also vibrate to some extent since its outer edge is fastened to the aluminum diaphragm and its center is held rigid by the back electrode. When the diaphragm vibrates, the pressure on the granular carbon of the front and rear carbon electrodes varies, and moreover the entire cell vibrates, thus keeping the granules well shaken up. The construction of the carbon chamber prevents tampering with that part of the instrument and at the same time makes the chamber sufficiently moisture-proof for most locations. The rear electrode may be clamped in any desired position by means of the screw *f*.

Two types of this transmitter are made, one for local circuit systems, and one for common-battery systems. The two types differ only in the size of the parts contained in the carbon chamber.

27. The **White solid-back transmitter** is used by all the licensees of the Bell Telephone Company. The back electrode, which is made of carbon, is carried in a heavy brass block mounted on a firm bracket attached to the front plate of the instrument. The chamber in which the back electrode lies also contains the front electrode, also made of carbon, which is attached to the diaphragm by a short piston rod. The space between the electrodes is nearly filled with very uniform, hard, granules of carbon, and the chamber is closed at the front end by a thin mica diaphragm through which the piston rod passes to the main diaphragm, which is of aluminum, .022 inch in thickness, and is heavily damped by means of rubber-tipped springs exactly as in the Kellogg transmitter. The White and Kellogg transmitters operate in exactly the same manner, and if the parts of the Kellogg transmitter were reversed in position, that is, if the aluminum diaphragm were fastened to the rear carbon electrode, and the front electrode rigidly fastened to the frame, the two instruments would be almost identical in construction.

28. Transmitters for Central-Energy Systems. With the advent of central-energy systems, telephone designers apparently returned to primitive methods, because in the first central-energy systems they connected the transmitter and receiver in series in the line circuit. However, they knew more about telephone apparatus, and so designed the transmitter and receivers to suit their conditions. They used complete metallic circuits of copper wire, low-wound receivers (from 4 to 20 ohms in resistance), and high-resistance transmitters (from about 20 to 90 ohms). Thus, the variable transmitter resistance was quite an appreciable part of the total resistance of the line circuit. The higher the resistance of the line circuit and the higher the E. M. F. of the central battery, the higher should be the resistance of the transmitter. The White solid-back transmitters vary in resistance from about 18 to 90 ohms for 24-volt central-energy systems and average from 35 to 50 ohms. However, the Kellogg transmitters now used in Buffalo on a 24-volt system, but with lines averaging rather longer than usual, are said to have a resistance of 320 ohms. The ordinary Kellogg transmitters have a resistance of about 200 ohms. On 48-volt systems the same manufacturers now propose to use 1,000-ohm transmitters. Transmitters may be designed to have almost any resistance and current-carrying capacity by suitably varying the cross-section of the carbon electrodes, the distance between the electrodes, and the amount, size, and resistivity of the carbon granules between the two electrodes.

_____

CALLING APPARATUS

_____

THE MAGNETO-GENERATOR

29. So far only the apparatus and methods employed in the actual transmission of speech have been considered; but of hardly less importance are the means for attracting the attention of a party at a distant station, with whom it is desired to converse. For this purpose the *magneto-generator* and the *polarized bell*, or *ringer*, have been extensively used.

The **magneto-generator** is in reality the simplest form of dynamo. It consists essentially of an armature of iron, wound with a large number of turns of fine, insulated, copper wire, and adapted to be rapidly revolved between the poles of a powerful permanent magnet. Its principles of operation can be readily understood by reference to the left-hand portion of Fig. 12. A represents the armature, around the shank of which is wound a coil of wire, as shown. S and N are the south and north poles of the permanent magnet, and P, P the pole pieces, usually of soft cast iron, attached to the magnets and bored out so as to afford a space in which

the armature may revolve.   The armature is in a rather
powerful magnetic field, the lines of force passing from the
north to the south pole of the magnet.

When the shank of the armature is horizontal, that is,
when turned 90° from the position shown, all the lines of
force pass through it, and therefore through the coils, as
they follow the path easiest for them to travel.   When
turned 90° farther, it reaches a point like that shown in the
figure, where practically no lines pass through the coils,
because their plane is parallel to the direction of the lines.
Thus, in a complete revolution of the armature, starting at
a point where the plane of the coils is vertical, the lines of

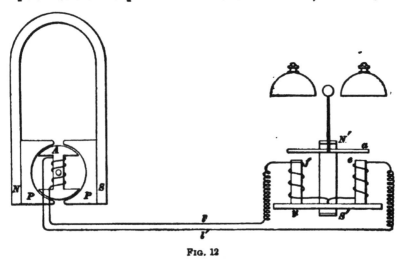

FIG. 12

force passing through the coils will decrease from a maximum to zero, and then to a maximum in the other direction,
and then decrease to zero and increase to the maximum in the
first direction again.   It is well known that a coil moving in
a field of force so as to include within itself more or less of
the lines will have an induced current set up in it, and that the
direction of this current depends on the direction of the
lines through the coil, and on whether they are increasing
or diminishing.   Now, since during each complete revolution of the armature the number of lines passing through the
coil will vary from a maximum in one direction through zero

to a maximum in the other direction, and then through zero again to the first maximum, it follows that for one revolution two impulses of current will flow in the coil, and these impulses will be in opposite directions. The current is therefore an alternating one.

30. A common form of magneto-generator is shown in Fig. 13, in which M, M, M are the permanent magnets, bent into horseshoe form from bar steel especially adapted for the purpose, then hardened flint hard, and magnetized. These magnets are clamped to their cast-iron pole pieces by screws S, S passing through washers and clamping plates P, P, and then between the magnets, as shown. The end plates B are of brass, secured to the ends of the pole pieces by screws. These end plates form bearings for the armature and also for the crank-shaft carrying the large gear-wheel G, which engages the small gear-wheel G' secured directly to the armature shaft. The pole pieces are bored out accurately and the armature turned to fit, so that the air gap between them will not be more than $\frac{1}{64}$ inch. The winding on the armature for an ordinary so-called series-generator is usually of No. 36 B. & S. gauge silk-covered wire, and the resistance about 600 ohms. The resistance of some generators is as low as 250 ohms. One terminal of the winding is soldered to a pin driven in the armature core, and thus makes contact with the frame of the machine through the bearings. The other end of the winding is fastened to a pin Q, driven into the end of the armature spindle, but insulated from it. Connection with this pin is made by the spring A, which therefore forms the other terminal of the machine.

31. The Automatic Shunt.—A very important adjunct to the magneto-generator is an automatic device that either short-circuits the armature when used in series-telephones or opens the generator circuit when used in bridging telephones. Series and bridging telephones will be explained later. It is found desirable on series-telephones, while the generator is not in use, to remove the resistance of the armature from the circuit by having a path of practically no

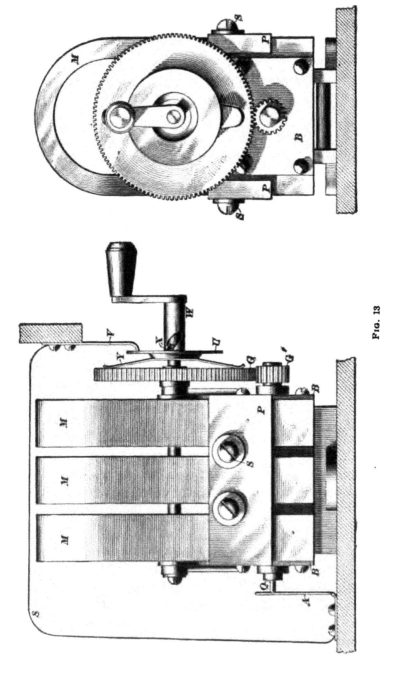

Fig. 13

resistance closed around it. This is accomplished in a variety of ways, but the one shown in Fig. 13 illustrates the principle involved. The plate or collar U normally bears against the spring V, which is connected by a short wire S with the spring A. This forms a low-resistance shunt, or short circuit, around the winding of the armature. When the crank of the generator is operated, the slotted sleeve W, which is loosely mounted on the shaft, is automatically pressed toward the generator by means of the pin X riding on the inclined sides of the slot in the sleeve. This presses the collar U from the spring V, thus breaking the shunt. This always happens before the generator starts to turn, and hence always removes the presence of the shunt when the generator is being operated. As soon as the hand is removed from the crank, the spring Y presses the collar U again into contact with spring V.

Similar devices are arranged to close the generator circuit only when the handle is turned. The use of such devices will be made clear in what follows.

The voltage given by a telephone generator depends on the strength of the magnetic field, on the number of turns of wire on the armature, and on the speed of rotation. At the usual rate of turning by hand, a machine for ordinary service will generate a pressure of from 65 to 75 volts at a frequency of about 15 complete cycles per second.

THE POLARIZED BELL

32. The **polarized bell**, or **ringer**, as it is more often called, is a device for receiving the alternating currents generated by a magneto-generator and causing them to produce audible signals. It is shown in Fig. 12 in connection with a magneto-generator. A polarized bell, as ordinarily constructed, is shown in Fig. 14, in which m, m are the coils of an electromagnet, y the yoke, and a the armature. The cores of the magnets, the yoke, and armature should all be of the best grade of annealed soft iron. The bracket b, on which the armature is pivoted, is of brass, and is carried on the brass standards g, g, which also pass through and support the yoke y. The hammer h, consisting of a slender brass rod and a ball for striking the gongs, is carried by the armature; while the permanent magnet NS, bent as shown in Fig. 14 (b), is carried on the bracket b. This magnet, by induction, gives the yoke y, and consequently the core ends, or poles, e, f, a positive polarity and the armature a a negative polarity. Each end of the armature will, therefore, be attracted with about equal force by the two poles e, f, and will adhere to the one to which it happens to be nearest. The two coils are wound so that a current traversing them in series

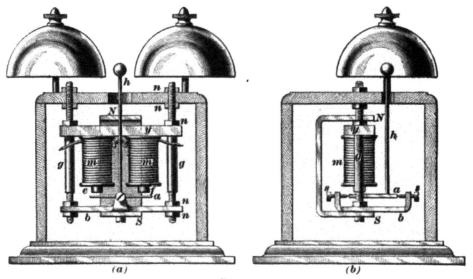

Fig. 14

will tend to make one of the poles positive and the other negative. Suppose that a current is sent through the coils in such a direction as to make e positive and f negative. This will strengthen e and weaken f, and therefore a will be attracted by e, causing the hammer to strike against the right-hand gong. If the current is reversed, f will be made stronger and e weaker, and the armature will be attracted by f, causing the hammer to strike against the left-hand gong. If the impulse of current is so strong as to entirely reverse the polarity of the pole that it weakens, that pole will tend to repel its end of the negatively polarized armature, thus aiding the attraction of the other pole.

For series-telephones the coils are wound with a No. 31 B. & S. gauge silk-covered copper wire to a resistance of 100 to 120 ohms, or with No. 30 wire to a resistance of 60 to 80 ohms. For bridging telephones, however, they are wound to a much higher resistance—1,000, 1,200, 1,600, and even 5,000 ohms.

33. Complete Magneto-Bells.—The term **magneto-bell** is usually applied to the combination of a magneto-generator and ringer mounted in a box, as in common use in telephone work. A fairly good machine will ring its own bell through a resistance of 10,000 ohms, which is the ordinary test applied to them. Thus, a 10,000-ohm magneto-bell, or generator, does not mean that either the bell or generator has a resistance of 10,000 ohms, but that the generator should be able to ring its own bell through a circuit whose total resistance is 10,000 ohms.

AUXILIARY APPARATUS

THE HOOK SWITCH

34. As the apparatus for sending and receiving both articulate speech and signals perform entirely separate functions, it has been found necessary, in order that they will not interfere with each other, to provide means for cutting one set out of circuit while the other is in use. It is evident that during the idle periods of the instrument it is necessary that the ringer should be left in circuit, in order that an incoming call may be received. As soon, however, as a call is received or sent, it is better to cut the calling apparatus out of circuit, so that the telephone, or *voice, currents*, as they are called, will not have to pass through the high resistance of the ringer and generator coils. The alternative switching of the talking and calling apparatus into circuit was at first accomplished by ordinary hand switches, but it was soon found that people could not be relied on to operate them, and the **hook switch,** now in universal use, was designed to accomplish these changes of circuit automatically, without the volition of the user of the telephone.

35. The **Kellogg hook switch** is shown in Fig. 15. The lever *l*, carrying the hook for the receiver on its outer end, is pivoted at the rear end to a bracket screwed to the inside of the bell box. The hook is held down by the weight of the receiver, which causes the contact spring *a* to make a firm contact with *b* and *b* with *c*; in this position, the three springs *a*, *b*, and *c* are electrically connected together. When the receiver is removed from the hook, the spring *n* lifts the hook, first separating the springs *a*, *b*, *c*

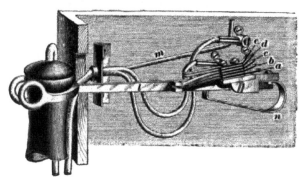

FIG. 15

from one another and then causing *c* to make a firm contact with *d* and *d* with *e*; in this position the three springs *c*, *d*, and *e* are electrically connected together. The rear ends of the springs are insulated from each other and firmly fastened to a fixed support. The springs are German silver with platinum contacts. This type of hook switch is used for both common-battery and magneto systems. Evidently the number of spring contacts above or below the spring *c* may be diminished or increased, or some of the springs may be left idle, or the middle spring *c* may have an insulating piece on the top or bottom of its front end so that it will make connection only with the springs below or above it; thus, it may be made suitable for almost any system.

There are about as many forms of switches as there are manufacturers.

The weight of the receiver, should it be dropped or left hanging, is born by the cord *m*, one end of which is firmly fastened inside the receiver and the other end inside the telephone box or case; therefore, no strain ever comes on the flexible receiver leads or on the terminal screws to which the latter are connected. It should also be noticed that there are no exposed binding posts or screws, either on the receiver or on the outside of the box casting. By this arrangement there are no exposed metal connections from which a subscriber can receive a shock while handling the receiver properly, as might otherwise happen if the line were crossed with a power or lighting circuit; neither are there any exposed screws, for persons so inclined, to meddle with.

The connections of hook switches to the various circuits of a telephone and the action of the switches in general will be more readily understood by considering the circuits of a telephone instrument.

___

BATTERIES

36. The **battery** used with the Blake transmitter was in nearly all cases one Leclanché cell. Two or more cells, as a rule, caused undue heating between the electrodes and a buzzing or cracking noise. With the various forms of solid-back transmitters, more battery power is needed. Two cells of any good sal-ammoniac battery will, under ordinary circumstances, give good results, and are considerably used. Dry cells are now being used extensively in place of (wet) Leclanché cells. In long-distance work two, and even three, Fuller bichromate cells have been used a good deal. The advantages of this cell are that it does not become exhausted so quickly as the sal-ammoniac cell, and that it has a higher E. M. F. and a lower internal resistance, and is therefore capable of giving greater current.

The adoption of the central-energy system is doing away with primary cells at the subscriber's instrument for local city service and even for quite long distances. No batteries are required for such systems at the subscribers' stations, while at the central office storage batteries are invariably used. The latter give excellent results on account of their exceedingly low internal resistance and their constancy of E. M. F. Modern solid-back transmitters work best with from .1 to .2 ampere, although one party recommends as high as .32 ampere.

COMPLETE TELEPHONE INSTRUMENTS

SERIES INSTRUMENTS

37. By a **series instrument** is meant one in which the ringer and the generator when in operation, and therefore not short-circuited by the automatic shunt device, are connected in series with each other across the line wires when the lever switch is in its lowest position.

In Fig. 16 are shown the various parts of a complete telephone connected in circuits as they are in actual practice in a subscriber's series-telephone set. G is the magneto-generator, C the ringer, or call bell, R the receiver, T the transmitter, H the hook switch, B the battery, and S and P the secondary and primary, respectively, of the induction coil. The hook lever is pivoted at its right-hand end, and is free to move between the contacts on its lower and upper sides.

While supporting the receiver, the lever is depressed, against the force of a spring, into engagement with the lower contact. When the receiver is removed, however, the spring immediately moves the lever into engagement with the upper contacts, in which position it is shown.

38. Suppose the receiver to be supported on the hook; an alternating current from the magneto-generator of some distant station may enter the instrument, we will say, by the line wire terminating at binding post A. It will then pass by wires a, b to the armature spring of the generator; two paths will then be presented, one through the armature winding to the frame of the generator and to wire g, and the other through the shunt wire s, shunt spring and collar

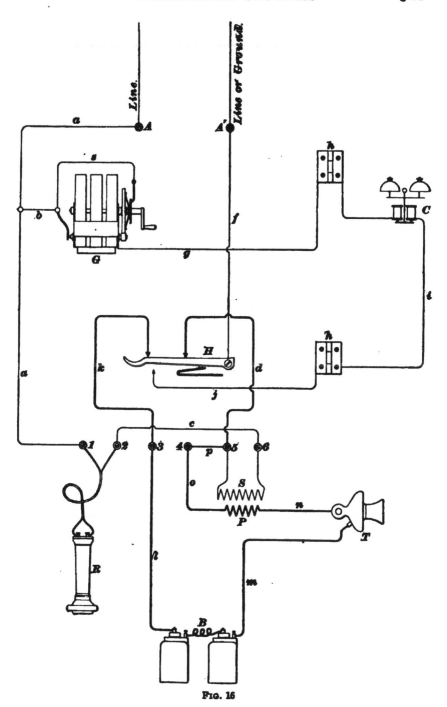

FIG. 16

on the generator shaft, to frame of the generator and wire g. As the latter path is a short circuit of practically no resistance, nearly all the current will pass through it. From the wire g, the current will pass through the upper hinge h on the lid of the box, thence through the coils of the polarized bell C, and thence by wire i, lower hinge h, and wire j to the lower contact of the switch. As the lever is depressed, the current will pass through the lever and by wire f to the return side of the line or to ground, according to whether a double or a single wire is used. This current will cause the gongs to sound, and thus give the desired signal. Let us now suppose that, instead of receiving a signal, the generator is operated to send one. The hook being down, the current from the generator will travel over exactly the same path as that just traced, but there will be no shunt circuit through wire s, because the operation of the generator will break the continuity of that circuit between the shunt spring and the collar on the shaft.

After having sent or received a signal, the receiver is removed from the hook and the apparatus is connected as shown in the figure. It will be noticed that the calling circuit is now open at the lower contact point of the hook switch. The circuits through the talking apparatus have, however, been closed at the upper contacts, and are as follows: Starting at binding post A, a current coming over the line will pass through wire a–receiver R–wire c–secondary winding S–wire d–upper right-hand contact–hook lever–wire f–binding post A'; this is called the secondary circuit, and contains only the receiver and secondary winding of the induction coil. Another circuit that is also closed by the rising of the switch lever contains the battery, transmitter, and primary winding of the induction coil. The wires forming the primary circuit are shown by heavy lines, and may be traced from the battery through wires l and k–left-hand upper contact–lever–right-hand upper contact–wires d, p, and o–primary winding P–wire n–transmitter T–wire m–battery.

In this position the transmitter, when spoken to, causes

47—29

fluctuations of the current flowing in the primary circuit that act inductively on the secondary circuit, causing corresponding alternating currents to pass out over the line.

BRIDGING INSTRUMENTS

39. In Fig. 17 is shown the arrangement and connection of apparatus forming a complete bridging telephone. The bell and the generator circuits are permanently connected across the line binding posts A, A', but the generator circuit is normally open between q and r, being closed only while the crank is being turned. When the switch is up the receiver R and secondary winding S are connected in series and to the binding posts A, A' through a–1–R–2–6–S–5–d–H–f and the transmitter circuit is closed through B–l–3–k–H–d –5–4–o–P–n–T–m–B.

The bell is wound so as to have a large number of turns and a high resistance and consequently may be left permanently connected across the line, because its inductance together with its resistance is so high that the very rapidly fluctuating voice currents do not pass through it to an appreciable extent, but pass into the line circuit, which offers much less opposition to them.

It is usual to mount the generator, ringer, and hook switch all in the same box. The terminals of the circuits leading from these parts are led to binding posts A, A' on the top of the box for the line terminals, and to posts 1, 2, 3, 4, 5, and 6 on the bottom of the box. The numbered posts on the bottom are for facilitating the connecting of the other parts of the telephone not properly included in the magneto-box.

CENTRAL-ENERGY INSTRUMENTS

40. Instruments for use on central-energy, or common-battery, systems can hardly be classed as series or bridging instruments. No generators are used, the central office being signaled by the mere removal of the receiver from the hook. They usually have an ordinary polarized bell, varying in resistance from 80 to 2,500 ohms (sometimes in series with a condenser of about 2 microfarads capacity) connected across the line circuit, or between one line wire

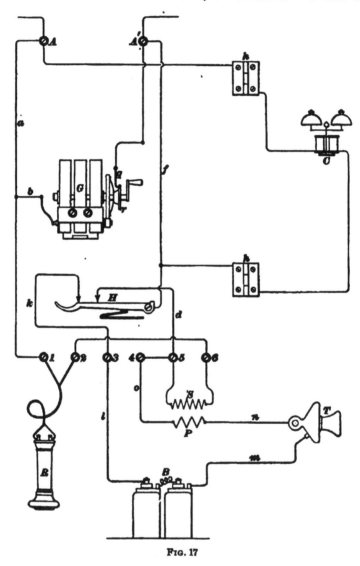

FIG. 17

and the ground, when the receiver is on the hook. The transmitters and receivers are arranged in various ways. In the simplest arrangement the transmitter and receiver are connected in series across the line circuit when the receiver is removed from the hook. Very satisfactory results can undoubtedly be obtained with the transmitter and receiver in series, if both are suitably designed. But the use of induction and impedance coils in connection with the subscribers' telephone instruments seems to give better results and is replacing the more simple arrangement, although the latter has given almost perfect satisfaction for local-exchange work for a number of years in several systems, including at least one of about 3,000 subscribers that was in operation in 1904.

_____

PARTY LINES

41. When several parties desire intercommunication by telephony, the simplest way is to place all the instruments on one line circuit. Then, when one party turns his generator crank, the bells of all the parties will ring. Lines of this description, which are called party lines, may be divided into two general classes: First, those on which the bells of all the subscribers ring when a signal is sent to any one. In this a code of audible signals is employed to enable the parties at the various stations to distinguish their calls from those of the other stations. These codes are usually made up of a various number of rings or various combinations of long and short rings, so that a party may at once by sound tell whether his attention is desired at the instrument. The second class embraces all those systems in which means are provided for ringing the bell of any subscriber on a line without disturbing any of the others. Systems arranged under this second class may be termed *selective signaling systems*, in order to distinguish them from the systems of the first class in which no means of selective signaling is employed.

_____

NON-SELECTIVE PARTY LINES

42. There are two general methods of arranging instruments on party lines. One of these is to place all the instruments in series in the line wire, and the other is to connect them in multiple across the two sides of the line circuit. The first of these systems is called the *series system*, and the second the *multiple*, or *bridging*, *system*.

43. Series Party Line.—When arranged in series, instruments with low-resistance ringer coils are used. The line wire is cut at each station, and the two ends so formed are connected with the two binding posts at the top of the instrument. The instruments are thus connected in series with the line that enters at one binding post and goes out at the other. Another reason will now be made apparent for the placing of the shunt around the generator armature in each instrument; for if the resistance of the armatures were left directly in the circuit of a series party line, it would be necessary for the voice currents of two stations in use to pass through the armature coils of all the stations not in

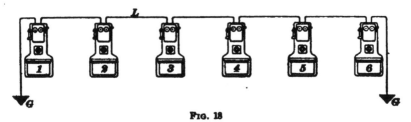

FIG. 18

use. Ringing currents would also have to pass through the same path, thereby greatly reducing their effectiveness. Fig. 18 shows a series party line, using the ground as part of the circuit. A line wire could, of course, be substituted for the ground return.

If two subscribers on a party line are talking, the ringer coils at all the other stations are included in the circuit. This is a necessary evil in the series system, inasmuch as these coils must at all times, when the instrument is not in use, be left in the circuit in order that each instrument may · receive a call at any time. The evil may be reduced to a minimum by winding the ringer magnets to a low resistance and by cutting out the generators, as already stated; but even then it is a serious defect, and is not capable of giving the best service.

So far as the actual ringing of bells is concerned, as many as forty instruments may be placed on a line, but this renders the talking exceedingly poor. Besides this, there is always an incessant ringing of bells, and a consequent annoyance and confusion of signals. As a rule of good practice, not more than ten instruments should be placed on a series line, though, of course, conditions sometimes render a greater number necessary.

44. The Bridged Party Line.—The other method of connecting telephones on a party line is to place them in

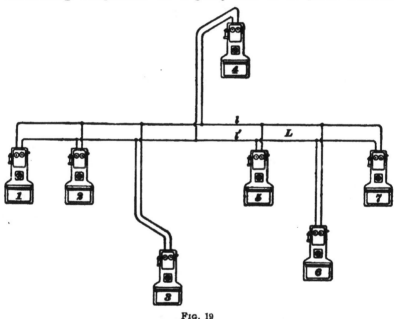

Fig. 19

multiple or bridge connection. Each instrument is in a separate bridge wire between the two sides of the line circuit, that is, either between the two line wires, where a double line or metallic circuit is used, as shown in Fig. 19, or between the line wire and ground, where a grounded circuit is used, as shown in Fig. 20. In this arrangement a bridging telephone is used. The ringer magnets are wound to a high resistance—usually 1,000, 1,200, or 1,600 ohms, and occasionally even as high as 5,000 ohms—and may be left permanently in the circuit. The great number of turns on the magnets prevents the rapidly alternating voice currents from leaking through, and forces them to pass along the line. The generator at each station is in a normally open circuit that is automatically closed and connected directly to the terminal binding posts of the instrument when the generator is operated. Current from any generator in operation on a line passes along the line wire and divides, passing through all the ringer magnets in multiple.

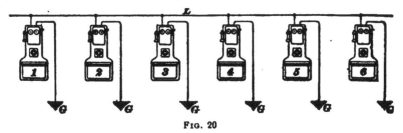

FIG. 20

These lines give better service in talking than series lines, and for other reasons are to be preferred. It is not advisable, however, to place more than twelve instruments on a single line, though, of course, it is possible for a much greater number to be used. A generator capable of producing a larger current at about the same voltage is required where many more bridging bells than usual are connected to the same line circuit. In the latter case, higher resistance bells may be used, thus requiring less current for each bell.

SELECTIVE-SIGNALING PARTY LINES

45. Selective-signaling party lines embrace all those systems in which means are provided for ringing the bell of any one subscriber on a line without ringing the bell of any other subscriber on the same line. They may be divided into three systems, as follows: First, those employing step-by-step mechanisms operated by impulses of current from the central station in such manner as to close the calling circuits at the subscribers' stations successively; second, those using the harmonic, or reed, system of selecting, wherein currents of various frequencies are employed for actuating the different signals; and third, those using current impulses flowing in different directions for operating the different bells and called polarity systems.

46. Step-by-Step Signaling.—The general plan used in step-by-step systems is that of employing a mechanism at each of the various stations on the line that will close a circuit at its station after a certain predetermined number of current impulses have been sent over the line. In order to call any particular station, for instance No. 10, the operator would send ten current impulses over the line, which would bring a contact lever at station 10 into engagement with its own button. Those levers at the stations having smaller numbers would have passed over their buttons, while those at stations having larger numbers would not yet have reached their buttons. After the completion of the circuit at the desired station, the operator then sends the ringing current over the line, which flows through the circuit completed at that station, and sounds the bell of that subscriber only.

This general method of selective signaling appears to be a very promising one, but nevertheless has come into but little practical use, because it is difficult to secure proper electrical contacts between the stationary and movable contact points. Besides this, the step-by-step mechanism is always rather complicated—an undesirable feature, especially when in the hands of inexperienced parties.

47. Harmonic Signaling.—Every vibrating pendulum or reed has a natural period of vibration out of which it is a comparatively difficult matter to make it vibrate with any great amplitude. This fact is made use of in the **harmonic selective systems** by placing a reed or pendulum at each subscriber's station, adapted in each case to be acted on by an electromagnet in the line wire. The reeds at all the stations are tuned to vibrate at different rates, and when any one is thrown into vibration of the desired amplitude it completes a circuit or circuits at that station, by which the signal is sounded. At the central office transmitting devices, or keys, are provided, each adapted to send over the line circuit rapidly pulsating or alternating currents, the frequencies of which correspond to that of the reed at one of the sub-stations. Thus, in order to call any particular station on a line, the operator, by means of one of the transmitting keys, sends a pulsating or alternating current of the proper fre-quency to line. The frequency of the impulses of this current being the same as the rate of vibration of the reed at the station that it is desired to call, will throw that reed into motion and cause it to complete the calling circuit at that station. As none of the other reeds at the other stations have a rate of vibration corresponding to that of the particular current used, their reeds will not be thrown into vibration, or at least not with sufficient amplitude, and their signals will remain inert.

The harmonic method of signaling has not proved very successful. It is very difficult to secure a good contact between a rapidly vibrating reed and a stationary contact, and furthermore the adjustment of the rates of vibration of the various stations is likely to be changed by variations in temperature or by other unavoidable causes. The Kellogg selective four-party line system operates on this principle.

48. Polarity Systems.—This class of systems relates to those in which selective signaling is accomplished by changes in the direction of the current flowing in the line. This general method is now extensively used in many exchanges. A relay with polarized armature and magnet cores or an ordinary polarized bell can readily be made to respond to current impulses in one direction only. Obvi-ously, this in itself on a single line wire affords means for signaling either of two stations exclusively of the other, for one of the relays or bells may be arranged to respond only to currents in a positive direction, and the other only to those in a negative direction.

49. The so-called *Genessee system*, which has been exten-sively used, is operated in the following manner: The ordinary magneto-bell shown in Fig. 14 has a spiral spring attached to one end of the armature *a* so as to normally keep the hammer *h* against one gong. If direct, but pulsating, currents flow through the coils in such a direction as to move the hammer against the other gong, the bell will ring because the current impulses cause the hammer to hit one gong while the spring draws it back against the other gong between each impulse while no current is momentarily flowing through the coils. A pulsating current in the opposite direction will not ring the bell because the magnetism produced in the cores merely tends to pull the armature in the same direction as the spring and there is no tendency to move the hammer against the other gong. By putting the spring on the opposite end of the armature of another similarly polarized and similarly wound bell and sending pulsating currents through the coils in the latter direction this bell will ring. By connecting two such bells, wound to a high resistance, across a line in parallel, it is possible to ring either bell by sending a pulsating current through the line in the proper direction.

By connecting two such bells between one wire of a metallic circuit and the ground and two similar bells between the other wire and the ground, it is possible to ring any one of the four bells desired by connecting between the ground and the proper line wire a generator that will send pulsating currents in the proper direction. The circuits are arranged so that all conversations are carried on over the two wires, that is, over a complete metallic circuit. It is not practical to describe here the more elaborate but less used selective-signaling systems.

_____

COMMON FAULTS AND THEIR REMEDIES

50. In telephones, as in all other systems employing electrical connections, certain defects, called **faults,** will from time to time develop, however well the apparatus may have been constructed and erected. The most common troubles are due generally to one of three causes: First, loose or dirty connections at the binding posts of the instrument, at the binding posts of the batteries, or in joints in the line wires; second, in exhausted, poor, or weak batteries; third, crossed, open, or defective wires. These troubles, of course, do not include those arising from inferior or defective instruments, which it is impossible to enumerate on account of the large number of different makes of instruments now in use. In the case of a defective instrument, the best thing to do is to return it to the dealer for repairs. If any trouble or poor service is noticed on a telephone line, look first for badly constructed joints, loose connections at binding posts, dirty or corroded contacts and connections, and defective batteries. If the connections are dirty, corroded, or greasy, scrape the wires and clean out the binding posts; then screw the wires firmly in place. If the telephone does not then work properly, examine the batteries and see whether they are run down, or if the zincs are eaten away. With wet batteries, it may be that the water has evaporated; in dry batteries, the zinc may be eaten through, or the batteries may be otherwise defective. The simplest way to test the battery is to try a new battery, and see if it will make the telephone work properly; if it does the trouble was with the old battery. If the trouble is present after changing the battery, examine the line connections and the line outside; if any loose connections are found, correct them at once. When inspecting the line outside, see that it does not touch anything except the insulators, and that it is neither crossed nor broken. On grounded lines, examine the ground connection first, and see if it is in good condition; and if a plate is used, see that it is in moist ground and below the frost line.

A frequent trouble with transmitters is a frying noise; this is usually caused by too much current or by loose connections. If any of the coils in the instrument have been damaged by lightning, the smell of the charred insulation can frequently be detected when the door of the telephone is opened. If this is the trouble, the only thing to do is to replace the coil that has been burned out. One thing that should be carefully avoided, is the placing of nails, screws, screwdrivers, scissors, or metallic instruments on the top of the telephone box. In a series-telephone this might cut out the instrument; on the bridging line, it might result in all the instruments on the line being thrown out of service. A very short list of only the most common faults that occur in ordinary series and bridging instruments, with suggestions as to their cause, will now be given.

51. Cannot Ring or Receive a Ring.—The line or generator circuit may be opened in a series instrument or short-circuited in a bridging instrument. If a series-bell, connect the two main binding posts at the top of the instrument together; if the bell will not then ring when the generator is operated, the trouble is probably a broken wire inside of the box. For a bridging instrument, remove the two line wires at the top of the instrument; if the bell will now ring, the trouble was due to a short circuit outside the instrument, probably between the line wires.

52. Can Be Called But Cannot Call Others.—This may be due to weak or defective generators, or to bell coils of different resistance on the same line.

·53. Cannot Be Called But Can Call Others.—This may be due to imperfect adjustment of the bell armature or gongs, or to bell coils of different resistance on the same line. In a bridging instrument it may be due to a defect in the automatic cut-in device of the generator.

54. Can Hear But Cannot Be Heard.—In such cases the trouble is usually with the battery or transmitter circuit. A careful examination of all connections therein should be made. It may be due to exhausted battery, cells improperly connected, a packed transmitter, or a broken wire in the battery circuit. The person using the instrument may stand too far away; one should stand so that the lips are about 1 inch from the transmitter.

55. Cannot Hear But Can Be Heard.—In such cases the trouble is usually in the receiver circuit, and is probably due to a defective or improperly adjusted receiver, or to a short circuit in the receiver coil or in the receiver cords. It may be due, however, to a defective transmitter or a weak battery at the transmitting station; or possibly, in a bridging instrument with a ground return, to an imperfect ground connection, the ground connections at the several other instruments on the same line being good.

56. Weak Ringing of Bells.—Loose connections, bad joints in the line, or imperfect ground connection at terminals in case a ground return is used. The bell adjustment may also be defective.

57. Instrument Receives and Transmits Rings But Nothing Can Be Heard at Either Station.—This may be due to loose connections or a broken wire in the receiver, in the receiver cord, in the secondary winding of the induction coil, or to poor or loose contacts in the switch hook, or to weak batteries at both stations. With a series instrument the following test may be made to determine whether the trouble is in the receiver or cord. Disconnect the cord from the box, but allow the receiver to remain on the hook. Remove the line wires from the binding posts and place the two ends of the receiver cord in the line binding posts and turn the generator handle. If the receiver or cord circuit is not broken the bells will ring. The wires in a cord may be broken, and yet the break may not be apparent if the cord is held in a certain position. The result produced is a scraping sound in the telephone, or it may interrupt the speech so that a word is only audible occasionally.

58. If Clapper Clings to One Gong.—Move that gong toward the other gong and against the clapper. A slight adjustment of this kind will usually remedy the difficulty.

59. Rasping and Squeezy Noises in Receiver. Loose connections, or excessive current in the battery circuit. This trouble may also be caused by a buckled diaphragm in the receiver, or by particles of foreign substance lodged between the diaphragm and pole piece of the receiver. Or, the position of the diaphragm may not be correct. In modern receivers no provision is usually made for adjusting the distance of the diaphragm from the magnet. Where such adjustment is possible, the diaphragm should be .015 inch from the magnet. Or, it may be due to a weak magnet; the magnet should be strong enough to at least hold the diaphragm by its edge.

60. Bell Rings Frequently Without Apparent Cause.—The line wire swings across telegraph or other wires in which a current is flowing.

TELEPHONE LINES

61. The construction of telephone lines is in general the same as that of telegraph lines. Certain peculiar conditions are found, however, in telephony that occur in no other class of work.

Telephone lines as first constructed used the earth as the return side of the circuit, in the same manner as is universally done in telegraph work. Such lines are called *grounded lines*, and the systems using them are called *grounded systems*. Such lines frequently form part of the return circuit for trolley systems and give considerable trouble for that reason, and also, because induction and cross-talk from neighboring parallel lines cannot be eliminated. These disturbances caused grounded systems in which the earth is used as a return circuit to be extensively superseded by a *common return system*, in which all the lines, after passing through their respective telephones, are connected to a common return wire instead of to the earth, the common return usually being a wire of larger size than the individual line wires. This eliminates the trouble due to trolley currents, referred to above, and is much more satisfactory than the earth as a return circuit. However, induction troubles are relieved but little, if at all, by this method.

Two separate wires for each line circuit is now considered the best practice and is replacing the other systems. Such lines are called *complete metallic circuits*. When such circuits are properly put up all disturbances due to induction may be eliminated.

INDUCTION ON LINES

62. When a single telephone-line wire runs parallel with another on the same poles for a considerable distance, a very peculiar phenomenon occurs. It is found that although the two wires are entirely separated, and highly insulated from each other, a conversation on one circuit can be heard in the other. Similarly, a telephone line running in the vicinity of electric-light, power, telegraph, or trolley lines will have currents set up in it that cause noises in the receivers, always exceedingly annoying, and sometimes so loud as to prevent conversation. These phenomena are, of

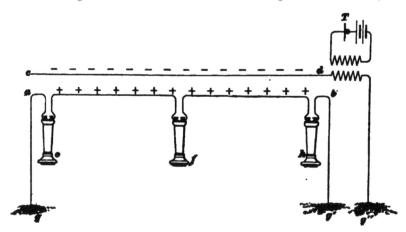

FIG. 21

course, due to induction, and it has been proved that this induction is electrostatic instead of electromagnetic, as was at first commonly supposed.

63. Cross-Talk.—The phenomenon of overhearing conversation on a circuit other than that over which it is originally conducted is called cross-talk. It is very confusing when two or more subscribers are talking at the same time and was at one time the main difficulty to be overcome in practical work, when a number of circuits were necessarily run in close proximity. The action by which induction is produced may be explained by Fig. 21, in which a telephone line *a b*, 200 or 300 feet long, contains three telephone receivers, one at each end and one at the center. The two ends of *a b* are connected to the earth at *g*, *g'*. A telephone line *c d* is supposed to run parallel to *a b* and at a uniform distance therefrom of ⅛ inch throughout its entire length. One end *c* is open, that is, it is left entirely disconnected, while the other end *d* is connected with earth at *g''*. A transmitter *T* in a local circuit with a battery is connected with this line through an induction coil in the usual manner.

A sound is produced in front of *T* by a tuning fork or some other source, and the line *c d* will receive a succession of charges that are alternately plus and minus. This induces charges on *a b* of opposite signs, the two conductors with the air between forming a condenser, the free charges passing off into the ground. If a minus charge be formed on *c d*, as

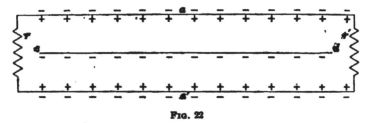

Fig. 22

in the figure, a plus charge will be held on the line *a b*. As the charge on *c d* changes to plus, the positive charge on *a b* is released, and two equal currents pass from the center of line *a b* in both directions to the earth at *g*, *g'*. A plus charge on *c d* produces the same effect, but the currents are in an opposite direction and meet at *f*. It will be observed that a sound is heard distinctly by the receivers *e*, *h*, while receiver *f*, for the reason that no current passes through it, is silent. This shows that a current is produced by electrostatic and not by electromagnetic induction, for, if it were produced by the latter, a sound could be heard in all the receivers.

64. Eliminating Cross-Talk.—When a complete metallic circuit is used, no cross-talk or other induction is developed if the disturbing wire is at equal distance from each of the circuit wires. This fact may be explained by

the aid of Fig. 22, in which a, a' represent the halves of a
complete metallic circuit; r, r', the coils of receivers placed
at opposite ends of the circuit; and $c\,d$, a disturbing wire
placed in such a position that all its points will be the same
distance from a that they are from a'. Alternating charges
are produced on $c\,d$ either by a telephone or by any other
source, and alternating charges with opposite signs to these
are produced, by induction from $c\,d$, on a and a'. When a
negative charge is produced on $c\,d$, as in the figure, a positive charge is induced on the sides of a, a' that are nearest
$c\,d$; while on the other sides negative charges are produced.
As the charge on $c\,d$ reaches zero, a current will flow across
each wire, but not lengthwise, for the potential is exactly
balanced on each side of the wires a, a'. The charges of
opposite sign simply come together across the wires. There
is no tendency for the free negative charge on a, for instance,
to flow away, as in the case of a grounded
wire, because there is an equal negative charge
on the entire circuit. If, however, the wire $c\,d$
is nearer a than a', the free negative charge
on a will be greater than that on a', and some
current will flow. In this case the lines are said to be *out of
balance*, and some cross-talk or induction will take place.

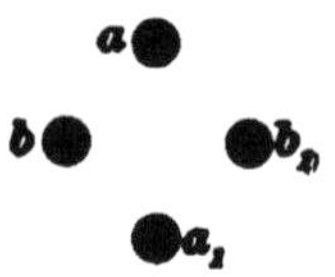

FIG. 23

It may be readily perceived that only four wires forming
two complete circuits can be placed so as to fulfil the condition shown in Fig. 22. Their position is shown in Fig. 23,
in which a, a_1 are the cross-sections of the wires of one circuit, and b, b_1 those of the other. It is, therefore, necessary
to resort to the method of transpositions shown in Fig. 24
for other cases.

TRANSPOSITIONS

65. In Fig. 24 the disturbing wire is $c\,d$ at unequal distances from the two circuit wires a, a'; the resistances r, r'
represent the subscribers' instruments. As $c\,d$ is at unequal
distances from a, a', the induced potentials will not be balanced; cross-talk would occur, were it not that the two wires
are transposed, as shown in the figure. This transposition

breaks up the charge, so that the resulting currents are too small to seriously affect the telephone. When $c\,d$ receives a negative charge, as in the figure, a positive charge is induced on the nearer ends of a, a', while a negative is repelled to the farther ends. As the potential on $c\,d$ reaches zero, currents flow from the points e, f and come together at e', f', where no effect will be produced on a

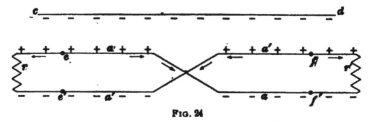

Fig. 24

telephone receiver; and if there were no retardation from resistance and self-induction in the apparatus, they would be in the center of each transposition. However, on account of this retardation, the greater part of the current will flow from the neutral points in a direction away from the apparatus through the part having less retardation. The neutral points, therefore, are nearer to the apparatus; this fact enables the transpositions to be the longer as the

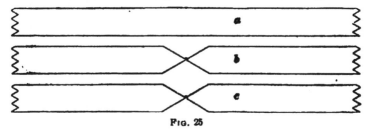

Fig. 25

retardation is increased. The greater the number of transpositions, the smaller will be the charges and, therefore, the resulting currents.

66. Where many wires are used, the transpositions must be carefully planned, since, although for two circuits one may be transposed so that no cross-talk shall occur, for one more circuit the same transposition will cause cross-talk between the second and third. This is shown in Fig. 25, where it may be seen that the average distances of the two wires of c from either wire of b are not equal to each other. In fact, the circuits b and c bear the same relation to each other as if they had not been transposed. To overcome

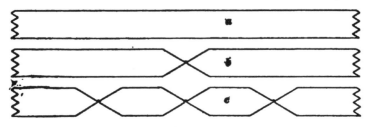

FIG. 26

this difficulty, twice as many transpositions as there are in the second must be made in the third, and so on, as is shown in Fig. 26. It is not necessary in practice, however,

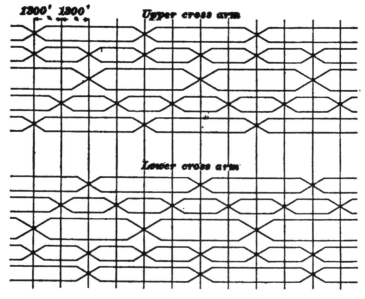

FIG. 27

to exactly balance the induction effect. Similar transpositions for every other circuit have been found to work very well.

67. Fig. 27 shows the scheme of transposition on the New York-Chicago telephone line, the poles on which transpositions are made being 1,300 feet apart. The same scheme of transposition is used on every other set of cross-arms. Thus, the first line of cross-arms at the top of the pole line will be arranged as shown in the upper part, of Fig. 27, the next set will be arranged as shown in the lower

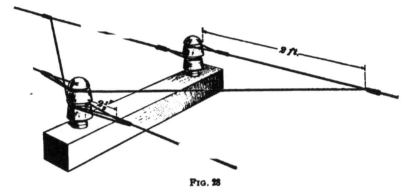

Fig. 28

portion of that figure, the third set will be like the first, and the fourth like the second, and so on throughout the entire number of sets of cross-arms.

68. An excellent way of making transpositions, where ordinary twisted wire joints are used, is shown in Fig. 28. In any case the cross-over wires should either be insulated

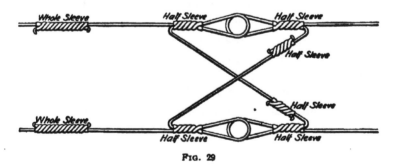

Fig. 29

or else so bent as to avoid the possibility of crosses, the latter plan being the more common.

Where the joints at transpositions are made with McIntire sleeves, the transpositions are usually made as shown in Fig. 29.

SIZE OF WIRE FOR VARIOUS PURPOSES

69. It is difficult to give any definite rules for the proper size of wire to be used in construction work, but the following sizes will ordinarily answer for the purposes mentioned.

In the country and small towns use: for distances not to exceed 8 miles, No. 14 B. B. galvanized-iron wire; for distances not to exceed 25 miles, No. 12 B. B. galvanized-iron wire; for distances from 25 to 100 miles, No. 10 B. B. galvanized-iron wire; and for distances of 100 miles and over, hard-drawn copper wire.

The size most generally used on farmers' lines is No. 12 B. B. galvanized-iron wire, weighing about 165 pounds to the mile, although No. 14 will answer up to about 8 miles.

For small city or town lines, No. 14 B. W. G. galvanized-iron wire of B. B. grade is extensively used, although in towns where cable forms part of the line, steel wire may be used. For lines connected with large city exchanges, hard-drawn copper wire (usually No. 12 B. & S.) is almost always used.

For toll lines not exceeding 75 miles B. B. galvanized-iron wire, generally No. 10 B. W. G., but in a few cases No. 8 B. W. G., is used; from 75 to 150 miles the E. B. B. grade, or hard-drawn copper, should be used. For good toll lines of any length, the best practice calls for complete metallic circuits of hard-drawn copper, No. 10 B. & S. up to about 500 miles and No. 8 B. & S. up to about 1,000 miles.

For interior wiring use No. 16 or No. 18 B. & S.; in dry places weather-proof office wire (copper), and in damp places rubber-covered wire.

TELEPHONE CABLES

70. When telephone lines enter cities, it is customary to bunch them into cables, containing from 26 to 406 pairs, or complete metallic circuits. These wires vary from Nos. 19 to 24 B. & S. gauge copper, and are insulated with paper, usually dry, but occasionally saturated with an insulating compound. The whole cable is usually covered with a lead sheath to keep out moisture and to protect the wires.

In telephone cables the two wires in each pair forming a complete metallic circuit are twisted together, thus successfully preventing cross-talk. The pairs in each concentric layer are given a twist in a direction opposite to that of the pairs of the next layer. Thus, if any two pairs be considered, and one wire of one of them be taken as the disturbing wire, it will readily be seen that the two wires of the other pair are at equal average distances from it. The method of twisting pairs in cables affords the most perfect transposition known.

If the cable gets wet or damp, leakage takes place and produces cross-talk in that way. The only remedy is to dry out or replace the damp portion.

Telephone cables have been sufficiently treated in connection with telegraph cables, hence it is unnecessary to further consider them here.

PUPIN LOAD-COIL SYSTEM

71. It is a well-known fact that an articulate voice current consists of a fundamental wave and a large number of overtone waves of different frequency, and furthermore that the greater the frequency of an overtone wave the more does inductance tend to make it lag behind the fundamental, whereas capacity has the opposite effect, so that the greater the frequency of an overtone wave the more does it tend to lead the fundamental wave. It has been known for some time that the proper amount of inductance inserted in a circuit will neutralize a given capacity for one particular frequency. The difficulty in telephone circuits is the fact that the capacity exists mostly between the line and the ground or other conductors and is distributed throughout its length and that the frequency varies from about 32 to 40,000 periods per second. To Prof. M. I. Pupin is due the credit of practically proving that by connecting properly designed inductive coils in series with the line and at definitely calculated distances apart, that the distributed capacity of a line could be neutralized by this inductance. He showed that by very nearly neutralizing the capacity by inductance for the highest frequency commonly occurring in conversation, namely, 750 periods per second, that the neutralization was better for all lower frequencies and sufficient for all others. The coils must have a definite inductance and be located definite distances apart, the proof and calculation of which are too complex to be given here. When this is properly done the overtone waves are so little displaced from each other and from the fundamental wave (for which reason it is called a distortionless circuit) that distinct conversation can be held not only over much greater distances but also over smaller sized wires. In the practical application of this system to long lines the line leakage may enter as a disturbing factor for which some allowance must be made. The *Pupin load coils*, as they are called, are now used on the New York-St. Louis, New York-Chicago, and other long circuits, and on some underground cable lines in large cities. On the New York-Chicago line it is claimed that their use has improved the transmission 100 per cent. This circuit consists of No. 8 B. W. G. hard-drawn copper line wires with load coils in series with the line every $2\frac{1}{2}$ miles. Professor Pupin is confident that a conversation could be held without difficulty between New York and San Francisco over a circuit properly equipped with these inductance coils.

TELEPHONE SYSTEMS
(PART 2)

SWITCHBOARD SYSTEMS

SWITCHBOARD DEVICES

1. The ordinary form of exchange for handling a large number of subscribers comprises a central office, from which the lines to the various telephone users, or *subscribers*, radiate. These lines terminate at the central office in what is termed a **switchboard**, which must contain: apparatus for attracting the attention of the operator when a subscriber desires a connection; means for the operator to connect the switchboard telephone with the line of a subscriber, in order to ascertain his wants; means for connecting his line with that of any other subscriber; means for calling the subscriber desired; and, lastly, means enabling either subscriber to signal the operator at the close of the conversation.

LINE DROPS

2. The apparatus for calling the operator may be an electromagnetic annunciator or a very small incandescent lamp, whose circuit is controlled by a relay in the line circuit. A simple form of annunciator, or **line drop,** as it is called, is shown in Fig. 1. *C* is an electromagnet mounted behind the front plate of the switchboard; *S* is a shutter pivoted below and held in its upper position by a catch carried on the lever *L*, rigidly secured to the armature *A*.

When a current is sent through the coil C, the armature A is attracted, thus removing the catch from engagement with the shutter and allowing it to fall. An auxiliary feature, known as a *night-alarm attachment*, is shown in connection with this drop. When the shutter falls, the small spring t is pressed into engagement with the insulated pin u, by the cam projecting from the bottom of the shutter. The spring t and contact u form the terminals of a circuit containing a battery and an ordinary vibrating bell, and the bell is therefore sounded whenever a drop falls. This arrangement is used only at night, when the exchange is not busy, or in small exchanges not requiring the constant attention of an operator.

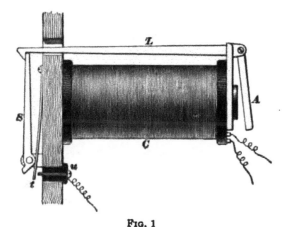

Fig. 1

3. A good switchboard drop should have all its parts rigidly mounted on the same framework, so that it can be readily removed, as a whole, from the switchboard. The framework should be of a metal or other material that is not affected by atmospheric changes, and where drops must be mounted close together, as is usually the case, they should be iron clad, or very nearly so; that is, the coil of each drop should be practically surrounded with a shell of iron, because if only an iron core is used, the drop is very liable to act, by electromagnetic induction, on neighboring drops and cause trouble. Cross-talk may be produced by the electromagnetic induction of one drop in whose line circuit voice currents are flowing, on a neighboring drop whose line circuit is at the time in no way connected with the first.

Induction between neighboring drops and other similar devices may be eliminated by placing the devices farther apart; by so connecting them—when associated with a line circuit—that the insertion of a plug in a jack disconnects them from the circuit, or by surrounding the coil more or less with iron. The latter construction confines the lines of force to an all-iron circuit, thereby preventing their spreading out through the air and thus cutting neighboring drops each time they increase or decrease in intensity. Moreover, a high resistance is often very desirable in a drop or relay, in order that it may be left connected directly across the circuit during a conversation without appreciably interfering in any way with the voice currents in the line

Fig. 2

circuit. This principle is very extensively used in telephone apparatus. When a coil having inductance must be left in a circuit through which voice currents must flow, it should be wound with as few turns as possible; and to further reduce its opposition to voice currents, it may be shunted with a comparatively high but non-inductive resistance, or with a condenser.

4. Tubular Relay.—Relays are now being extensively used in all central-energy telephone systems. In Fig. 2 is shown a **tubular,** or **iron-clad, relay,** illustrating the good features mentioned in connection with drops. The coil and iron core of this relay are held in the iron shell, which is made of one piece, by means of a nut and projecting brass screw *e,* by which it is also fastened to a frame. The terminals of the coil are brought out by means of the brass rods b, c, which are rigidly fastened to the coil and easily slide through holes in the iron shell without touching it. The iron armature a has a projection d. The weight of the piece d, the springs m, n, and the way in which the armature is balanced on the front edge of the piece o, tend to hold the armature a slight distance from the core and shell. The screw s merely holds the armature from falling off when the relay is handled, and does not in any way bind it. When current flows through the coil, both the core and shell attract the armature, but even then a copper rivet or screw prevents it from touching either the core or shell. This is very essential, as otherwise residual magnetism would cause the armature to stick to the core after the current ceased to flow through the coil. When the armature is not attracted, the springs m, n touch the lower contact strips; but when attracted, the projection d lifts the springs m, n so that they part from the lower, and touch the upper, contacts. As these relays are mounted close together and are connected across the circuit during a conversation the iron shell is a very desirable feature.

JACK AND PLUG

5. In Fig. 3 is shown a spring jack and a double contact plug, the two metallic portions of which are adapted to engage the contacts in the spring jack. In this jack, s is the

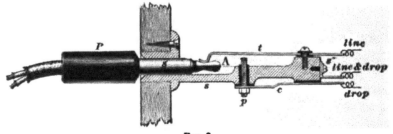

FIG. 3

metallic framework on which the various parts are mounted. Carried on, but insulated from this framework is the spring t that normally rests on the insulated pin p, to which is connected a metallic strip *c*. One side of the line circuit is connected with the metallic frame *s* by means of the screw *s''*. The other terminal of the line is connected with the rear projection of the spring *t*, while the rear projection of the strip *c* is connected with one terminal of the winding of the line drop, the other terminal of which is connected to the frame of the jack at *s''*. The plug *P* has an insulated handle and two metallic contacts *A* and *s'*, to each of which is connected, on the interior of the insulated handle, a terminal of one of the conductors of a flexible cord. The contact *A* is termed the *tip*, while the contact *s'* is termed the *sleeve*, of the plug, the latter surrounding the former, but carefully insulated from it by means of a tubular bushing of hard rubber or similar material. The conductor of the cord circuit that makes contact with the tip of the plug is usually designated as the *tip strand* of the cord, and the conductor making contact with the sleeve of the plug is termed the *sleeve strand*. For similar reasons, the contact spring *t* in the jack is termed the *tip spring*, while the contact *s* with which the sleeve of the plug engages, in this case the metallic frame of the jack itself, is called the *sleeve contact*, and the two sides of the line are frequently termed the tip and sleeve conductors, respectively. When the plug is inserted into the jack, its tip makes contact with the spring *t* and raises it from engagement with the pin *p*, thus breaking the circuit through the line drop, and connecting the tip spring *t* to the tip *A* of the plug, and the sleeve contact *s* to the sleeve *s'* of the plug. Other arrangements of plug and jack contacts and connections will be illustrated in the diagrams. of switchboard systems.

SIMPLE MAGNETO-CALL SWITCHBOARD

6. The Line Circuit.—The circuits of a simple magneto-call switchboard are represented in Fig. 4. The line conductors *l*, *l'* are connected, respectively, with the tip spring *t* and sleeve contacts *s* of the jacks *J'*, *J''*, while the drops *D'*, *D''* are connected between the anvil *p* and the

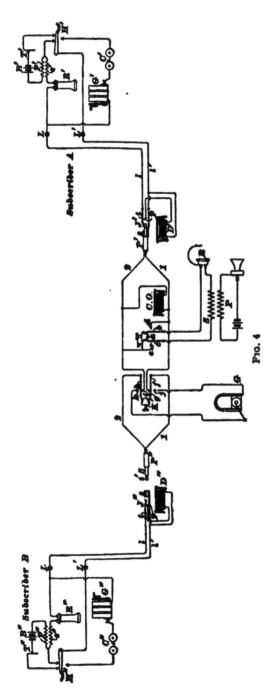

FIG. 4

sleeve side of the line. Under normal conditions, as shown at J'', the drop is connected between the tip and sleeve sides l, l', respectively, of the line. The path, therefore, of a calling current coming in from a subscriber A, when there is no plug inserted in the jack J', may be traced from the tip side l of the line to the tip spring t, anvil p, coil of the drop D', belonging to A's line, and to the opposite side l' of the line.

7. The Cord Circuit. — P', P'' are termed, respectively, the *answering* and *calling plugs*, because P' is used to answer an original call and P'' for calling the subscriber wanted. The tips of the two plugs are connected together through the tip strands 1 of the flexible cords, while the sleeves are similarly connected by the sleeve strands 2 of the same cords. The clearing-out drop CO is permanently bridged across the tip and sleeve strands of the cord of the answering plug. In order that this drop shall not form a circuit of too low resistance between the two sides of the circuit, it is wound to a high resistance, from 500 to 1,000 ohms. The impedance of the drop is still further increased by enclosing the entire winding in a tube of soft annealed iron, which forms the return portion of the magnetic circuit of the core, furnishing a path of low magnetic reluctance for the lines of force; therefore, with a given iron core, a given number of turns in the coil, and a given current, the total number of lines of force, and consequently the inductance, will be a maximum.

From the formula, the impedance $= \sqrt{R^2 + (2\pi n L)^2}$, it is evident that the impedance will be much larger than the simple resistance R when both the inductance L and the frequency n are large, as they will be with this tubular form of drop during a conversation. While talking, the average value of n is about 300, but for the current from a telephone generator it is only about 15 to 20. The impedance of the drop is therefore much larger for the talking than for the generator current, and consequently sufficient of this latter current will flow through it to drop the shutter; but practically none of the talking current can pass through it.

8. Ringing and Listening Keys.—A key K when depressed will connect the secondary circuit of the operator's telephone set across the two strands of the answering plug P', the action of this key being as follows: The conical wedge a, when pressed by the button, presses the two springs b, c, which form the terminals of the secondary of the operator's circuit, against the stationary anvils d, e, which are connected, respectively, with the two strands of the answering cord.

K' is the ringing key by which the operator may connect the terminals of the ringing generator G with the two strands of the cord of the calling plug P''. Normally, the tip strand 1

of the calling plug is connected with that of the answering plug by means of the spring f resting against the anvil g of the ringing key, and in a similar manner the circuit between the sleeve strands 2 of the pair of plugs is maintained continuous by the spring h resting against the anvil i. When the key K' is pressed, however, the springs f, h are forced out of engagement with the anvils g, i and are pressed against the contact anvils j, k, forming the terminals of the generator G. Thus, the operation of the key K' disconnects the calling plug P'' from the answering plug P', and at the same time connects the calling plug with the terminals of the generator G.

9. Calling Central.—In this figure the circuits are shown corresponding to the position of the apparatus when the operator has inserted the answering plug P' into the jack J' of a calling subscriber, for the purpose of ascertaining his wants. The subscriber A, for the purpose of sending a call, operated his generator and caused his drop D' at the central office to display its signal. The operator, in answer to this signal, has inserted the answering plug P' into the jack J', thereby continuing the circuit of the subscriber's line to the cord circuit, and at the same time cutting out the drop D' by virtue of the lifting of the spring t from the anvil p. It is in this position that the apparatus is shown.

10. Listening In.—The next step on the part of the operator is to depress the key K, thereby forcing the springs b, c into engagement with the anvils d, e, and bridging the secondary circuit containing the head-receiver R and the secondary S of the induction coil across the cord circuit. The operator is able to converse with the calling subscriber in the ordinary manner, the circuit over which this conversation takes place being traced as follows: Secondary winding S of the operator's induction coil, spring c of the operator's key, anvil e, sleeve strand 2, sleeve of plug P', line wire l', binding post L' of the telephone at station A, lever H' of the hook switch, secondary coil s', receiver R', binding post L, line wire l, tip spring t of the jack, tip of plug P', tip strand *1* of the cord, anvil *d* of the operator's key, spring *b*, operator's receiver *R*, and back to the secondary *S*. This circuit is acted on by the operator's transmitter through the primary coil *P*, in the ordinary manner.

11. Making Connection.—Having learned that the subscriber *A* desires to be connected with subscriber *B*, the operator inserts the calling plug P'' in the jack J'' of subscriber *B*, thus cutting out his line drop D'' and establishing connection with his line wires l, l'. The operator then depresses the ringing key K', thereby connecting the generator *G* directly across strands *1* and *2* of the calling cord, and thus ringing the bell C'' at the station of subscriber *B*. This calling current from the generator *G* does not pass to the station of the subscriber *A*, because in its action the key K' disconnected the plug P' from the rest of the circuit by breaking the contacts *h, i* and *f, g*. It is not desirable to send the calling current from the switchboard generator to the station of the subscriber calling, because he is holding his receiver to his ear, and the generator current, in passing through it, will produce a violent noise, sometimes very painful to the ear drum, and always annoying. For this reason, in good switchboards provision is made whereby the answering plug is cut off when the calling generator is connected with the calling plug.

12. Talking Circuit.—When subscriber *B* removes his receiver from its hook, the talking circuit between the two subscribers is complete, and may be traced as follows: secondary s'' of the induction coil at the station of subscriber *B*, receiver R'', binding post *L*, line wire *l*, tip spring *t* of jack J'', tip t' of the plug P'', tip strand *1* of the calling cord, spring *f*, anvil *g*, tip strand *1* of the answering plug, tip of plug P', tip spring *t* of jack J', line *l* of subscriber *A*, binding post *L*, receiver R', secondary s', hook lever H', binding post L', line l', sleeve *s* of jack J' and of the answering plug P', sleeve strand *2* of answering plug, anvil *i*, spring *h*, sleeve strand *2* of calling plug, sleeve *S* of calling plug, sleeve *s* of jack J''', line *l*, binding post L', hook lever H'', and back to the secondary s'' of his induction coil.

13. Clearing Out.—After the conversation is completed, one or both of the subscribers should operate their generators or ring off, as it is usually termed. This will send a generator current over the line of the two connected subscribers, a part of which current will find a circuit through the clearing-out drop CO, and cause its shutter to fall. The clearing-out drop is used to signify when a disconnection or other service is desired. The operator, seeing the signal, should again depress the key K and inquire if the subscribers are through talking; if she receives no response, she will disconnect the lines by removing the plugs. If, however, one of the subscribers desires to be connected with still another subscriber, the operator may receive orders as before, and complete the second connection desired.

CENTRAL-ENERGY, OR COMMON-BATTERY, SYSTEMS

14. The idea of replacing all the transmitter batteries and the signaling generators at the subscribers' stations by a single source of current located at the central station has proved so successful that most of the new large exchanges are now operated on this plan. They are called **central-energy**, or **common-battery**, **systems** because all the electrical energy is supplied from the central exchange.

15. Battery in Series in Line Circuit.—The most obvious solution of the problem was to do away with the induction coil at the subscribers' stations, placing the transmitter and receiver at each station in series in the line wire, the talking battery being placed at the central office in series in the circuit formed by the two connected lines. This arrangement would require a separate battery for each pair of cords and plugs. In this case the transmitter at one station serves simply to vary the resistance of the entire circuit and therefore to cause corresponding variations in the current flowing through the two receivers, which are connected directly in the same circuit. It is desirable when such an arrangement is used, that the resistance of the circuit shall be as low as possible, and to this end the receiver coils should be wound to a lower resistance than would otherwise be desirable and the transmitter designed to have a high resistance, a large proportion of which is variable. The simple arrangement at the subscriber's station, consisting of a transmitter and receiver in series, has been used with considerable success even in large exchange systems, but it is now being installed only in house, hotel, and other small systems.

THE STONE SYSTEM

16. The principles of one of the successful systems by which a common battery is made to serve for all the transmitters of an exchange is illustrated in Fig. 5, which shows the actual talking circuit between the subscribers when they are connected at the central office by the cord circuits, the receivers at each station being removed from the hook. A, A', A'', and A''' are subscribers' stations connected with the central office by the metallic circuit lines as shown. The portion of the circuit included between a, a' may be considered as the tip strand of one cord circuit and that portion between the points b, b' as the sleeve strand of the same circuit. The two metallic-circuit lines form, when thus connected, one continuous circuit, including the subscribers' receivers and transmitters connected in series. Bridged across this circuit at the central office is a battery B connected between two impedance coils I, I'. A'', A''' represent two other stations similarly connected together and bridged across the same battery B through two other impedance coils I'', I'''. Current from the positive pole of the battery flows through the impedance coil I to a point c, where it divides, and, having passed through the line and instruments of stations A, A', returns to the point c' where the two portions of the current unite and pass through the

impedance coil I' to the negative pole of the battery. In a
similar manner current flows from the same battery B
through the two stations A'', A''' that are connected together
by the strands d, d' and e, e' of another cord circuit. The
impedance coils are made of rather coarse copper wire
wound about heavy iron cores so that, while their ohmic
resistance is very low, their impedance to rapidly varying
currents is very high.   From this it follows that they will
allow steady currents to pass through them with comparative
ease, but will form a practical barrier to rapidly varying

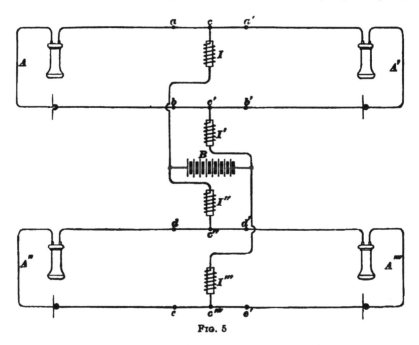

FIG. 5

currents—such as are produced when talking. While the
transmitter and receiver at each subscriber's station are connected in series, it is clear that these two subscribers' lines
are connected in multiple with respect to the battery B and
the impedance coils I, I'. Assuming that the resistance
of the two lines is the same, equal portions of current will
pass through each subscriber's station.   If, however, the
transmitter at A is so operated as to increase the resistance

of the line through that station, a greater portion of the current from the battery B will then be forced through the line of station A' because the impedance coils placed in the bridged circuit with the battery tend to prevent any fluctuations in the current through them. The current flowing through the coil is therefore maintained practically constant, and an increase in resistance in one of the lines tends to cause less current to flow through that line, and more through the other. In a similar manner, if the transmitter at station A is so operated as to decrease its resistance, the greater portion of the current flows through the line leading to station A, and the smaller portion through the line leading to station A'. Whatever changes take place in the resistance of the circuit of one line wire cause corresponding changes in the current flowing in the other line wire, and these changes in current act on the receiver at the other station in the ordinary manner. By this arrangement the transmitter at one station has only to vary the resistance of the circuit leading from the central office to that station.

17. Prevention of Cross-Talk in Stone System. The battery B is connected across the cord circuit of each pair of plugs. The use of the impedance coils renders it possible to make a single battery B serve for a large number of cord circuits, the various leads from the points c, c', c'' and c''' on each cord circuit being led through separate impedance coils to the terminals of the common battery.

It might seem at first that this arrangement would produce cross-talk—that conversation being carried on between subscribers A and A' would also be heard on the lines A'' and A''', by virtue of the fact that the two circuits are connected together through the impedance coils. This, however, is not usually the case for the impedance coils I, I' confine nearly all the fluctuations of current to the circuit of the lines leading to stations A and A', and whatever fluctuations do find their way through the impedance coils I, I' can complete their circuit through two paths, one of which is through the battery B and the other through the impedance coils I'', I''' and the combined lines of subscribers A'', A'''. The circuit through the impedance coils I'', I''' contains a large amount of impedance and considerable resistance, while the circuit through the battery B contains no impedance and practically no resistance; therefore, the fluctuations in current that find their way through the impedance coils I, I' will be short-circuited by the battery B and will not pass through impedance coils I'', I''' to the lines of subscribers A'', A'''.

This method of supplying current to the transmitter batteries was devised by Mr. John S. Stone, of the Bell Company, and the principle involved is extensively used.

THE HAYES SYSTEM

18. A system devised by Mr. Hammond V. Hayes is shown diagrammatically in Fig. 6, in which A and A' are sub-scribers' stations, and C the central office; J, J' are repeat-ing coils, each having two equal windings j, j'. One end of the winding j is connected to one end of the winding j' and these two ends are also connected to the battery B. The other terminal of each winding of the repeating coils is con-nected, as shown, with the line wire leading to a subscriber's station. These repeating coils are merely induction coils,

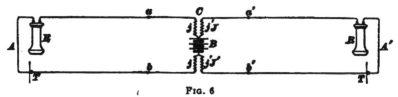

Fig. 6

having windings of equal resistance, and the same number of turns. They are wound on heavy, soft-iron wire cores, and in practice, the two coils J, J' are wound on the same core.

As in the preceding figure, a, a' may be considered the terminals of the tip strand of the cord conductor of a pair of plugs, and b, b' the terminals of the sleeve strand. Current flowing from the positive pole of the battery B will divide and pass through the separate windings j, j' of the repeating coil J to the line wires leading to stations A, A'. The two portions of the current will then proceed through the instruments at this station, and back to the central office through the positive side of the line, and through the windings j, j' of the repeating coil J' to the negative pole of the battery.

The action of one line circuit on the other is probably easier to understand in this case than in the Stone system. The two circuits may be considered as entirely separate, connected inductively only through the windings of the repeating coil. If a variation takes place in the transmitter at station A, the current flowing from the battery B through the circuit of that line will be correspondingly varied. These

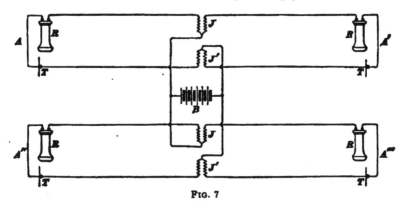

FIG. 7

variations will pass through the windings j, j of the repeating coils, and will therefore act inductively on the windings j', j', which are in circuit with the station A'. It is easy to see that any fluctuations in current in the circuit of station A will induce similar currents in the circuit of station A', and vice versa. When station A is transmitting, the coils j, j act as primary coils, j', j' being the secondary coils. When section A' is transmitting, the functions of the coils are reversed, j', j' serving as the primaries, and j, j as the secondaries. As in the Stone system, the battery B may be made to serve an unlimited number of cord circuits, a separate repeating coil being placed between each pole of the battery and each side of the cord circuit, as shown in Fig. 7.

19. Automatic Signaling.—One of the great advantages of common-battery systems is their adaptability to automatic signaling from the subscriber to the central office. The arrangement is such that no separate action on the part of the subscriber is necessary in order to transmit a signal to the central office, this result being accomplished automatically by the removal of his receiver from its hook.

COMPLETE SYSTEM FOR SMALL EXCHANGE

20. Arrangement of Circuits.—In Fig. 8 is shown the circuits of a common-battery system for exchanges of moderate size, embracing a feature not heretofore considered. This feature consists in the use of supervisory signals in the cord circuit, by the operation of which the operator is at all times made aware of the condition of any two lines connected together for conversation. An electromagnetic signal E is used in the line circuit and so arranged that the target is lifted and thus displays the line number when the magnet coil is traversed by a current, the target automatically dropping into its normal position when the current ceases. The cord circuit is arranged according to the Hayes system for supplying current to the subscribers' transmitters, the coils j, j' being wound on one iron core, and j'', j''' on the other, the magnetic circuit of the cores being completed through the two iron yoke pieces y, z. This makes the impedance to alternating currents much higher than it would be if the iron circuit was not completely closed, as would be the case were the connecting yokes y, z omitted. The terminals of the coils j, j' are connected together and to the negative pole of the battery B; the remaining terminals are connected with the tip strands of the cord circuit. In a similar manner, the coils j'', j''' are connected together and to the positive pole of the battery B, while the remaining terminals are connected with the sleeve strands of the cord circuit. Connected in the tip strand of each cord are the electromagnetic signals O, O', adapted to be operated by the passage through the subscribers' talking circuits of the

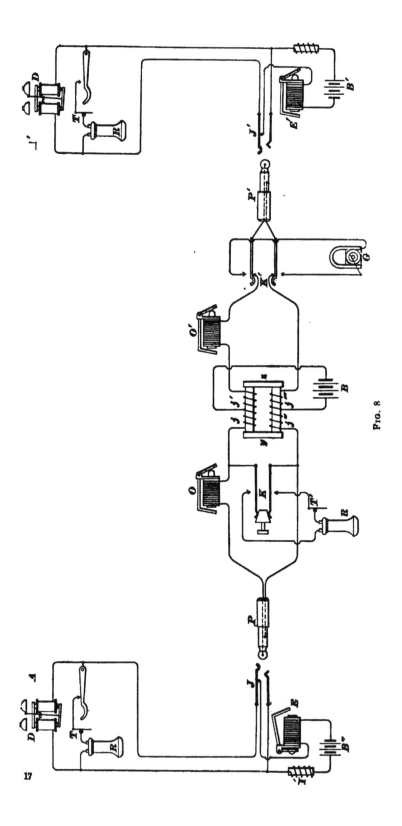

Fig. 8

current from battery B. By a key K the operator's circuit, including the transmitter T' and the receiver R, may be bridged across the circuit; and by a similar key K' the generator G may be connected with the calling plug P'.

21. Operation.—To understand the operation of the signals, assume that subscriber A desires a connection with subscriber A'. On removing his receiver from its hook, the signal E is automatically displayed by the passage of the current from battery B'' through the circuit. The operator answers with the plug P, thus opening the circuit through the signal E, which is restored by gravity. Current from the battery B now flows through the line and cord circuit, thus causing the magnet of signal O to raise its target. Having ascertained the subscriber desired, the operator inserts plug P' into the jack of the line leading to subscriber A', and operates the ringing key K'. The insertion of the plug P' will not cause the signal O' to operate because of the high resistance of the bell D at that station. As soon, however, as the subscriber A' lifts his receiver, a low resistance path will be formed between the two sides of the line circuit, through the receiver and transmitter, thus allowing a comparatively large current to flow from the battery B, which will cause the target of signal O' to rise. This will inform the operator that subscriber A' has responded. If the signal O' is not raised, she will know that the subscriber has not responded and will ring him up again.

22. Supervisory Signals.—The two subscribers converse as described in connection with Fig. 6, the coils j, j'' acting as primaries and j', j''' as secondaries when subscriber A is talking. When subscriber A' is talking the functions of the coils are reversed. Both supervisory signals O, O' remain up as long as the subscribers remain in conversation. When, however, either one of them hangs up his receiver, the low resistance path through his transmitter and receiver will be broken and replaced by one of high resistance through his bell, and the signal O or O' will fall, owing to the reduction in current through it. When both signals are down the operator knows that the conversation is finished. If only one falls, the other remaining up, the operator concludes that one of the subscribers desires another connection, and inquires of him what it is. A rather high non-inductive resistance or a condenser connected in parallel with each supervisory signal O, O' would improve the talking qualities of the circuit.

23. The Lamp Signal.—Hand in hand with automatic signaling and common-battery systems came the adoption of the incandescent lamp for signaling purposes in telephone exchanges, the lamp presenting many advantages over the electromagnetic signals so far considered. The signals O, O', Fig. 8, may be replaced by lamp signals, which, however, would not be connected directly in the cord circuit but in separate circuits controlled by relays placed directly in the cord circuit in positions corresponding to the magnets of signals O, O'. In a similar manner, the line signal may be a lamp, whose circuit is controlled by a line relay connected in place of the signal E. This arrangement will be fully described in connection with the multiple-board common-battery system. Although three batteries are shown for the sake of clearness, only one battery would be used for all circuits.

LARGE SWITCHBOARD SYSTEMS

TRANSFER SWITCHBOARDS

24. Thus far only those switchboards for use in com-paratively small exchanges, that is, those having not over four or five hundred subscribers, have been described. Under ordinary circumstances, the number of subscribers allotted to each operator during the busy portion of the day does not exceed one hundred and sixty. It is obvious that any one of the subscribers whose lines terminate on the section of a board before a certain operator may desire a connection with any other subscriber on that or any other section. Where not more than three or possibly four sections are used, the plan usually adopted in making a connection between subscribers on different sections is to reach across the face of the boards with the calling plug of the pair used in answering the call, and inserting it into the jack of the called subscriber. The extent to which this can be done is limited by the length of the cords, the sizes of the boards, and the reach of an average operator.

When the number of sections becomes so great that it is not practical to reach across with a pair of cords in order to complete a connection, other means have to be provided. One of these is to provide a system of auxiliary circuits, usually termed *trunk*, or *transfer*, lines, running between the various sections of the switchboard, and to provide these trunk lines with means by which an operator may connect one end of one of these trunk lines with the line of a subscriber, the operator at the other end, when instructed to·do so by the first operator, completes the connection between her end of the trunk line and the line of the subscriber called for. The trunk lines thus serve as an auxiliary connecting circuit between two subscribers' lines that cannot be well connected by a single pair of plugs and cords. A switchboard arranged to operate on this general principle is termed a **transfer**, and sometimes an **express, switchboard**. It is obvious that in transfer switchboards two operators are required to complete a connection between two subscribers, unless the line of the subscriber called for is within easy reach of the operator who answers the call. The term transfer switchboard is used in distinction to *multiple switchboard*, which will now be considered.

THE MULTIPLE SWITCHBOARD

25. Main Features.—The primary object of the multiple switchboard is to so arrange the apparatus that any operator can connect the line of a calling subscriber with that of any other subscriber in the entire system without the aid of another operator. The entire board is divided into sections, each usually containing working room for three operators; on each section is placed, besides the line signals and jacks of the lines, the calls of which are to be attended to at that section, the jacks connected with every other line in the exchange. Each operator can reach over one operator's position on each side of her own.

On some of the common-battery multiple switchboards that the Bell Company has installed, each operator attends to the lines of one hundred and sixty subscribers, a great increase over the number per operator's position customary with approved types of annunciator boards, which runs from sixty to seventy-five. With the common-battery board the supervision over the lines in use and the number of movements required of each operator is considerably reduced, thus reducing the cost of operation, as well as the initial cost of the switchboard, by being able to reduce the total number of sections almost one-half.

In order to have the arrangement of the multiple jacks the same at each section, and also to have the answering jacks conveniently located near their respective line signals, it is usual to have the answering jacks duplicated at each operator's position. So there are in the case cited above one hundred and sixty jacks more at each position than there are subscribers' lines entering the exchange, in addition to other jacks required for trunk lines to other parts of the exchange.

26. An idea of the general plan of a multiple switchboard may be obtained from the simple diagram shown in Fig. 9, in which L, L', and L'' represent three subscribers' lines, each passing through jacks j, j', and j'' on the various sections A, B, and C of a multiple board. The drop D and answering jack J of line L are located at section A, and in like manner the lines L', L'' are equipped with answering jacks J', J'' and drops D', D'' at sections B, C, respectively. If the subscriber L sends in a call, his drop D will attract the attention of an operator at section A, who will answer it by inserting one plug of a pair into the answering jack J at that section, and finding that the connection desired is with line L' or L'', will insert the remaining plug of the pair into the jack j' or j'' of the line wanted. It is not difficult to understand that inasmuch as every line throughout the whole exchange is provided with a jack on section A and on every other section also, the operator at A will have means within her reach of connecting the line L of the calling subscriber with any subscriber's line coming into that exchange.

27. Busy Test.—Suppose the lines L, L' to be connected together at the central office, and that while they are so connected subscriber L'' sends in a call. His drop D'' at section C will be thrown, and the operator at that section will place herself in communication with him by

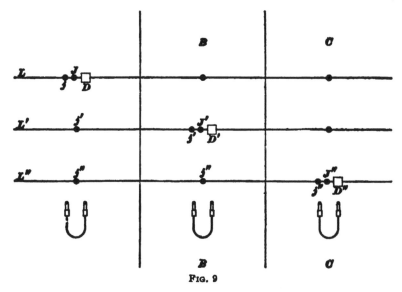

Fig. 9

inserting the answering plug of a pair into the answering jack J'' of his line. Suppose, further, that this subscriber desires to be connected with line L. It is evident that the operator at section C will, unless special means are provided, have no means of knowing that the line L is already connected with another line at another section of the board. Should she therefore connect the line L'' with the line L, the three subscribers would be connected together and much confusion would result. In order to prevent an operator from making connection with a line that is already in use at another section of the board, there is employed the so-called **busy test,** which forms an essential feature in every multiple switchboard.

The test for a busy line is usually performed by the operator applying the tip of the calling plug of a pair to the sleeve of the jack of the subscriber called for. If the line is busy, she will hear a click in her telephone, while if it is free or disengaged, silence will inform her to that effect. This order could be reversed, a click indicating a free line and silence a busy one, but this is unusual. The details of the busy test can only be understood in connection with the complete circuits of a multiple switchboard, which will be next considered.

Although magneto-call multiple switchboards have been extensively used and are still in use, nevertheless they are being rapidly crowded out by central-energy multiple systems, because the latter are more economical to operate and give much better satisfaction to both subscribers and owners of telephone systems, although the magneto-call switchboard may be somewhat less expensive in first cost. The simplicity of operation and construction and the low first cost of the subscribers' telephone instruments is a very advantageous feature of central-energy systems and partially equalizes the more expensive and complicated switchboard circuits required. To describe more than one multiple switchboard would require more space than can be devoted to the multiple system here; hence, a central-energy multiple switchboard that is extensively used will be explained. The central-energy systems developed by the various telephone manufacturing companies differ more or less from one another, but if the principles of the systems and the central-energy switchboards described in this Section are thoroughly understood there should be no great difficulty in understanding any system.

CENTRAL-ENERGY MULTIPLE SWITCHBOARD

28. The central-energy, or common-battery, system that is here described is used by the licensees of the American Bell Telephone Company and employs the Hayes principle of utilizing one battery for all the line circuits. **Fig. 10** shows two subscribers' telephone instruments connected to the common battery B; all exchange apparatus, except what is actually in the circuit while a conversation is being held, is omitted. Two repeating coils are represented, each having two coils of 40 ohms and 2,000 turns each. In practice, the two repeating coils (four coils) are wound on one core of soft-iron wire, so arranged as to form a practically closed, or complete, magnetic circuit. Connected in the cord circuit are two supervisory relays $A R$, $C R$, each of 20 ohms resistance and shunted by non-inductive resistances $A r$, $C r$ of 30 ohms each. The non-inductive shunts offer much less opposition to the voice currents than the relay windings. However, the shunt is of sufficiently high resistance to compel enough of the battery current to flow through the relay coil to operate it.

29. Subscriber's Instrument.—At station A, the subscriber's circuits are shown in a customary and convenient way, while at station D the same circuits are arranged to facilitate an explanation of the theoretical action of the instrument. The coil s, having a resistance of 15 to 17 ohms and about 1,700 turns, is usually termed the secondary and p, having a resistance of 25 to 30 ohms and about 1,400 turns, the primary. The ends i, i' are the inside, while o, o' are the outside; the two coils are wound in the same direction, as indicated.

When the transmitter is at rest, it will have a fixed resistance and the current will flow from the positive terminal of the battery through l–o–s–i–c–T–a–l' to the negative terminal of B; arrow 1 indicates the direction of this current through s. The number of turns in s and the normal current strength are about such that the iron core I is probably magnetized to a point where it is most sensitive to the slightest change in the current in either s or p; that is, a given change in the strength of the current produces a greater change in the magnetism of the core than would be the case were it magnetized to a much less or much greater degree.

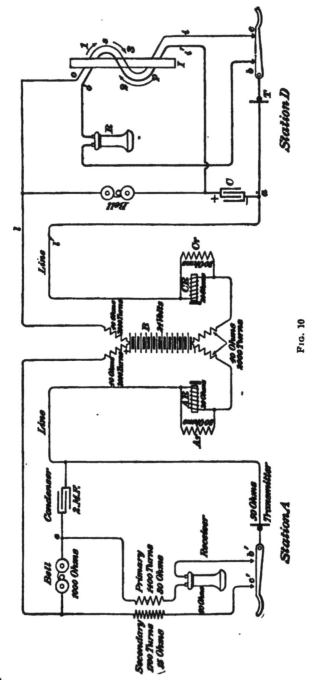

FIG. 10

The following explanation of the action of the subscriber's instrument only is based on one given in "The American Telephone Journal," by Mr. W. W. Dean. The difference of potential between the points a, c will be equal to the current times the transmitter resistance. The condenser C, being connected from a to b, will be charged to the potential that exists between a and c, the $+$ and $-$ signs on the figure indicating the polarity of the charge. Suppose that the resistance of the transmitter is suddenly lowered by speaking into it. The difference of potential between a, c will suddenly be reduced, because the current will not be proportionally increased in strength, and this potential being lower than that of the condenser, will cause the latter to discharge some of its electricity through $i'-p-d'-R-b-T-a$, the direction of this discharge current through the primary being shown by the arrow 2. This current will induce a current in the secondary in the opposite direction; that is, in the direction of arrow 3. Since s has more turns than p, this current 3 will be of higher potential than the inducing current; and being in the same direction as the battery current 1, which is also increasing in strength because the transmitter is decreasing in resistance, will cause a considerable increase in the current in the line circuit. When the transmitter increases in resistance, the potential between a and c rises and current flows into the condenser from b through $R-d'-p-i'$; that is, in the opposite direction through the primary to the arrow 2. This induces a current in the secondary s in the opposite direction to the arrow 3. The battery current through s at the same time decreases in strength because of the increase in the transmitter resistance, while the induced current in s will still further reduce the strength of the current in the line circuit. Thus, the changes in current strength in the line are much greater than would be the case were no induction coil used. This can be proved by reversing the secondary coil in the circuit, which will cut down the loudness about one-half.

The fluctuating current produced by talking into the transmitter at station A flows through $l-o-s-i-c-T-a-l'$. This induces in p a current that flows through $d'-R-b-T-a-C-i'$,

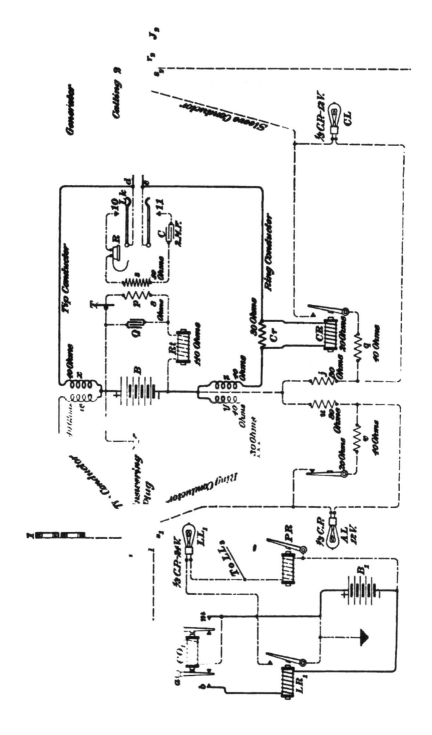

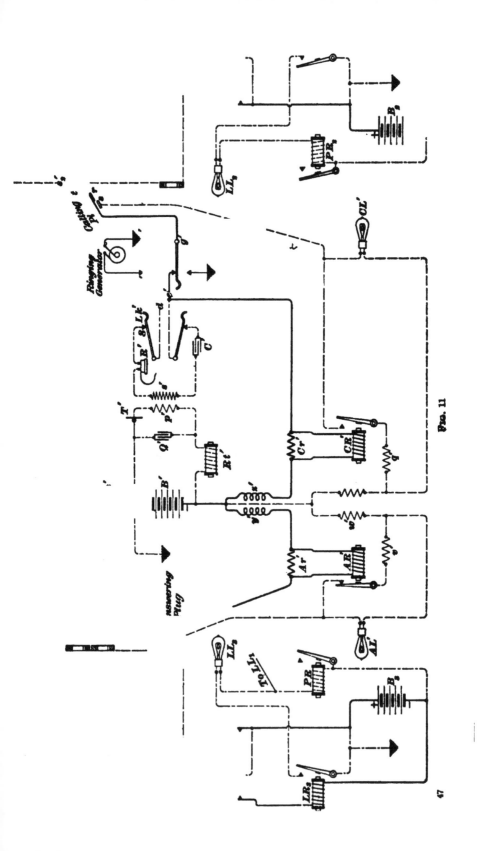

Fig. 11

47

the condenser offering little opposition to this very rapid alternating current. Or, it may be considered that the alternating potential induced in p causes the condenser to be charged alternately in opposite directions, or the normal charge on the condenser to be varied in amount, in either case the charges flow in and out of the condenser and produce the desired effect on the receiver R.

30. Line Circuit.—Two operators' cord circuits and three subscribers' line circuits are shown in Fig. 11. Normally, the subscriber's receiver rests on the hook switch, and the circuit between line wires l, l' is closed only through the bell and condenser. On account of the condenser, no battery current can then flow through any part of the subscriber's instrument. At the frequency of the alternating current (about 15 to 20 periods per second) used for ringing the subscribers' bells, the impedance due to the 1,000-ohm bell and the condenser is small enough to allow sufficient ringing current to pass through them to properly ring the bell. The 24-volt storage battery B, at the central office is normally connected through the *line relay* $L R_1$ and the contact points b, m of the *cut-off relay* $C O_1$ to the two line wires l', l.

31. Operation of Line Circuit.—When the subscriber removes the receiver R_1 from the hook, the transmitter T_1 and the secondary winding of the induction coil form a circuit of sufficiently low resistance to allow enough battery current to flow from $+ B_1$ through m–i–l–c_1–T_1–l'–a–b–$L R_1$ to $- B_1$ to cause the line relay to attract its armature and close the local circuit containing the line lamp $L L_1$ and the line-pilot relay $P R$. This causes the line lamp to light, thereby notifying the operator that her attention is desired on that line. The circuits of all the line lamps at any one operator's position are collected together at one point, from which a single wire is connected to the pilot relay for that one position. Thus, the lighting of any one line lamp will cause the line-pilot relay to attract its armature and close the local circuit containing another lamp called the line-pilot lamp. To avoid making the diagram complicated, the local circuit containing the line-pilot lamp, which is connected between $+ B_1$ and the relay contact screw, is not shown. The line-pilot lamp enables the supervising operator to observe how the operator at any one position is attending to her work, for the line-pilot lamp will light up and go out every time a signal is received and answered, respectively. Hence, the supervising operator knows that if the line-pilot lamp remains lighted an undue length of time that her assistance is needed at that position. There is one line relay and line lamp for each line circuit, but only one line-pilot relay and one line-pilot lamp for each operator's position. The line-pilot lamp is usually placed in some convenient position on the face of the board where it can be seen by the supervising operator. The pilot lamps are usually somewhat larger and of higher candlepower than the line lamps. The lamps used as signals usually have over them glass caps, called opals, of different colors so as to be easily distinguished; thus, the line lamps may be white, the supervisory lamps, red, and the line-pilot lamps blue.

32. Cord Circuit.—In each cord circuit there is one repeating coil having four windings w, x, y, z; in the answering side of the circuit there is one supervisory relay $A R$ controlling the answering supervisory lamp $A L$; and in the calling side, one supervisory relay $C R$, controlling the calling supervisory lamp $C L$. As these two relay coils are directly in the talking circuit, they are shunted by a non-inductive resistance that is sufficiently high, however, not to interfere with the operation of the relay by the battery current, the relay coil having a sufficiently low resistance. The resistance u (80 ohms) plus that of the cut-off relay $C O_1$ (30 ohms) associated with the line, is about equal to the hot resistance of the lamp $A L$, and hence with the two in series across 24 volts, the lamp must light with 12 volts across its terminals. When, however, the 40-ohm coil v is connected by the supervisory relay $A R$ in parallel with the lamp $A L$, the 40-ohm coil takes such a large portion of the current flowing through u that the lamp goes out. The 140-ohm retardation coil Rt in the operator's transmitter circuit prevents cross-talk between the various operators' circuits because the very rapid fluctuations in current caused by the transmitter are practically confined to the circuit Q-T-p. The listening key is normally open, but will remain either open or closed; that is, the operator does not have to hold it in either position. A 2-microfarad condenser C in the operator's listening circuit prevents the battery current from flowing through the operator's receiver when the listening key is closed. The ringing generator supplies current through the calling plug when the ringing key Rk is pressed. The plugs have three contacts, known as the *tip t*, *ring r*, and *sleeve s*; they make contact with corresponding parts of the jacks.

The supervisory lamps are mounted in the horizontal plug shelf in front of the plugs with which they are associated; about $\frac{1}{10}$ ampere is required to properly light a supervisory or line lamp. There are usually fifteen such cord circuits for each operator's position, or forty-five to a section.

33. Operation of Cord Circuit.—Assume all circuits to be in their normal position and that subscriber *1* has just lifted the receiver off the hook in order to call for subscriber *2*. For convenience, several batteries and ringing generators are shown, but there is only one battery and one ringing generator. Normally, no current flows through the cut-off relay and consequently a rests against b and i rests against m. The removal of receiver R_1 from the hook connects the transmitter T_1 and one winding of the induction coil across the two line wires, thus allowing sufficient current to flow from $+B_1$ (through the circuit m-i-l-c_1-T_1-l'-a-b-LR_1 to $-B_1$) to close the line relay LR_1 and cause the line lamp LL_1 and the line-pilot lamp to light. The operator then inserts an answering plug, usually the rear plug, into the answering jack belonging to line *1*, which is located on the switchboard immediately above the line lamp LL_1. This produces two results: a current flows from $+B$ through w-t_1-l-c_1-T_1-l'-r_1-$\begin{Bmatrix} AR \\ Ar \end{Bmatrix}$-$y$ to $-B$. Enough of this current flows through the answering supervisory relay AR to close it, thereby shunting the answering supervisory lamp AL by the resistance v, which prevents the lamp AL from lighting. Current for the subscriber's transmitter now comes through the repeating coil and the cord circuit. Current also flows from $+B$ through ground–CO_1–s_1–$\left\{{AL \atop v}\right\}$–$u$ to $-B$ and closes the line cut-off relay CO_1, thereby breaking the circuit at b through the line relay LR_1, which in turn causes the line lamp LL_1 and also the line-pilot lamp to go out. The operator closes her listening key Lk, thereby bridging her receiver circuit R–s–C across the tip and ring conductors, which enables her to converse with the subscriber. The retardation coil Rt not only limits the current in any one transmitter circuit, but it also enables current to be supplied to all the operators' transmitters from one battery without producing the cross-talk and interference that would otherwise result. The condenser Q improves the transmitting qualities by being charged and discharged through the primary winding p in much the same manner as already explained in connection with the subscriber's telephone. Variations in current strength through the winding p of the operator's induction coil induces an E. M. F. of variable intensity in the secondary winding s, thereby causing a fluctuating current to flow in and out of the condenser C through the circuit s–R–10 (the listening key Lk is now supposed to be closed as shown at Lk') –d–x–B–z–Cr–e–11. This fluctuating current induces, through the repeating coil $w\,x\,y\,z$, a similar current in the subscriber's circuit, thereby affecting the receiver R_1. In a very similar manner, fluctuating voice currents, due to the subscriber talking into his transmitter, are repeated in the operator's receiver circuit. The non-inductive shunts across the supervisory relays allow the voice currents to readily flow through either ring conductor.

Suppose that subscriber 2 is wanted, that the usual busy test, which is explained later, has been made and the line found to be not busy, and that the operator has therefore inserted the calling plug into a multiple jack of line 2, opened the listening key $L k$, and closed the ringing key $R k$ as here shown. Current now flows from $+ B_2$ through $C O_2$-s_2-$C L$-j to $- B$, thereby lighting the calling supervisory lamp $C L$ and closing the cut-off relay $C O_2$. The closing of $C O_2$ prevents the operation of the line relay $L R_2$ when subscriber 2 presently removes his receiver from its hook. Current also flows from the ringing generator through 5-f-t_2-l_2-bell and condenser in the subscriber's telephone-l_2'-r_2-g-4 back to the generator, thereby ringing the subscriber's bell, because this alternating current can readily pass through the condenser and bell. It will be seen that, when the operator rings the desired subscriber, the ringing key breaks the circuit leading to the answering side of the cord circuit, and hence, does not ring back in the waiting party's ear. The operator then releases the ringing key. When the called subscriber takes down the receiver R_2, current flows from $+ B$ through x-d-6-f-t_2-l_2-c_2-T_2-l_2'-r_2-g-7-e-$C R$-z to $- B$, thereby operating the calling supervisory relay $C R$, which connects the resistance q in parallel with the calling supervisory lamp $C L$ and causes the latter to go out. Hence, the going out of lamp $C L$ signifies that the called subscriber has answered the telephone. Subscribers 1 and 2 can now converse, their circuits being practically the same as shown in Fig. 10.

When the conversation is finished a supervisory lamp will light as each subscriber hangs up his receiver. For, suppose that subscriber 2 hangs up his receiver, current can no longer flow from $+ B$ through x-d-6-f-t_2-l_2-c_2-T_2-l_2'-r_2-g-7-e-$C R$-z to $- B$, because this circuit is open at c_2. Hence, $C R$, being deprived of current, releases its armature, thereby opening the circuit through q and allowing the lamp $C L$ to get enough current to light it. Similarly, the lamp $A L$ will light when subscriber 1 hangs up his receiver. If only one lamp lights, it signifies that the subscriber whose supervisory lamp has not lighted desires another connection. When both lamps are lit the operator removes both plugs from the jacks, and all circuits return to their normal conditions.

Evidently both supervisory lamps go out because their circuits are both open at the sleeves of the withdrawn plugs.

34. Busy Test.—To explain the **busy test**, assume that the circuits are exactly as shown in this figure; that is, subscribers *1* and *2* are connected together, subscriber *2* not having yet removed his receiver from the hook, and that the call of subscriber *3*, for connection with subscriber *2* whose line is now busy, has just been answered by an operator at another section of the switchboard. The latter operator makes the busy test as follows: Her listening key $L k'$ is already closed because she has just been conversing with subscriber *3*. Leaving this key closed, she touches the tip t of her calling plug to the sleeve contact s_s' of the multiple jack in her section belonging to the line of subscriber *2*. Current flowing from $+ B$ through the ground–$C O_s$–s_s'–s_s–$C L$–j to $- B$, makes potential of the sleeve contacts of all jacks belonging to the line of subscriber *2* different from that of the ground and produces a click in the operator's receiver R'. For current then flows from $+ B'$ through x'–d'–f'–t–s_s', where it unites with current that is flowing from $+ B$ through $C O_s$–s_s', and they flow as one current through s_s–$C L$–j to $- B$. Before touching t to s_s', the points d', e', and hence the terminals of the condenser C', had exactly the same potential difference as the terminals of the battery B', but now the current flowing from $+ B'$ through x'–d'–f'–t–s_s'–s_s –$C L$–j–B–ground to $+ B'$, has suddenly lowered the potential of the point d' and hence the charge on condenser C' has suddenly decreased, thereby producing a momentary current and click in the receiver R'. If the subscriber is listening at R_s, he will hear this busy test, because the winding x' of the repeating coil induces a momentary current in w', which flows through the subscriber's circuit and back through y' and B' to w'. If the test shows the line to be busy, the operator so informs subscriber *3*, and if no request for another connection is immediately made, she returns both plugs to their normal positions.

If the receiver R_s has been removed from the hook, the

test will be practically the same; the only difference being that the resistance q is now in parallel with CL. If the line is not busy no click will be heard in the operator's receiver R', because, there being no plug in jack J_s, s_s has the potential of $+ B_s$, that is, the same potential as the ground, and t also has the potential of $+ B'$, which is identically the same battery as B_s, and hence the difference of potential between points d' and e' does not change at all when t is touched to s_s'. Although a subscriber may have removed his receiver from the hook, his line will not test busy until a plug is inserted in one of the jacks belonging to his line.

INTERIOR SYSTEMS

HOUSE SYSTEMS

35. Two general plans may be followed in installing telephone systems in buildings: one is to run a separate circuit from every station to a switchboard placed at the central office, and any of the switchboard systems adapted for small exchanges may be used for this purpose. The other is to run at least one more wire than there are stations through all the stations, and to provide a switch at each station whereby the instrument at that station may be connected with any other station. This system is known under several names, among which are the *intercommunicating system*, the *house system*, and the *speaking-tube system*. This system, which will be here called the *house system*, has the distinct advantage, for a compact system, not exceeding about twenty telephones, over the use of a central switchboard, that it does not require the service of an operator to make the connections between the various stations.

MAGNETO-BELL SYSTEM

36. A good and simple system using ordinary magneto-generators, polarized bells, the necessary talking apparatus and switches is shown in Fig. 12. Only three subscribers' stations, numbered *1*, *2*, and *3*, are shown, but more can be added. Below each telephone is an intercommunicating switch and an ordinary push button. Both of these are usually mounted on the backboard of the telephone set. The lever of the intercommunicating switch is adapted to slide over and make contacts with any of the buttons *1*, *2*, and *3*. Three line wires run through each station, each bearing a number corresponding to the number of the station to which it particularly belongs. Each line wire is permanently connected with the buttons bearing the same number at each station. On each switch the button bearing the same number as the station to which that switch belongs is placed at

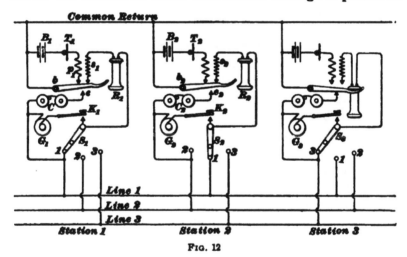

FIG. 12

the left-hand end of the row, and is called the *home button.* Thus, at station *3* the button marked *3* is placed at the left-hand end of the row of buttons, instead of occupying its regular place in the sequence of numbers. The system of electrical connections is in nowise affected by this, the button being merely shifted from the position it would ordinarily occupy to the first position, because it is more convenient to have it there. The common return wire runs through all of the stations with the other line wires.

37. Operation.—The operation of house systems is very simple. All of the levers normally rest on the left-hand or home button. When, however, one station desires to call another, the party at the calling station moves the lever on his telephone to the button bearing the number of the station he desires to call. This act connects his telephone to the line belonging to that station. He then presses the push button and turns his generator handle, and after the subscriber has responded the two converse over the line wire bearing the number of the called station and the common return wire. As shown in this figure, the lever at station 2 has been moved to button 1, in order to call station 1, and, by tracing the circuit, it will be found that these two stations are now connected in a metallic circuit formed by line 1 and the common return.

With house systems it frequently happens that the user will forget to return the switch to the home position, which should always be done. However, in this system the wiring is such that the bell at any station can be rung from any other station, no matter on what button the switch lever has been left, but before a conversation can be carried on the switch at the station called must be returned to the home contact if not already there. This is the most desirable arrangement.

For instance, suppose that station 2 has called up station 1 whose switch lever has been carelessly left on button 3. The bells at stations 2 and 1 will ring; the ringing current flows from generator G_2 through K_2–S_2–1–line 1–bell C–e–b–common return–b_2–e_2–bell C_2–G_2. It cannot flow through any other circuit, hence no other bell will ring. But, when the receiver R_1 is removed from the hook, while S_1 rests on 3, no current can pass from line 1 through any apparatus at station 1, because the circuit is open at 1, e, and K_1. As soon, however, as the switch S_1 is placed on button 1 the talking current flows through s_2–R_2–S_2–line 1–S_1–R_1–s_1–b–common return–b_2–s_2. The transmitter T_1, transmitter battery B_1, and primary winding p_1 of the induction coil are connected in a local circuit when the receiver is off the hook. If a bell rings, the party answering the call, if nothing can be heard when the receiver is taken down, will naturally look at the switch and if he finds it off the home button will return it to that button, after which the conversation may be held. This compels the return of the switch to the home button each time a station is called, if it is not already there, and tends to cultivate the habit of returning the switch to its home position. By this arrangement, the desired bell can always be rung, and neither the ringing nor talking currents are subdivided through idle stations where the switches may not have been returned to their home positions.

If station 2 should want to hold a consultation with both stations 1 and 3 simultaneously, station 2 would first call up station 3, request that the switch there be turned to button 1 and left there; station 2 would then call up station 1 and the three stations would be connected in parallel between line 1 and the common return. The talking current originating at one station would then subdivide between the other two stations. This would reduce the loudness but probably not enough to seriously interfere with the conversation. At the end of the consultation all switches should be returned to the home position. This system requires one more line wire than there are stations, and a regular series magneto-generator telephone, intercommunicating switch, and push button at each station. The generator need not have an automatic shunt device, however. If it is provided with an automatic cut-in device that normally leaves the generator on open circuit the push button will not be required.

COMMON SIGNALING-BATTERY SYSTEM

38. The common signaling-battery system is one in which one battery located in any convenient position is used in common by all stations in the system for signaling purposes only. The diagram shown in Fig. 13 is about the best way to wire the system, for the ringing and talking currents flow only through the stations properly connected; neither are subdivided through any other stations no matter where the switches at the other stations may be left. There is the usual transmitter battery, talking apparatus, and

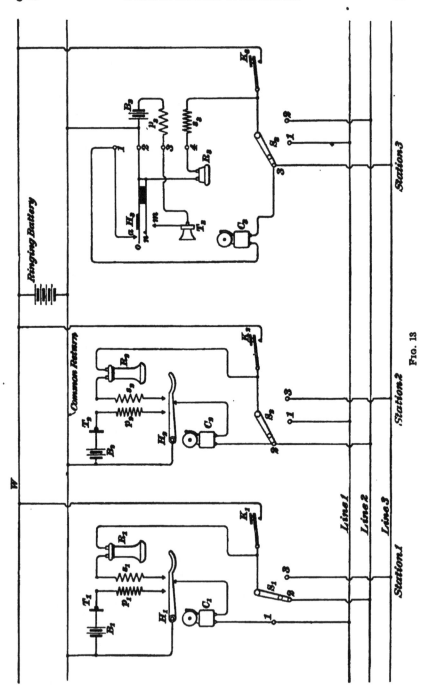

Fig. 13

intercommunicating switch at each station. The means for signaling consist‧ of an ordinary vibrating bell and push button at each station and one conveniently located battery for ringing the bells.

With S_1 resting on 2 and K_1 closed, current from the ringing battery flows through K_1–S_1–button 2–line 2–button 2 at station 2–bell C_1–hook switch and common return back to the ringing battery. The voice currents flow through s_1–R_1–S_1–button 2–line 2–button 2 at station 2–S_2–R_2–s_2–H_2–common return–H_1–s_1. The bell at the station making the call does not ring. It is necessary to arrange the bells so that only one bell will ring, because two vibrating bells do not work well in series, unless the vibrating contact of one is removed, which would not be applicable to a system having over two stations. This system requires two more line wires than there are stations.

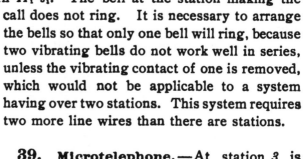

FIG. 14

39. Microtelephone.

—At station 3 is shown the connections for a so-called **hand microtelephone,** a good idea of which may be readily obtained from Fig. 14. The hand microtelephone consists of transmitter and watch-case receiver, both a little smaller and lighter in weight than the ordinary transmitters and receivers, and a switch, all mounted in one handle that can be conveniently held by one hand with the receiver to the ear and the transmitter mouthpiece in front of the mouth. The switch in the handle can easily be held closed by the same hand that holds the microtelephone. When the switch H_3, Fig. 13, is pressed, o parts from a and o, n, and m are all connected together; hence, this switch replaces the ordinary hook switch. The induction coil p_3, s_3, the bell or buzzer C_3, the push button K_3, and the intercommunicating switch S_3 may all be mounted in a little box and located in a convenient place, while the microtelephone may be hung where most convenient, or placed in the pigeon hole of a desk.

The numbers *1, 2, 3, 4* in Fig. 14, and in the upper part of station *3* in Fig. 13, represent the four wires running out of the microtelephone handle and are here numbered exactly as on the tags fastened to the microtelephone, which is now made by several companies.

CENTRAL-ENERGY HOUSE SYSTEM

40. One Wire for Each Station and Three Common Wires.—Where it is desirable to have no batteries at the subscribers' stations, the central-energy system shown in Fig. 15 may be used. The objection to most all central-energy house systems is the liability to cross-talk when

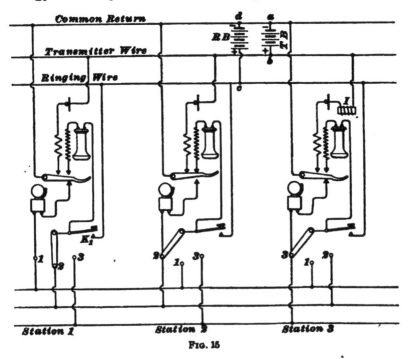

Fig. 15

several pairs of telephones are in use simultaneously. The fact that all the wires for a house system are usually run in one cable tends to increase the cross-talk if the wires are not run in pairs. In the arrangement shown, two centrally located batteries are used, one set of cells *a b* supplies the various transmitters with current when the receivers are removed from the hook switches, and the cells dc are used for ringing the bells. More, cells are generally required for ringing the bells than for the transmitters, and it is preferable to use separate batteries for ringing and talking purposes. However, on a small system and where the same number of cells will answer for both ringing and talking purposes, one wire less may be used. In this case the various transmitters may be connected to the ringing wire and the transmitter wire and battery TB may be omitted. In the latter arrangement two, and in the one shown in the figure, three more line wires are required than there are stations. The necessary number of cells may be connected between b and c in series with TB and all the cells between a and c used as a ringing battery.

Cross-talk may be reduced by connecting a 25- to 50-ohm impedance coil I in the transmitter circuit at each station, as shown only at station 3. In a system of any size or importance, at least this much should be done to reduce the cross-talk. To practically eliminate cross-talk requires not only these impedance coils, but also a twisted pair of line wires for each station in addition to the wires used in common by all stations for signaling purposes. For systems having less than six telephones or where the line wires do not run along in one cable for more than about 100 feet, the systems shown should be sufficiently free from cross-talk. A little cross-talk may often be endured, considering the fact that systems not entirely free from it usually cost less to install.

41. A Pair of Wires for Each Station.—A more desirable central-energy house system than the one just described and one that is suitable for a larger system because it is freer from cross-talk is shown in Fig. 16. There is one battery RB for signaling purposes and another TB for supplying current to all the transmitters. From RB two wires W, X run through all the stations. There are in addition to W, X as many twisted pairs of line wires running through all the stations as there are stations. Each wire of each of

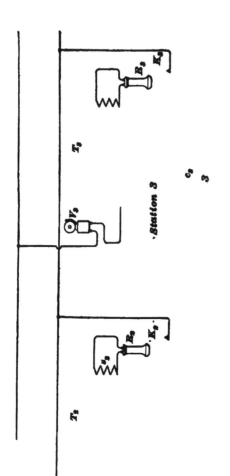

Station 3

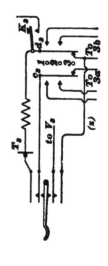

Fig. 16

the latter pairs is connected through an impedance coil to the talking battery TB. The two batteries and all the impedance coils are located at some one convenient place. A two-arm strap switch, or its equivalent, is necessary at each station. The switch shown in the figure consists of two metallic arms c_1, d_1 insulated from each other, but mechanically connected by a bar and handle so that one handle moves both arms. Ordinary vibrating bells are used. The transmitter is connected in series with the primary of an induction coil and the receiver, in series with the secondary and in a permanently closed local circuit. At station 3, the bell V_s is entirely disconnected from the circuit while conversing. This prevents the possibility of interference while talking, due to charging, discharging, and leakage currents through the bell circuit. However, it is very doubtful if this would be serious enough to warrant the use of the extra contacts between e and n, o, and the two extra contacts n and o thereby required.

To call up station 1 from station 2, turn the switch so that c_2 and d_2 rest on contacts $1, 1$ and close the push button K_2. Current then flows from $+ RB$ through K_2–d_2–wire $1b$–q–V_1 –W to $- RB$. When the receivers are removed from the hooks the battery TB supplies both transmitters with current through the impedance coils fa and Ib, and the fluctuating voice currents flow through T_1–p_1–d_1–$1b$–d_2–p_2–T_2–H_2–c_2–$1a$– c_1–H_1–T_1. These fluctuating currents induce currents in s_1 and s_2, thereby affecting the receivers R_1 and R_2.

The bell at any station can be rung from any other station, even if the switch has not been returned to its home position, but the conversation cannot be held until the switch at the station called is returned to its home position, which is a desirable feature.

As in practically all house systems, there is nothing to prevent a third station from listening to a conversation of two others, if the switch at the third station is turned to the proper contacts. However, the parties conversing can probably tell that such is being done from a click or a decrease in the loudness of the tone.

The wiring used by the Couch and Seeley Company in one of their systems is shown for one station at (*z*). It is practically the same as the wiring shown at station *3*. The switching connections are made by means of push buttons *1, 2, 3*.

AUTOMATIC SWITCH SYSTEMS

42. With the original house systems it was absolutely necessary for the calling party to return the intercommunicating switch to the home position when through with a conversation in order to avoid leaving the station cut out. With the systems as now connected this is not necessary, although desirable. In the older systems considerable trouble was caused by people, not accustomed to its operation, constantly forgetting to return the switch to its home position. This led to the development of the automatic return switches and systems. An automatic intercommunicating switch is one that automatically restores all connections to their normal positions when the receiver is hung on the hook switch.

43. The **DeVeau automatic house system** is shown in Fig. 17. At station *1* is shown the automatic switch mechanism and the wiring of a wall set; the general appearance of the same set with the back open is shown at station *2*. Since the principle of the automatic switch is practically the same for all sets it is not shown in detail at stations *3* and *4*, which represent desk sets. Inside the telephone box are two insulated plates (see station *1*), s_1 called the *sliding plate* (same as s_s at station *2*) and in front of s_1 (in this view) a fixed plate r_1 (same as r_s at station *2*), called the *ringing plate*. Behind these plates in this view are push buttons, the handles of which project through the front of the box. The push-button part of the mechanism is shown separately and enlarged in Fig. 18, in which the sliding plate *s* is drawn toward the reader, that is, upwards by a spring f_1, which is shown only in Fig. 17. When *a*, Fig. 18, is pressed, the shoulder *b* draws the sliding plate *s* to such a

VB_2

VB_1 TB

position that b can pass through the hole in it. If pressed in as far as it will go, the phosphor-bronze spring 2 will touch the ringing plate r. When released, the flat shoulder of b will be pressed upwards by spring 2, against the surface of s, and thus keep 2 in contact with s through b. This is what happens when any button is pressed, whether the receiver is on or off the hook. At station 1, Fig. 17, is shown the position of the mechanism when the receiver is on the hook.

When the receiver is placed on the hook, m presses the sliding plate s_1 down and the pin e causes the piece m to move to the right until the end n is released by the shoulder in m; then the sliding plate s_1 is drawn up by the spring f_1. As the sliding plate moves down and all the holes in it come

just opposite the shoulders on all the push-button rods, all the push buttons, that may have been pressed in, are pushed out by their springs (2 and 5, Fig. 18) to their normal positions as the hook switch moves down. At g, Fig. 17, is shown a convenient ar-

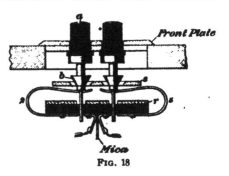

FIG. 18

rangement for adjusting the tension of spring f. No induction coils are used and, when talking, the two transmitters, two receivers, and one talking battery are connected in series between the common-return and one of the other wires. Good results can readily be obtained by this arrangement for ordinary house systems.

To call station 3 from station 1, press in push button 3 as far as it will go. Current then flows from the positive terminal of the common ringing battery through $y-r_1-3-$line $3-z-k-b-a-VB_3-l$ to the negative terminal of the ringing battery. But very little of the ringing current will flow from k through T_3-R_3-l, because the path through $k-b-a-VB_3$ has much less resistance. The bell has a resistance of only about 3 ohms, while the receiver has a resistance of about 10 ohms, and the transmitter anywhere from 25 to 85 ohms. The spring *3*, when released, returns part way and rests against the sliding plate s_1. When both receivers R_1 and R_2 are removed from their hooks, current flows from TB_1 through s_1–*3*–line *3*–*z*–*k*–T_2–R_2–*l*–common return–*g*–R_1–T_1–H_1–*j* to $-TB_1$, thus enabling the two parties to hold a conversation. One station can call up and hold a consultation with several stations by simultaneously pressing in the buttons corresponding to the stations wanted. The several stations called will be in parallel with each other, and in series with the station that called them. The talking efficiency will be diminished somewhat, but not enough to prevent a consultation between three or four parties.

HOTEL TELEPHONE SYSTEMS

44. In a hotel almost any switchboard system can be used, but it is customary to install only central-energy systems with very simple instruments in the guests' rooms. In most cases, communication between the substations and office only is required, but sometimes communication between the substations is afforded by means of simple jacks and plugs at a central-office switchboard. Very often a telephone system is to replace or to be added to an annunciator system, the wiring for which is already installed; in such cases the same wires between the various stations and the office can and must, generally, be used for the telephone system. Most annunciator systems have one or two wires that run through all stations, and one wire that runs between the office and each substation.

PLUGBOARD ANNUNCIATOR SYSTEMS

45. Communication Between Office and Substations Only.—In Fig. 19 is shown a system that admits of communication between the office and substations, but does not admit of cross-communication between any two stations. The connections are practically the same as in a

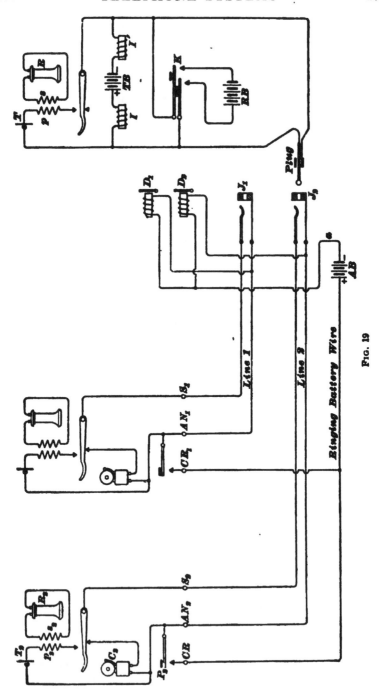

FIG. 19

system made by the Couch and Seeley Company. All batteries are centrally located. D_1, D_2 represent the annunciator electromagnets of an ordinary hotel annunciator, and J_1, J_2 the spring jacks of a simple plug switchboard. One ringing-battery wire runs from the office to all substations and a pair of wires from the office to each substation.

Any substation calls the central office by merely pressing a push button, and it makes no difference whether the substation receiver is on or off the hook. For instance, the pressing of P_2 operates the annunciator D_2, the annunciator battery AB supplying the current; the office attendant inserts the plug in the correspondingly numbered spring jack J_2 and takes the office receiver R off its hook. Conversation can then be carried on between the central office and substation 2. The talking battery TB is connected, through two impedance coils I, I', to the cords of the plug and supplies all the current for talking purposes.

To call up any substation, as 2, the office attendant inserts the plug in jack J_2 and closes the ringing key K, thereby ringing the ordinary battery bell C_2. When both receivers R_2, R are removed from their hooks, the two parties can converse. RB and AB may be one and the same battery and, if desired, a bell or buzzer may be included in the circuit at a, so that it will ring whenever any annunciator drop is operated. The substation instrument can be simplified and cheapened by connecting the receiver and transmitter directly in series, omitting the induction coil. The spring jacks and plug can be replaced by a simple, or automatic, two-wire intercommunicating switch.

46. Communication Between All Stations.—The simple system shown in Fig. 20 allows communication not only between the office and substations, but also between the substations through the office plugboard. This is a system designed by the Couch and Seeley Company for ordinary hotels. It requires one wire from the office to each substation and two wires common to all stations. All batteries are located at the central office, where an ordinary annunciator,

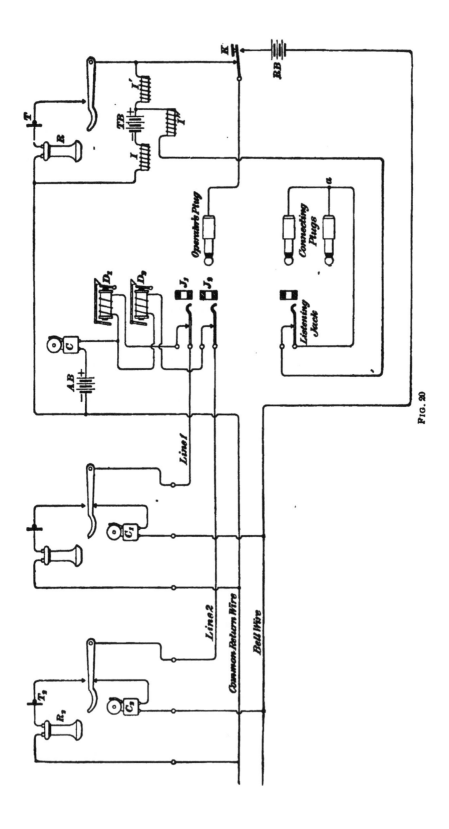

FIG. 20

plugboard, connecting cords, and plugs are also provided. This is not intended for use where more than two substations would need to be connected together at any one time. Only three impedance coils I, I', I'', one talking battery TB, one ringing battery RB, one annunciator battery AB, one office bell, or buzzer, C, one listening jack, one operator's plug, and one pair of connecting plugs are required, no matter how many substations there may be on a small system, for which this arrangement is only intended.

The removal of a receiver at a substation is all that is necessary to call up the office. For instance, if R_2 is removed from the hook, current flows from $+AB$ through C–D_2–J_2– line 2–T_2–R_2–common-return wire to $-AB$; thus operating the annunciator D_2 in that line and ringing the bell C. The office attendant inserts the operator's plug in the correspondingly numbered jack J_2. The plug, when inserted, connects with the line spring only and opens the circuit running to the annunciator electromagnet. A conversation can then be held between substation 2 and the office. In this case the two telephone sets are directly in series, the battery TB and impedance coils I, I' being bridged across the wires connecting them. If substation 2 desires to talk to substation 1 the operator's plug is withdrawn from jack J_2, inserted in jack J_1 and the ringing key K closed, thereby causing current from the ringing battery RB to flow through and ring the bell C_1. As soon as substation 1 replies, the operator's plug is removed and the connecting plugs inserted in jacks J_1 and J_2; then the parties can converse. While substations 1 and 2 are conversing, current from $+TB$ flows through the impedance coil I'' and the listening jack to the point a, where it subdivides, flowing through the line wires 1 and 2 and substations 1 and 2 to the common-return wire, where the currents reunite and flow back through I to $-TB$. For the voice currents the two substations are in series through the wire connecting the two plugs, and also that portion of the common-return wire between the two substations. The office attendant must insert the operator's plug in the listening jack to determine, by listening, when the conversation is completed. When listening, the coil I' is used. When the conversation is complete all plugs are withdrawn and the annunciators restored to their normal position.

---

47. The Couch and Seeley Company's **hotel switch-board system**, shown in Fig. 21, allows communication not only between the office and substations, but also between the substations through the office switchboard. This system being more elaborate and complete is suitable for a larger hotel than the one last described. It is a central-energy system, the talking battery TB being connected through two impedance coils m, n across each cord circuit. The two impedance coils in each cord circuit are wound on one iron core and so arranged as to constitute a clearing-out drop CD for that cord circuit. There is one listening key Lk, and two ringing keys Rk and Rk' for each cord circuit. There is a separate battery RB for ringing the subscribers' bells and operating the line drops D_1, D_2, and D_3. From the office there are two line wires (one pair) running to each substation and one common-return wire running through all the stations.

If the receiver R_2 is removed from the hook, current flows from $+RB$ through g–h–D_2–i–R_2–T_2–j to $-RB$, thereby dropping the shutter of D_2. The operator inserts one plug, say P', into the corresponding line jack J_2, and closes the listening key Lk; this bridges the battery TB, through the coils m, n of the clearing-out drop, across the cord circuit and also across the operator's receiver R and transmitter T, thereby supplying both the substations and operator's set with current from TB and enabling the operator to converse with the substation. Furthermore, the shutter of the clearing-out drop is raised. If substation *1* is desired, the operator will insert the other plug P of the same pair in jack J_1 and close the ringing key Rk, thereby causing current from $+RB$ to flow through g–o–t–v–H_1–C_1–common-return to $-RB$ and ring the bell C_1.

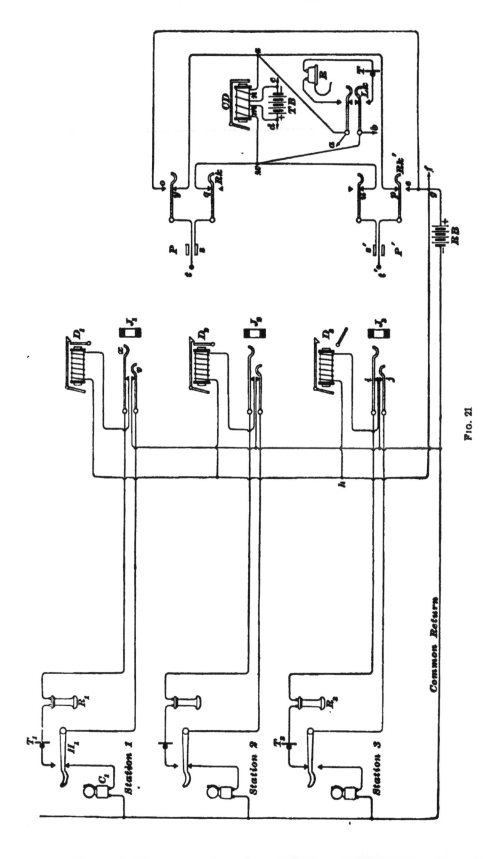

Fig. 21

When the receiver R_1 is removed from the hook, current flows from $+ TB$ through m to w where it divides, part flowing through q–s–x–R_1–T_1–H_1–v–t–y–z and part through u–s'–i–R_2–T_2–j–t'–p–z, where the two currents reunite and flow through n to $- TB$. The voice currents flow through T_1–R_1–x–s–q–w–u–s'–i–R_2–T_2–j–t'–p–z–y–t–v–H_1–T_1. When both receivers are hung up no current can flow through the coils of CD, and the drop is so constructed that the shutter then falls. The shutter remains up only while one or both receivers are off the hook. The clearing-out signals are in a retired position when the plugs are disconnected from the jacks and when the plugs are in the jacks, but both receivers of the connected stations are on their hooks. While this would not be a very satisfactory arrangement for a large switchboard, it is found to be no great inconvenience for a small switchboard for which this system is only intended.

There may be as many cord circuits and clearing-out drops as required. The same operator's set may be connected to each listening key by leads a, b; leads f run to contacts o, e on each ringing key, and leads c, d to each clearing-out drop. No one plan of wiring will meet all conditions, and it is usually found that each party requiring a telephone' system, whether it be an exchange, hotel, or house system, has his own ideas as to just how it should operate.

TELEPHONE ACCESSORIES

PROTECTION OF TELEPHONE CIRCUITS

48. The way in which the telegraph circuits and apparatus are protected by means of static lightning arresters and tubular fuses has been fully explained. The same protecting devices are extensively used in telephone systems. It is customary to use tubular fuses at each end of each conductor running through a telephone cable and to use static arresters where the conductors, whether bare or cable lines, enter the exchange building. The static protectors used are generally two rectangular-shaped pieces of carbon separated a very small distance by mica or silk, one carbon being grounded and the other connected to the line.

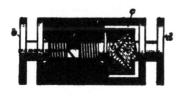

FIG. 22

In addition to the static arresters, a so-called *sneak-current arrester* is connected in each line circuit. The Sterling sneak-current arrester is shown in Fig. 22. It consists of brass terminals that screw into a cylindrical piece of hard rubber r, and a thin shell of hard rubber e fitting snugly over the latter, but somewhat longer, so as to leave a small space at c. The cone-shaped end v, forming part of the piece d, fits into a corresponding hollow portion of o, the cone v being fastened to o by a solder that melts at a low temperature. The brass piece o forms a small spool in which is wound fine insulated German-silver wire; the spool and coil about fills the space c. One end of the coil is soldered to the brass spool and, hence, is in contact with d and the other end runs through a hole in r and in the center of b and is soldered to the piece b after the latter has been screwed into place. The resistance of the coil varies from about 4 to 30 ohms, depending on the system with which it is to be used. The end pieces b, d are held between strong flat springs that tend to pull them apart. If ¼ ampere flows through this coil for 15 seconds sufficient heat is generated to soften the solder and allow the springs to separate the pieces v, o, thus producing a wide air gap in the circuit and usually grounding one of the springs. Sometimes the springs are arranged to ring a bell when the coil opens the circuit. Such devices are called **sneak-current arresters** because they will interrupt a very small current, called a sneak current, that is usually caused by leakage between the telephone line and some lighting or power circuit. Such a current is sufficient to damage coils or devices in the switchboard, but still too small to melt a fuse wire of sufficient size to be reliable.

Protecting devices are sometimes used at each telephone. They usually consist of a carbon static arrester, and either a fuse or a sneak coil. One of the carbons is, of course, grounded and a fuse or sneak coil is connected in each line. Tubular fuses are preferable, but fuses stretched on a fiber strip are sometimes used.

POLE CHANGER

49. A **pole changer** is a device employed in telephone systems that will rapidly interrupt or reverse a current from a battery in such a manner that it may be used to ring ordinary polarized bells. It is really an automatically operated interrupting or reversing switch, the speed being such as to give about fifteen or thirty current impulses per second, respectively. Pole changers are used in small telephone exchanges where it is inconvenient or impossible to use a power-operated generator or where the pole changer would be the more economical to install and operate.

The diagram of connections of the pole changer made by the Illinois Electric Specialty Company is shown in Fig. 23; R is an ordinary pony relay, M an ordinary two-coil (one coil above the other) electromagnet whose iron armature n is fastened to a vibrating hard-rubber lever $h\,k$ that is supported by quite a wide flat spring s securely fastened at l. The spring i and contacts a, b are fastened to the vibrating lever $h\,k$, the connections being brought out so as not to interfere with the vibrations of the lever. For clearness, a and b are placed alongside of each other in this figure, but in the device itself one is above the other. When the lever is at rest and the spring r is properly adjusted, a should make contact with e and b with g and a, b should clear c, d, respectively, by $\frac{1}{16}$ inch.

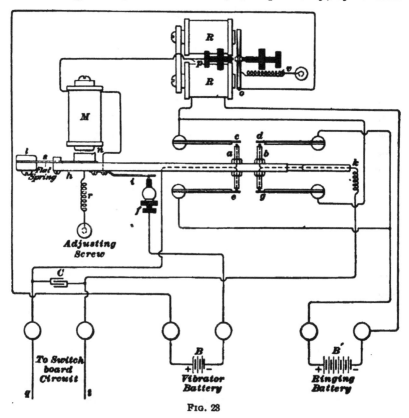

FIG. 23

When the circuit is closed by connecting a telephone line, in whose circuit is an ordinary polarized bell, across $q\,t$, current flows from $+ B'$ through e–a–q–t–k–b–g–R to $- B'$. This causes R to attract its armature, so that current can now flow from $+ B$ through o–p–M–i–f to $- B$, thereby causing M to attract its armature n and thus break this circuit at i and to also break the circuit containing B' at e and g. However, the latter circuit is almost immediately closed again at c and d, but the connections are so arranged that the direction of the current between q and t is reversed, for it now flows from $+B'$ through d–b–k–t–q–a–c–R to $-B'$. When the circuit is broken at i, the magnet M no longer attracts its armature, and hence the spring r pulls the lever back to its starting position, reversing the direction of the current between q t in doing so. This operation is repeated about fifteen times per second.

The contact points are so adjusted that the circuit between o, p is closed just an instant before that between i, f is closed and i, f separate an instant before o, p. This is accomplished by adjusting f so that one circuit is opened at i before the other circuit is opened at e g. The momentum of the lever, assisted by the spring s, is sufficient to cause the lever to vibrate a short time, even though current ceases to flow through M. The tension of the relay spring v should be adjusted lightly. Thus the circuit of M will always be closed when i and f come together, provided the circuit between q and t is closed by a bell or otherwise. The condenser c is used to diminish the sparking. It is connected directly across the switchboard circuit, but as this circuit receives current only when a calling plug is in the jack and the ringing key is closed, it does not form a continuous load upon the ringing battery. Therefore, the condenser need not be removed from the switchboard ringing circuit when the latter is not in use. Four dry cells are required at B, just enough to properly magnetize the low wound magnet M, but sixty-five or more dry cells connected in series are required at B' in order to furnish a strong enough ringing current through the highest resistance line and bell in use in any one exchange.

The Warner pole changer is very similar in construction, but the circuit containing the coils of the electromagnet does not pass through the contacts of the relay R (M is connected between i and $+B$) and the relay R, which is connected in series with ringing battery, is used to connect the condenser directly across the line wires for just a moment as the circuit is broken at e, g or c, d.

47–34

APPLIED ELECTRICITY

ELECTROPLATING

1. Introduction.—Under the title *Applied Electricity* will be considered a number of applications of the electric current that are of more or less practical importance, but of which an extended treatment is not warranted in a Course of this character. The object here is simply to give general information regarding these matters, so that the principles and processes involved will be understood.

2. Under the subject *Chemistry and Electrochemistry* it was explained that when a current is sent through a solution of metallic salt, such, for example, as copper sulphate, metal is deposited on the electrode toward which the current flows through the solution. If two plates of copper are suspended in a solution of copper sulphate, and current is allowed to flow between the plates, copper will be deposited on the plate (the cathode) toward which the current flows, and it will be removed from the plate (the anode) which the current leaves to flow through the liquid. In other words, metal is deposited on the plate that is attached to the negative side of the source of current; and is removed from the plate connected to the positive side. The plate that is connected to the positive side and from which metal is removed is called the *anode;* that on which the metal is deposited is called the *cathode*. Instead of using a copper plate, the cathode may be replaced by any other article that is capable of conducting current, and this article will have deposited on it a thin coating of copper. This process, by means of which objects may be coated with a thin coat of metal, either for ornament or protection, is known as **electroplating**. After the deposit, or coating, has been made by means of the current, the objects are polished and lacquered, or finished in any other manner desired.

3. The solutions used in plating always contain a salt of the metal to be deposited, and the anode or plate of metal from which the deposit is to be made is suspended in the solution and connected to the positive pole of the battery or dynamo furnishing the current. When the object to be plated is suspended in the bath and connected to the negative pole, the passing current decomposes the metallic salt and deposits the metal in a layer, the quality and thickness of which depend on the metal used, the strength of the current, and the time during which it flows. The metal that is thus taken from the solution is immediately replaced with metal taken from the anode, because the free acid radical at once combines with the metal of the anode that is gradually eaten away and has to be replaced from time to time.

4. It is sometimes desired to plate articles with a layer of metal of varying thickness, as, for example, on silverware at those parts that are subject to the greatest amount of wear. This process is called **sectional plating**. One method of producing this result is to bring the anode close to the cathode, and the deposit at that point becomes heavier. It follows from this, that, when a uniform coating of metal is to be deposited, a certain minimum distance must be maintained between the anode and the cathode. Also, the cathode may be moved occasionally, in order to present another side to the anode, in case the plating bath is very small.

5. Source of Current.—For small plating operations, primary cells may be used as the source of current, but where plating is done on a large scale, dynamos are much cheaper and more convenient. Special dynamos are used for plating work, that is, they are special inasmuch as their voltage and current output are different from that of ordinary machines, and this necessitates certain changes in the construction; otherwise, plating machines are the same as any other type of dynamo. Most modern plating dynamos are compound wound, so that they will maintain their voltage when handling a large amount of work, i. e., when delivering a large current. Machines for this work are wound for a low voltage and large current output; hence, the armature has but a few turns of large conductor. Small machines are frequently wound for 5 volts, and larger machines for 7½, 10, or 12 volts. The commutator has a small number of broad heavy segments, and the whole construction of the current-collecting part of the machine is much larger and heavier than is usual for ordinary dynamos of similar output at higher voltage. The shunt field of the machine is usually provided with a rheostat in order to allow variation in the voltage. Large electroplating dynamos are very often separately excited because the low voltage is not well adapted to shunt excitation.

In case it is desired to use direct current from city lighting or power mains for plating purposes, either a motor-dynamo or dynamotor can be used. The former consists of a low-voltage plating dynamo directly coupled to a motor wound for the supply voltage. In this outfit there are two distinct armatures and fields and the voltage of the plating dynamo can be regulated by varying the excitation of its field. In the dynamotor there is but one field magnet, the armature being provided with two windings and two commutators. One winding is designed to receive current from the source of supply and operate the machine as a motor; the other winding delivers current at low voltage to the plating tanks. The dynamotor is simpler and cheaper than the motor-dynamo but it has the disadvantage that the delivered voltage cannot be adjusted by varying the field excitation. A variation in the field strength changes the speed of the dynamotor but does not alter the voltage because both sets of conductors on the armature revolve in a common magnetic field. In case alternating current is the only source of power, it must, of course, be changed to direct current, and it is necessary to use a motor-generator set consisting of an

alternating-current motor coupled to a direct-current generator, or a rotary converter designed for low voltage.

If it is necessary to use batteries as a source of current
they should be of a type that can furnish a current continuously without running down to any great extent; or, in other
words, they should be closed-circuit cells. The following
types will be found suitable: Edison-Lalande, Gordon, Fuller
bichromate, Daniell, Gravity (crowfoot, etc.), Smee, Bunsen,
and Grove. Open-circuit cells, such as the Leclanché, and
various forms of dry cell are not suitable as they soon run

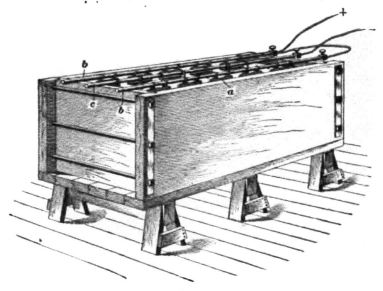

Fig. 1

down if called on to deliver current for any length of time.
Storage batteries are, of course, suitable, but batteries are
usually employed only where no other source of current
is available, and as some other source is necessary to charge
storage batteries, they are but little used in connection with
electroplating.

6. Plating Tanks.—The tanks in which plating operations are carried on may be made of glass, earthenware or
stoneware, and wood. For large tanks, wood is used almost
exclusively. They must be substantial and well made,

otherwise there may be considerable loss of solution on account of leakage. The tanks are lined with a coat of asphaltum or similar material in order to make them tight. Fig. 1 shows a wooden plating tank with the anodes in place. The anodes a, which are shaped as shown in Fig. 2, are hung in two rows from the copper rods b, b. The work to be plated is hung by means of copper wires from the central rod c, which is attached to the negative pole of the dynamo; rods b, b are joined together and connected to the positive pole. Since a large current at low voltage is required for electroplating, the current is usually conducted from the dynamo to the tanks by means of heavy bare copper rods supported in such a way that

Fig. 2

there will be no danger of anything falling across them. The bars used for this purpose should be of ample cross-section, otherwise their resistance will have an appreciable effect on the amount of plating that can be handled by the dynamo. It must be remembered that the current is large, and that a very small resistance may cause a considerable drop in voltage. About 1 square inch cross-section of copper should be allowed per 1,000 amperes.

ELECTROPLATING SOLUTIONS

COPPER SOLUTIONS

7. Acid Solution.—The electrolyte for use in copper plating may be a simple acid solution of copper sulphate, or a cyanide of potassium solution with carbonate of copper. The acid solution is very easily made. The necessary ingredients are: Sulphate of copper, 1 pound; sulphuric acid, 1 pound; water, 1 gallon. The sulphate of copper is dissolved in hot water, after which cold water is added until the quantity named is made up. The sulphuric acid is then added, and after the solution has become cool, it is poured into the depositing vessel. This may be of glass or of wood, and should be thoroughly cleaned before using.

This acid copper solution will deposit fast. It is not, however, as generally useful as the cyanide solution. It is liable to give a rough deposit unless used with a very low current density, and, moreover, it cannot be used for coating steel or iron articles with copper. When steel or iron articles are to be plated with, say, nickel or silver, they are first given a light coat of copper and the final plating is then deposited on the copper. This gives a better looking deposit, and one that wears better than a deposit made directly on the steel or iron.

8. Cyanide of Copper Solution.—This may be made as follows: Carbonate of copper, 1 pound; carbonate of potash, 6¼ ounces; cyanide of potassium, 2 pounds; water, 3 gallons. The cyanide of potassium is dissolved in the greater part of the water, and the carbonate of copper, dissolved in a portion of the water, is added. The carbonate of potash, also dissolved in water, is then added, and the whole stirred up thoroughly. If, on trial, the solution does not deposit freely, more cyanide or carbonate, or both, should be added until the desired result is obtained.

9. Brass Solution.—By adding carbonate of zinc to this solution, a brass deposit of varying color, depending on the amount of zinc, can be obtained. The zinc solution may be made by dissolving two parts by weight of cyanide of potassium and one part of carbonate of zinc in sufficient water. This solution is then added to the bath until the required color of deposit is obtained.

In using cyanide of potassium, great care must be exercised in handling, as this chemical is a strong poison and very deadly. If the hands are cut or scratched, blood poisoning will probably result from contact with cyanide, so that the greatest precautions should always be adopted in preparing solutions of which it forms an ingredient.

NICKEL SOLUTION

10. In electrodepositing nickel, it is necessary to thoroughly clean the articles to be plated, because the bath is nearly neutral and has no power to remove oxide from the surface of the metal. The first operation is to remove any grease, such, for example, as may be due to handling. A hot potash bath is used for this purpose, made by dissolving ½ pound of caustic potash in 2 gallons of water. This solution is kept hot by means of a steam-pipe coil at the bottom of the galvanized-iron tank. The strength of the potash gradually lessens while in use, so that, while a few minutes only are necessary to cleanse an article in a fresh bath, a longer time will be required later on. On being removed from this bath, the article should be well scoured with pumice and water, and rinsed. It may then be placed in the nickel bath, but, since an almost imperceptible film of oxide may have formed on the surface, it is advisable to dip brass or copper articles for a few seconds in a solution composed of ½ pound cyanide of potassium dissolved in 1 gallon of water. This will insure the deposit adhering firmly. For steel or iron, it is better to use ½ pound hydrochloric acid in 1 gallon of water, contained in a wooden tank. For cast iron the hydrochloric is replaced by sulphuric acid, the solution being used cold. An immersion of 15 minutes to ½ hour will be required in this case, and the casting should be thoroughly scoured. When a bright finish is wanted on cast iron, the proportion of sulphuric acid should be doubled, and 2 ounces of granulated zinc added. After this dissolves, add ½ pound nitric acid and mix thoroughly.

The plating bath is made as follows: Double sulphate nickel and ammonia (pure), 12 to 14 ounces; water, 1 gallon. The nickel salts are put into a wooden tank and hot water is poured on them until they dissolve, this being hastened by stirring with a clean wooden stick. The liquid may be poured off into the plating tank after a short time, without waiting for all the salts to dissolve; then more hot water is poured on the salts, this being repeated until the salts are completely dissolved. When this is accomplished, cold water is added to bring the solution to the required measure.

This plating bath has a rather high resistance, and good results may usually be obtained by adding to the solution 10 per cent. of common salt. This has the effect of whitening the nickel deposit, making it more tough, so that the film of metal will not readily strip, and decreasing the resistance of the bath.

The nickel bar or plate used as an anode must be pure, in order to secure good results. Cast anodes are more readily dissolved by the action of the current than are rolled anodes, and are therefore to be preferred. For small work, however, rolled anodes are more generally used, as they are more easily obtained.

All the water used in plating baths should be distilled, or rain water may be employed if desired.

In order to obtain a good color, the bath should be slightly acid; but if too acid, the deposit will peel off. In working the bath, the anode should present as large a surface as the object to be plated. If it is too small, the solution will be robbed of metal and become too acid, in which case sulphate of nickel (single salts) should be introduced, by dissolving first in water and adding to the solution in the required quantity. If the bath has the full amount of metal, the acidity may be corrected by the addition of carbonate of ammonium dissolved in water. When the bath is alkaline, add double salts, even if it is necessary in consequence to reduce the exposed surface of the anode for the time being.

GOLD SOLUTION

11. The exact composition of the bath used for gold plating varies greatly, depending on the class of work, the color desired, etc. In plating the cheaper metals, anodes of sufficiently large surface are used to keep the solution up to full strength, but with gold plating this is not usually the case on account of the high cost of the anodes. They are, therefore, usually much smaller in surface than the article being plated, and the solution is reduced in strength as the plating progresses. On this account it is necessary to watch the bath carefully and strengthen the solution from time to time. Gold baths are usually made by dissolving chloride of gold in cyanide of potassium solution. Chloride of gold may be prepared by dissolving five parts, by weight, of gold in about six parts, by weight, of nitromuriatic acid (*aqua regia*), and evaporating to obtain the chloride so formed. Nitromuriatic acid is made by mixing nitric acid and muriatic (hydrochloric) acid in the proportions of one part of the former to two of the latter. In most cases, however, it is cheaper in the end to purchase the gold chloride ready prepared from a dealer in platers' supplies than to attempt to make it. An ounce of chloride of gold contains 312 grains of gold.

12. As stated above, gold baths are used for such a great variety of work and in so many different ways that it is impossible to state any general formula giving the proportions to be used. Some baths are worked hot, while others are worked cold, and the proportion of gold necessary depends on which method is used. With a hot bath, the temperature determines the color of the deposit; the higher the temperature, the darker the color. Hot baths are usually worked at temperatures varying from 90° F. to 140° F. With cold baths, the color of the deposit can be regulated by adjusting the current, the stronger currents giving darker deposits than light currents. For hot baths, the quantity of gold varies from about 11½ to 20 grains per quart of solution, but for cold baths it must be at least as high as 54 grains, and in some cases it may run over 300 grains. The following proportions for a cold bath are recommended by Roseleur: Fine gold (as chloride), .35 ounce; potassium cyanide (98 per cent.), .7 ounce; water (distilled), 1 quart.

13. Cold baths are generally used for large objects; whereas, for small, cheap objects, where a thin, close-grained layer is desired and where the articles have to be plated cheaply and quickly, hot baths are generally used. Too much cyanide in the bath makes the deposit a pale color. For hot gilding, the following bath is recommended by Roseleur: Sodium phosphate (pure), 2.11 ounces; neutral sodium sulphide, .35 ounce; potassium cyanide, 30.8 grains; fine gold (as chloride), 15.4 grains; distilled water, 1 quart.

Electroplating with gold requires an unusual amount of skill, because the metal is so valuable that the plater must use every means possible to produce the required results with the least possible amount of metal.

SILVER SOLUTION

14. The solution used for silver plating consists of chloride of silver dissolved in cyanide of potassium. The chloride can be prepared from metallic silver by first dissolving the silver in dilute nitric acid and then precipitating the chloride by adding a saturated solution of common salt. It is much better, however, for the plater to buy the chloride ready prepared. It costs no more in the end, and if the plater prepares it himself, it is more likely to be impure. Having obtained the silver chloride, the bath may be made as follows: Silver chloride, 3 ounces; cyanide of potassium, 10 to 12 ounces; water, 1 gallon. The chloride is first made into a thin paste with some of the water, and then added to a solution of 9 or 10 ounces of the cyanide in the remainder of the water. The solution can then be tried, and if it does not work freely the remainder of the cyanide can be added as required.

ELECTROTYPING

15. **Electrotyping** is a process by which reproductions are obtained of printer's type and cuts, to be used on the printing press in place of the original, which may thus be preserved in good condition. The forms of type, after having been prepared in the usual manner, are placed in a press and forced down on a sheet of wax that has previously been dusted over with fine graphite. On removing the type an impression in reverse will be found on the surface of the wax, which is again coated lightly with graphite, all superfluous particles being blown off. The wax sheet, or case, as it is called, is then coated with a solution of iron filings and copper sulphate, which forms a thin and smooth face of copper on the mold, insuring a uniform deposit. A connection is now made between the wax sheet and rods leading to the terminals of the dynamo. This serves to make electrical contact between the conducting surface and the dynamo. The wax mold is immersed in a plating bath, which deposits a coating of copper on the wax. The thickness of this coating of copper, or shell, as it is termed, depends on the length of time the mold is acted on by the plating solution. A sufficient thickness of copper deposit may be obtained in some cases after about half an hour, but a longer time is usually required, depending of course on conditions of operation. On removal from the bath, the shell is separated from the wax sheet and filled in with electrotype metal to give a firm backing. Careful manipulation is necessary in preparing the wax sheet and in making the impression. Current is supplied from a plating dynamo and during the first part of the depositing process the current is regulated so as to give a fine dense deposit.

CURRENT DENSITY, WEIGHT OF METAL DEPOSITED, AND VOLTAGE

16. It is important that the current density be kept at the correct value when articles are being plated. If the current density (amperes per unit area of surface) is low, the surface deposit will be hard and close grained, but

TABLE I
CURRENT DENSITIES FOR ELECTROPLATING
Crocker-Wheeler Co.

Metal and Character of Deposit	Amperes per 100 Square Inches
Copper electrotyping, best quality, tough deposit	1.5 to 4
Copper electrotyping, good and tough (for stereotypes)	4 to 10
Copper electrotyping, good solid deposit . . .	10 to 25
Copper electrotyping, solid deposit, sandy at edges	25 to 40
Copper electrotyping, sandy and granular deposit	50 to 100
Copper (cyanide bath)	2 to 3
Zinc (for refining)	2 to 3
Silver	1 to 3
Gold	.5 to 1
Brass	3 to 3.5
Iron (steel facing)	.5 to 1.5
Nickel, at first deposit 9 to 10 amperes per 100 square inches, diminishing afterwards to . .	1 to 2

the deposit will be made slowly. If, on the other hand, the current density is too high, the deposit will be softer and more open. If the density is forced very high, the deposit will become granular or even powdery, in which latter case it may not adhere at all. Table I gives the current densities required for various kinds of deposit.

17. Amount of Metal Deposited.—One ampere in 1 hour will deposit 1.186 grams of copper, 4.025 of silver, 2.451 of gold, and 1.095 of nickel. To deposit 1 pound per hour requires 382.4 amperes for copper, 112.6 for silver, 185.1 for gold, and 414.2 for nickel.

18. Voltage Required for Various Baths.—The voltage, like the current, is variable within certain limits for a given class of work. In any case the voltage should be carefully adjusted so as to give the current best suited to the work in hand. Table II shows the voltages commonly used for a number of different kinds of plating.

TABLE II
VOLTAGES FOR VARIOUS PLATING BATHS
Crocker-Wheeler Co.

Metal	E. M. F. Volts
Copper, acid bath	.5 to 1.5
Copper, cyanide bath	3 to 5
Silver	.5 to 1
Gold	.5 to 4
Brass	3 to 5
Iron, steel facing	1 to 1.3
Nickel, on iron, steel, copper, with nickel anode, start deposit with 5 volts, diminishing to	1.5 to 2
Nickel, on iron, steel, copper, with carbon anode	4 to 7
Nickel, on zinc	4 to 7
Platinum	5 to 6

EXAMPLE.—An electroplating dynamo is to be used for nickel plating. The machine must be capable of depositing 1.5 pounds of nickel per hour. Required: (*a*) the current output of the machine; (*b*) the voltage for which it should be wound.

SOLUTION.—(*a*) From Art. **17**, a current of 414.2 amperes is required to deposit 1 lb. of nickel per hr.; hence, the machine should have an output of, at least, 414.2 × 1.5 = 621.3 amperes. Ans.

(*b*) From Table II, it is seen that the maximum voltage required for nickel plating is 5 volts at the beginning of the deposit. Ans.

ELECTRIC HEATING

19. The heating effect of the electric current is shown whenever a current flows through a conductor. It may not be large enough to be noticeable, but whenever a current I flows through a conductor of resistance R, the number of watts expended is I^2R and the number of joules work done is I^2Rt, where t is the time in seconds during which the current flows. All this work reappears in the form of heat and raises the temperature of the conductor. The operation, therefore, of producing heat by means of electrical energy can be carried out with an efficiency of 100 per cent., so far as the mere conversion is concerned. Of course, if the heat is utilized for some particular purpose, all the heat developed in the conductor may not be available for the purpose intended; some would likely be lost through radiation to surrounding objects, so that, looked at in this way, the efficiency of the electric heating device might be considerably less than 100 per cent.

The applications of electricity to heating purposes have so far been comparatively limited on account of the cost. Its principal use for this purpose is in street cars, where electric heaters have been adopted quite largely. In this case the cost of current is offset to a considerable extent by the fact that the passenger space that would otherwise be occupied by a stove is saved, and, moreover, on street-railway systems the cost of current to the company may be only from 1 to 2 cents per kilowatt-hour; whereas, if a private consumer purchased current for heating purposes from an electric light or power company, it would probably cost him at least 5 to 10 cents per kilowatt-hour. Electricity is, therefore, comparatively little used for the heating of dwellings, but it is well suited to special heating appliances where the heat has to be applied locally. Owing to the fact that the heat can be so easily controlled and can be applied just where it is wanted, it often happens for work of this class that it may prove cheaper than other methods. Examples of applications of this kind are to be found in light cooking utensils, such as coffee pots, teakettles, etc., glue pots, soldering irons, laundry irons, heaters for embossing presses, small electric ovens, and a number of other similar appliances.

20. Heating of Air.—It requires an expenditure of about 18 joules to raise the temperature of 1 cubic foot of air 1° F.; or an expenditure of 18 watts (18 joules per second) would raise the temperature of 1 cubic foot of air 1° F. per second. This gives a means of estimating the amount of power that must be supplied to raise the temperature of a given volume of air a given amount in a given time, but the amount of power required to maintain a room at a given temperature cannot be predetermined unless some data are at hand showing the rate at which heat escapes from the room. It is evident that if a room is tightly closed and its walls are of non-conducting material, it will take much less heat to maintain it at a given temperature than if the walls were of good heat-conducting material or if the windows were partly open; also, the amount of heat required will depend on the outside temperature.

EXAMPLE.—It is desired to heat the air in an electric oven for baking armatures, by means of electric heaters. The oven is 6 ft. × 10 ft. × 8 ft. inside dimensions. Current is to be supplied to the heaters at 500 volts. The oven is to have its temperature raised from 60° F. to 175° in ¼ hour: (a) What will be the total current required to heat the oven? (b) What will be the resistance of each heater, assuming that two heaters are used in parallel? Neglect loss of heat due to radiation, etc. in working the example.

SOLUTION.—(a) The cubical contents of the oven are 6 × 10 × 8 = 480 cu. ft. Each degree rise is equivalent to 18 joules per cu. ft. The total rise in temperature is 175° − 60 = 115°. The total number of joules expended must, therefore, be 480 × 18 × 115 = 993,600. The heating is to be effected in ¼ hour or 1,800 seconds; hence, the number of joules expended per second, i. e., the number of watts, must be $\frac{993,600}{1,800}$ = 552. The current taken is $\frac{552}{500}$ = 1.104 amperes. **Ans.**

(*b*) The current taken by each heater will be $\dfrac{1.104}{2} = .552$ ampere,

and the resistance of each heater is $R = \dfrac{E}{I} = \dfrac{500}{.552} = 906$ ohms, nearly.

Ans.

It should be noted particularly in the above example that the power estimated is the amount required to raise the temperature of the air only. It does not take into account the heat absorbed by the walls of the oven, the loss of heat through cracks or openings, etc. The actual amount of heat that would be required would be considerably greater than that estimated above. The best plan in such a case is to install a number of heaters and have each heater controlled by a switch so that the temperature can be regulated.

21.   Heating of Water.—It requires 778 foot-pounds of work to raise the temperature of 1 pound of water 1° F.; 1 gallon of water weighs 8.34 pounds, so that the amount of work required to raise 1 gallon of water 1° F. would be $778 \times 8.34 = 6{,}488.5$ foot-pounds; 1 joule is equal to .737 foot-pound, hence the number of joules required to raise 1 gallon of water 1° F. is $\dfrac{6{,}488.5}{.737} = 8{,}803$, approximately. To raise a gallon of water from 50° F. to the boiling point, 212° F., but not to convert it into steam, requires an expenditure of $8{,}803 \times (212 - 50) = 1{,}426{,}086$ joules. 1 kilowatt-hour is equivalent to 3,600,000 joules, and if power cost, say, 8 cents per kilowatt-hour, the cost of bringing the gallon of water from 50° F. up to the boiling point would be $\dfrac{1{,}426{,}086}{3{,}600{,}000} \times 8 = 3.17$ cents, assuming that all the heat supplied went into the water. As a matter of fact, considerable heat is always lost; the vessel in which the water is placed has to be heated, and some heat is' always lost to the surrounding air or objects with which the heated surfaces may be in contact.

The amount of power required to raise the temperature of the water will depend on the time allowed for heating. If the gallon of water were to be raised to the boiling point in

20 minutes, 1,200 seconds, and if, say, 70 per cent. of the total heat supplied were effective in heating it, the joules per second, or watts, would be $\dfrac{1,426,086}{.7 \times 1,200} = 1,698$, approximately. If current were supplied at 220 volts, the heater would be designed for a current of $\dfrac{1,698}{220} = 7.7$ amperes, and would have sufficient wire to give a resistance of $\dfrac{220}{7.7} = 29$ ohms, nearly.

22. Power Consumption of Heaters.—As examples of the power consumption of heaters required for various kinds of work, the following list may prove useful: Coffee pot (about 1 quart), 500 watts; teakettle (boil 1 quart in 10 minutes), 700 watts; chafing dish (1 quart), 500 watts; stew pan (2 quarts), 700 watts; glue pot (1 quart), 700 watts; glue pot (1 pint), 500 watts; laundry irons (medium size), 250 to 300 watts.

The greatest arguments in favor of the use of electricity are its convenience and cleanliness. For a small amount of localized heating, it may prove almost as economical as coal or other fuels, but where heat is required in considerable quantity, it is bound to be more expensive unless the circumstances are such that current can be obtained at exceedingly low rates, as is sometimes the case where large water powers are available. Where the current is generated by means of steam power, only about 10 per cent. of the heat value of the fuel is obtained as electrical energy, and even assuming that all of this could be utilized in an electric heater, it is easily seen that the direct utilization of the coal in the ordinary way would give much cheaper results. _____

HEATING OF WIRES

23. The amount of current that a given wire will carry depends on the resistance of the wire, which in turn involves its material and dimensions, and the rise in temperature that is allowable. When current is sent through a given wire, the heat developed raises the temperature, and the temperature will continue to increase until a point is reached where the wire gets rid of its heat to the air or surrounding objects as fast as it is developed. When this occurs, the temperature becomes stationary, and it can only be further increased by increasing the current or fixing the wire in some way so that it cannot get rid of its heat as easily as before.

The amount of current that a given wire will carry with a given rise in temperature, or the rise in temperature with a given current, is a quantity that is variable, and by no means easy to calculate even when the circumstances under which the wire is used are specified. The temperature at which a wire is worked depends on the class of service for which it is used. In electric transmission the size of wire is nearly always selected so as to limit the loss in it to a certain specified amount, and the heating effect is usually of secondary importance. In the case of underground ducts, however, where large currents are transmitted, the heating effect should be considered, but for overhead lines the rise in temperature is usually quite small and the drop in the line is the main thing to be taken into account.

As already stated, the rise in temperature of a wire depends on the ease with which the wire can get rid of heat. A wire can give off heat in three ways: By conduction, i. e., by imparting its heat to objects with which it may be in contact; by radiation, i. e., by the heat passing out from the body, and by convection. For heat loss to occur by convection the wire must be situated in a gas or liquid. Small particles of these become heated and are carried off by currents set' up by the unequal heating. If a wire is strung in the air and heated by means of a current, the peculiar wavy appearance of the air above the wire can easily be seen. This is caused by convection currents of heated air rising from the wire.

24. Heating Effect In Copper Wires Suspended In Air.—The rate at which a wire loses its heat depends on the location of the wire; evidently a covered wire placed, say, in molding or iron-pipe conduit will heat to a different degree than the same wire bare, strung in the open air and carrying the same amount of current. Again, if a bare wire is strung in air, the dissipation of heat from it will depend on the nature of the surface of the wire, i. e., whether it is bright or dark, rough or smooth. It will also depend on whether the wire is strung vertically or horizontally, and whether or not it is exposed to air-currents or drafts. Table III shows the heating effect in bare copper wires suspended outdoors, both for wires having a bright surface and those having a black surface. This and other data here given are from Roebling's Handbook, and are based on experiments made by A. E. Kennelly. Table IV shows the heating effect for wires suspended in still air.

25. Heating of Wires in Molding.—Table V gives Kennelly's results on the heating of insulated wires in moldings. In this case the heat is dissipated almost entirely by conduction because the wire is in contact with the molding. It will be noticed that for a given size of wire and a given rise in temperature, the current is larger than in the case of bright wire suspended in still air. One would naturally expect the opposite to be the case, but still air is not as good a conductor of heat as the materials surrounding a wire placed in molding. An insulated wire strung in the air will not heat as much, for a given current, as a bare wire because the insulation increases the radiating surface. Loss of heat by radiation increases directly with the surface of the wire, while loss by convection appears to depend only on the length and is independent of the size of the wire.

26. Loss by Convection.—The amount of energy that is dissipated per second by convection from a wire strung in still air is about .053 joule per second, per foot of length, per degree centigrade rise in temperature. For example, if a current heated a wire 40° C. above the temperature of the surrounding air, heat energy would be lost by convection at the rate of .053 × 40 = 2.12 joules per second, or the loss by convection would be 2.12 watts per linear foot. The total loss for a given length of the wire would be 2.12 multiplied

TABLE III

HEATING EFFECTS OF CURRENTS

Bare Copper, Suspended Outdoors

Amperes	Rise in Temperature. Degrees Centigrade							
	5°		10°		20°		40°	
	Bright	Black	Bright	Black	Bright	Black	Bright	Black
	Diameters of Wires. Mils							
1,000			962	932	771	745	620	594
950			928	897	744	720	595	572
900			894	865	715	692	574	552
850			868	843	689	665	550	530
800			839	810	672	649	537	512
750		975	804	775	643	620	515	495
700	963	933	767	739	613	591	491	472
650	916	889	729	703	582	561	467	449
600	869	837	690	665	554	532	442	426
575	845	813	671	647	538	517	429	414
550	820	789	650	627	522	501	417	402
525	795	764	630	609	506	487	404	389
500	770	740	610	589	489	470	390	376
475	745	719	589	569	473	455	377	363
450	719	693	568	548	453	438	363	350
425	690	667	546	526	436	422	349	336
400	661	638	524	504	418	406	334	322
375	632	610	502	484	399	377	319	309
350	601	581	478	462	380	360	304	295
325	571	552	453	439	362	342	289	279
300	540	522	428	415	342	326	273	264
275	509	492	404	392	321	309	257	249
250	477	460	378	367	300	290	240	222
225	445	430	351	343	280	270	223	215
200	410	399	324	316	259	250	205	198
175	373	365	296	289	235	227	186	180
150	334	329	267	258	211	202	166	161
125	295	290	235	226	185	177	145	144
100	254	248	202	193	157	152	123	120
90	236	230	186	178	145	140	114	111
80	216	212	171	164	132	128	104	102
70	198	192	155	150	120	116	94	91
60	177	170	137	132	107	104	83	80
50	155	147	119	115	92	87	72	70
40	130	124	100	96	77	73	62	59
30	104	100	78	75	61	58	50	45
20	73	70	54	53	43	40	34	30
10	40	38	27	26	20	18	16	14

TABLE IV

HEATING EFFECTS OF CURRENTS

Bare Copper in Still Air

Amperes	Rise in Temperature. Degrees Centigrade							
	10°		20°		40°		80°	
	Bright	Black	Bright	Black	Bright	Black	Bright	Black
	Diameters of Wires. Mils							
1,000						968	911	750
950						930	878	723
900						893	844	695
850						858	809	666
800					1,000	823	771	638
750					950	785	734	610
700				960	900	748	696	580
650				910	850	708	660	550
600				858	800	668	621	518
575				833	775	648	603	503
550		995	980	808	750	628	583	488
525		978	948	780	725	607	563	461
500		960	913	751	700	584	543	455
475		925	880	723	675	563	523	439
450		895	843	696	• 648	541	501	421
425		860	808	669	620	520	479	406
400	1,000	820	770	641	592	498	457	387
375	950	783	731	612	564	475	435	369
350	900	745	690	581	536	452	413	350
325	850	708	654	550	506	428	390	331
300	800	668	615	519	475	403	366	312
275	750	628	575	487	444	377	341	292
250	696	586	534	453	412	351	317	272
225	642	545	494	419	379	323	291	252
200	586	500	453	384	345	296	265	229
175	530	454	406	349	310	266	239	208
150	470	404	360	311	274	226	210	194
125	408	352	308	270	235	206	182	161
100	343	300	258	226	195	170	150	135
90	315	272	237	208	178	158	137	123
80	286	246	'214	196	161	· 143	124	112
70	259	220	190	170	143	127	110	100
60	226	194	167	150	125	112	97	87
50	' 191	167	142	130	106	95	82	74
40	156	140	117	108	86	78	68	61
30	120	111	90	85	66	60	54	48
20	82	76	63	60	45	44	40	36
10	40	38	37	35	30	28	26	24

TABLE V

HEATING EFFECTS OF CURRENTS

Carrying Capacity of Insulated Wire in Moldings

Amperes	Rise in Temperature. Degrees Centigrade								
	5°	10°	15°	20°	30°	40°	50°	60°	70°
	Diameters of Wires. Mils								
300					446	411	386	367	354
280					427	393	369	350	338
260				450	409	375	352	333	321
240				430	390	356	333	315	304
220	.		436	408	370	337	315	298	285
200		448	414	386	350	317	295	280	268
190		437	403	375	339	308	286	270	258
180		425	391	364	328	298	277	260	249
170		411	378	352	317	287	266	250	239
160		398	364	340	305	276	256	241	229
150	445	383	351	326	293	265	244	230	218
140	431	370	338	312	281	253	232	220	206
130	417	354	322	300	269	240	220	208	195
120	400	339	308	285	255	228	208	195	182
110	383	322	292	270	240	214	195	182	170
100	362	302	276	253	223	200	182	168	158
90	343	284	259	237	208	185	168	154	143
80	322	264	240	218	192	169	153	139	130
70	300	242	220	198	174	152	139	123	116
60	275	220	195	175	155	135	122	108	101
50	250	195	175	152	132	118	104	91	86
40	217	169	144	128	110	95	85	75	70
30	178	136	115	100	85	73	66	58	54
20	132	100	71	69	59	50	45	40	37
10	78	58	42	35	30				

by the length in feet. The convection loss thus appears to be independent of the diameter of the wire.

27. Loss by Radiation.—As before stated, the loss by radiation depends directly on the surface of the wire and also on the character of the surface. A square inch of bright copper surface will radiate about .004 watt per degree centigrade rise in temperature. A black copper surface will radiate heat over twice as fast or at the rate of about .009 watt per square inch per degree centigrade rise. A japanned cast-iron surface can dissipate about .015 watt per degree centigrade rise in temperature by radiation and convection combined. In other words, an expenditure of 1 watt per square inch will maintain a japanned cast-iron surface at a temperature of about 66° C. above the surrounding air.

28. Fusing Effects of Currents.—The current required to fuse a wire of given size and material depends considerably on the length of the wire, its location, method of mounting, etc. For example, very short fuses of a given size of wire require a larger fusing current than long wires of the same size. If the wire is very short, the terminals have an appreciable cooling effect, so that fuses should never be shorter than 1 inch between terminals. The wire should also be protected from air-currents and kept from contact with other bodies. For these reasons enclosed fuses have largely replaced the older open type of fuse in places where fuses are used to protect electrical apparatus against excessive currents. Table VI shows the results of experiments by W. H. Preece on the fusing point of wires made of various metals. The table shows the current that will fuse wires of various diameters, all wires being tested under the same conditions.

29. Heating of Wire in Rheostats and Resistance Boxes.—The temperature at which the resistance wire in a rheostat can be worked depends on how the wire is arranged. In some rheostats the wire is completely covered by enamel, cement, or other material that excludes the air. Wires so arranged can be worked at a higher temperature than those placed in the open air, because there is not the same chance

TABLE VI

DIAMETERS OF WIRES OF VARIOUS MATERIALS THAT WILL BE FUSED BY A CURRENT OF GIVEN STRENGTH

W. H. Preece, F. R. S.

Current Amperes	Diameters. Inches								
	Copper	Aluminum	Platinum	German Silver	Platinoid	Iron	Tin	Tin-Lead Alloy	Lead
1	.0021	.0026	.0033	.0033	.0035	.0047	.0072	.0083	.0081
2	.0034	.0041	.0053	.0053	.0056	.0074	.0113	.0132	.0128
3	.0044	.0054	.007	.0069	.0074	.0097	.0149	.0173	.0168
4	.0053	.0065	.0084	.0084	.0089	.0117	.0181	.021	.0203
5	.0062	.0076	.0098	.0097	.0104	.0136	.021	.0243	.0236
10	.0098	.012	.0155	.0154	.0164	.0216	.0334	.0386	.0375
15	.0129	.0158	.0203	.0202	.0215	.0283	.0437	.0506	.0491
20	.0156	.0191	.0246	.0245	.0261	.0343	.0529	.0613	.0595
25	.0181	.0222	.0286	.0284	.0303	.0398	.0614	.0711	.069
30	.0205	.025	.0323	.032	.0342	.045	.0694	.0803	.0779
35	.0227	.0277	.0358	.0356	.0379	.0498	.0769	.089	.0864
40	.0248	.0303	.0391	.0388	.0414	.0545	.084	.0973	.0944
45	.0268	.0328	.0423	.042	.0448	.0589	.0909	.1052	.1021
50	.0288	.0352	.0454	.045	.048	.0632	.0975	.1129	.1095
60	.0325	.0397	.0513	.0509	.0542	.0714	.1101	.1275	.1237
70	.036	.044	.0568	.0564	.0601	.0791	.122	.1413	.1371
80	.0394	.0481	.0621	.0616	.0657	.0864	.1334	.1544	.1499
90	.0426	.052	.0672	.0667	.0711	.0935	.1443	.1671	.1621
100	.0457	.0558	.072	.0715	.0762	.1003	.1548	.1792	.1739
120	.0516	.063	.0814	.0808	.0861	.1133	.1748	.2024	.1964
140	.0572	.0698	.0902	.0895	.0954	.1255	.1937	.2243	.2176
160	.0625	.0763	.0986	.0978	.1043	.1372	.2118	.2452	.2379
180	.0676	.0826	.1066	.1058	.1128	.1484	.2291	.2652	.2573
200	.0725	.0886	.1144	.1135	.121	.1592	.2457	.2845	.276
225	.0784	.0958	.1237	.1228	.1309	.1722	.2658	.3077	.2986
250	.0841	.1028	.1327	.1317	.1404	.1848	.2851	.3301	.3203
275	.0897	.1095	.1414	.1404	.1497	.1969	.3038	.3518	.3417
300	.095	.1161	.1498	.1487	.1586	.2086	.322	.3728	.3617

for the wire to become oxidized by the action of the air. The wire of any heating device or rheostat is never worked as high as a red heat because continuous operation at such high temperature leads to deterioration. Resistance boxes and similar apparatus are designed nearly altogether from experimental data relating to the particular construction adopted.

TABLE VII
CARRYING CAPACITY OF GERMAN-SILVER WIRE

Number B. & S.	Circular Mils	Maximum Current Amperes	Feet per Ohm
10	10,381	8.5	60.90
11	8,234	5.4	47.60
12	6,529	4.6	37.80
13	5,178	3.8	29.90
14	4,106	3.2	23.70
15	3,257	2.7	18.80
16	2,583	2.3	14.90
17	2,048	1.9	11.80
18	1,624	1.6	9.40
19	1,288	1.2	7.25
20	1,021	.99	5.91
21	810	.88	4.69
22	643	.66	3.72
23	509	.55	2.95
24	404	.488	2.33
25	320	.438	1.85
26	254	.384	1.47
27	201	.343	1.16

It is difficult to give any general data applicable to such devices, though Tables VII, VIII, and IX will be found useful. The safe carrying capacities here given assume that the wire is wound in an open spiral so that the air has free access to it.

30. Carrying Capacity of German-Silver Wire. Table VII gives the resistance and current-carrying capacity of German-silver wire containing 18 per cent. nickel. The currents given assume that the wire has to carry the current continuously. If it is used to carry the current for a few seconds only, as, for example, in motor starting boxes, the current could be doubled without danger.

31. Carrying Capacity of Galvanized-Iron Wire. Table VIII gives values of the carrying capacity of galvanized-iron wire. This wire is frequently used for large

TABLE VIII

CARRYING CAPACITY OF GALVANIZED-IRON WIRE

Number Washburn & Moen Gauge	Circular Mils	Maximum Current Amperes	Feet per Ohm
3	59,536	63.80	645.0
4	50,625	55.60	549.0
5	42,849	47.50	463.0
6	36,864	34.80	398.0
7	31,329	30.10	337.0
8	26,244	26.60	283.0
9	21,904	23.20	236.0
10	18,225	19.70	196.0
11	14,400	16.20	155.0
12	11,025	13.90	119.0
13	8,464	11.60	91.4
14	6,400	9.28	69.1
15	5,184	6.96	56.0
16	3,969	5.80	42.8
17	2,916	4.29	31.4

rheostats where a considerable current-carrying capacity is desired. The wire is supposed to be mounted so that the air has free access to it.

32. Carrying Capacity of Tinned-Iron Wire.—In Table IX is given the carrying capacity of tinned-iron wire. This kind of wire has been used considerably in rheostats because the coating of tin prevents rust. In the table the maximum safe current is given both for coils mounted in a wooden frame and in an iron frame. Wooden frames are now seldom, if ever, used in resistance boxes on account of the fire-risk. However, it is often convenient to mount coils on wooden frames for temporary experimental purposes. The last column gives the amount of resistance that can be

TABLE IX
CARRYING CAPACITY OF TINNED-IRON WIRE

No. B. & S.	Area Circular Mils	Maximum Safe Current With Wooden Frame Amperes	Maximum Safe Current With Iron Frame Amperes	Safe Current for 1 Minute	Feet per Ohm	Pounds per Foot	Ohms per Inch of a Spiral Wound on .4-Inch Mandrel
8	16,509	17.40	20.30	43.6	250.00	.04000	.0050
9	13,094	14.60	17.10	36.6	173.00	.03300	.0066
10	10,381	12.30	14.30	30.8	137.00	.02751	.0095
11	8,234	10.30	12.00	25.8	108.00	.02182	.0131
12	6,529	8.70	10.10	21.7	86.40	.01730	.0182
13	5,178	7.30	8.50	18.3	68.50	.01372	.0245
14	4,106	6.10	7.10	15.3	54.30	.01089	.0353
15	3,257	5.10	6.00	12.9	43.10	.00863	.0492
16	2,583	4.30	5.00	10.8	34.10	.00685	.0690
17	2,048	3.60	4.20	9.1	27.10	.00543	.0960
18	1,624	3.00	3.50	7.6	21.40	.00430	.1345
19	1,252	2.50	2.90	6.3	16.50	.00341	.1963
20	1,021	2.20	2.50	5.4	13.50	.00271	.2636
21	810	1.80	2.10	4.5	10.70	.00231	.3725
22	643	1.50	1.77	3.8	8.49	.00184	.5220
23	509	1.30	1.49	3.2	6.73	.00146	.7350
24	404	1.08	1.20	2.3	5.34	.00116	1.035

obtained per inch length of spiral wound closely on a mandrel .4 inch in diameter. The column giving the safe carrying capacity for 1 minute shows values of the current that may be allowed for starting boxes or other resistances that are used for short intervals only.

ELECTRIC WELDING

33. A special application of the heating effect of the electric current is found in the **electric welding** of metals. Many welding operations can be brought about by means of the electric current that would be very difficult by the ordinary process of welding, which consists in heating the two pieces to be welded, and when the proper heat has been attained, hammering them together before they have had time to cool appreciably. In the Thomson welding process, which is the one most widely used, the two pieces to be welded are brought together when cold and a current of very large volume sent through the junction. The local resistance at the joint causes a large heating effect, and the two ends are soon brought up to a welding temperature. The clamps, holding the work, are then pressed together and the weld thereby effected. The principle of operation of the Thomson welder will be understood by referring to Fig. 3. Alternating current is employed for this work, because it is easier to obtain a large current at low voltage by means of alterna-

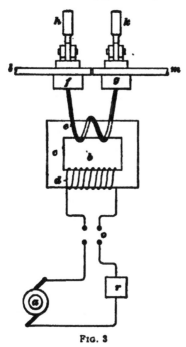

Fig. 3

ting current than by direct. In some small welders for light work, current is used directly from a special alternator provided for the purpose, but in most cases the current is generated at fairly high pressure and is stepped-down by means of a transformer. In Fig. 3, *a* is the alternator connected to the primary coil *d* of the welding transformer *b*; *c* is the laminated transformer core on which the windings are placed. The secondary coil of the transformer consists of very few turns, in many cases one turn only, of heavy conductor, and the terminals of the coil are attached to the clamps *f, g*. The two pieces *l, m* to be welded are securely clamped by means of cams operated by handles *h, k*, this method of clamping allowing the work to be quickly clamped or unclamped. A switch *o* is provided in the primary circuit so that the current can be readily cut off, and a regulating device *r* is also included to allow adjustment of the welding current; *r* is usually an adjustable reactive coil, but an adjustable resistance could be used, though it would not be as economical. Where only one welder is operated from an alternator, a convenient method of regulating the welding current is to change the field excitation of the dynamo by means of an adjustable field rheostat. One of the clamps *f, g* is arranged so that it can be moved so as to force the ends of the work together when the weld is being made. In some welders this is accomplished automatically by means of weights or hydraulic pressure. In other machines it is done by hand, either by turning a hand wheel or pulling on a lever. The clamps *f, g* are kept cool by means of water circulation.

The action of the welder is as follows: After the work is in place, the current is turned on, and since the ends of *l* and *m* are in contact, a large volume of current is sent across the joint. The voltage required is low, but the volume of current is very large; in the case of copper welding, the current may be as high as 60,000 amperes per square inch. The current soon heats the junction up to a welding heat, and the two pieces of metal are then forced together. The alternating current used for this work should be of low frequency, at least as low as 50 cycles per second, and better if lower than this for heavy work.

34. Fig. 4 shows a welder made by the Thomson Electric Welding Company for welding miscellaneous work having a cross-section of iron or steel up to .6 square inch. The welder is here arranged for welding flat iron hoops, but by using different forms of clamps a variety of work can be handled.

The transformer is contained in the case a, which forms the base of the machine. The work is held in the copper or bronze clamps at bc, which are operated by the handles de; f is the lever for forcing the ends together after the metal has been brought to a welding heat. When the handle is pulled to the left, the togglejoint at g straightens out and moves clamp c to the left, thus applying the pressure. Pipe connections hh are provided so that water can be circulated through the clamps in order to keep them cool. At h' the

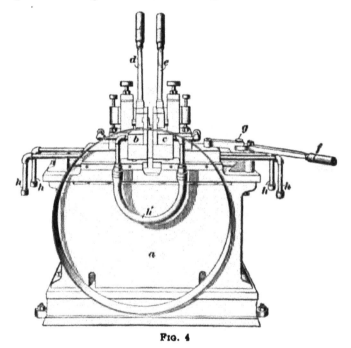

FIG. 4

pipe connection is of rubber hose in order to give flexibility and also to prevent current from flowing between b and c through the pipe. This welder requires a maximum of 18 horsepower for 35 seconds to effect the largest weld that it is capable of handling. Where hoops or rings are welded, a portion of the current, of course, flows around the ring, but this path is of such high impedance compared with that between the abutting ends that only a very small part takes the longer path. The welder shown in Fig. 4 is one of the simpler types.   For special lines of work, welders have been designed that operate to a large extent automatically; that is, the pressure is applied and regulated automatically, and in some cases the clamping is also effected by hydraulic pressure or some other method that permits of rapid handling of the work.

35.   Welding Transformers.—Fig. 5 shows one style of transformer used for a Thomson welder.   The transformer is double; one of the laminated iron cores is shown at a; there is a similar core on the other side.   Linked with the iron core is the massive copper casting b that forms the

Fig. 5

secondary coil of one turn.   This casting is divided by a slit c between the clamps d so that the current set up is compelled to pass through the work held in the clamps. The primary coil is not shown in the figure, but it is placed in the groove in the casting b which forms the secondary, so that it is linked by the core a in the same way as the secondary.

Fig. 6 shows another style of transformer designed for very heavy currents.   In this case the secondary $a\,a$ is made of two heavy copper castings of channel section, which, when bolted together, form a box entirely closing the primary coil. In some cases the enclosed space is filled with oil, thus thoroughly insulating the primary from the secondary. The laminated iron core that carries the magnetic flux passes through the opening *b*. The object in both these transformers is to provide a secondary of very low resistance and large current capacity, so that very large currents at low voltage can be applied to the work.

36. Power Required for Electric Welding.—The greater the amount of power supplied, the shorter will be the time for making a weld, and conversely, the less the power, the greater will be the time. These matters are regulated according to the material to be welded. Copper, brass, tool steel, and other metals that are deteriorated by being heated must be welded rapidly. Thus, no time is given for the metal to change its nature. That part of the metal where deterioration, if any, has occurred is pushed out from the weld by the pressure applied.

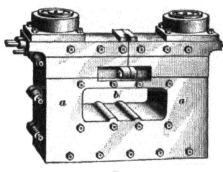

Fig. 6

Table X, given by the Thomson Electric Welding Company, shows the power required for welding iron, copper, and brass of varying cross-sections. The distance between clamps was twice the diameter of the iron pieces to be welded, three times the diameter for brass, and four times the diameter for copper. Tests have shown that from 70 to 75 per cent. of the power supplied is actually used in making the weld, so that there is comparatively little heat wasted. Although there is a great loss of heat in the steam engine, and also some loss in the dynamo, it has been found that the fuel cost for electric welding is but little if any more than for welding by the ordinary process, because in the electric process, the heat is applied just where it is wanted

TABLE X
POWER REQUIRED FOR ELECTRIC WELDING

Copper

Area Square Inches	Watts in Primary of Welders	Time in Seconds	Horsepower Applied to Dynamos	Foot-Pounds
.125	6,000	8	10.0	44,000
.250	14,000	11	23.4	142,000
.375	19,000	13	31.8	227,000
.500	25,000	16	42.0	369,000
.625	31,000	18	51.9	513,000
.750	36,500	21	60.6	700,000
.875	43,000	22	72.1	872,000
1.000	49,000	23	82.1	1,039,000

Brass

Area Square Inches	Watts in Primary of Welders	Time in Seconds	Horsepower Applied to Dynamos	Foot-Pounds
.25	7,500	17	12.6	117,000
.50	13,500	22	23.2	281,000
.75	19,000	29	31.8	508,000
1.00	25,000	33	42.0	760,000
1.25	31,000	38	52.0	1,087,000
1.50	36,000	42	60.3	1,390,000
1.75	40,000	45	67.0	1,659,000
2.00	44,000	48	73.7	1,947,000

Iron and Steel

Area Square Inches	Watts in Primary of Welders	Time in Seconds	Horsepower Applied to Dynamos	Foot-Pounds
.5	8,550	33	14.4	260,000
1.0	16,700	45	28.0	692,000
1.5	23,500	55	39.4	1,191,000
2.0	29,000	65	48.6	1,738,00
2.5	34,000	70	57.0	2,194,00
3.0	39,000	78	65.4	2,804,00
3.5	44,000	85	73.7	3,447,00
4.0	50,000	90	83.8	4,148,00

and is not wasted between welds. The electric process is therefore especially adapted to intermittent work. It also allows welds to be made that would be exceedingly difficult if not impossible by ordinary methods. It has not been used on very heavy work, but for light work it has found wide application. In the electric process the central parts of the metal become hot first, while in the ordinary method the heat is conducted in from the outer part and the center may be cooler than the outside. The electric process therefore gives a sound weld throughout; in fact, the weld is even stronger than the other part of the metal.

37. Rail Welding.—A special application of the Thomson welding process is the joining of steel rails, thus making the track one continuous piece. When rails are surrounded by paving, it has been found that they can be joined in this way without the expansion and contraction, due to heat and cold, throwing the tracks out of line. A special form of welder is suspended from a boom carried by a car designed for the purpose, the contacts are brought against opposite sides of the rail and by means of the current two pieces of iron are welded on at the joint, one piece on either side. When the pieces have been heated to a welding heat, pressure is applied by means of a hydraulic jack. A joint made this way on a 70-pound rail will stand a strain of 279,000 pounds, whereas, the maximum strain placed on the rail on account of variations in temperature is 150,000 pounds. The current for welding is obtained from a transformer, the primary of which is supplied from a rotary converter that takes direct current at 500 volts from the trolley line and transforms it to about 300 volts alternating. The average current supplied to the primary of the transformer during a welding operation is about 650 amperes. The surface of the rail at the weld is thoroughly cleaned either by grinding or by means of a sand blast before the weld is made. The electrical conductivity of the joint is as great as that of the rail itself, and under proper conditions four joints per hour can be made.

ELECTRIC ANNEALING

38. Another application of the heating effect is **electric annealing.** This process is used to soften parts of steel plates or castings on which it is desired to perform machine work. Fig. 7 shows the principle of a Thomson annealer as used for local softening on armor plates so that they can be drilled. The transformer is arranged so that it can be lowered to the plate and the heavy secondary terminals made to rest on the spot to be softened. A large current flows between the terminals, through the steel plate, as indicated by the dotted lines, and the heating caused thereby anneals the hardened surface so that it can be drilled or otherwise operated on

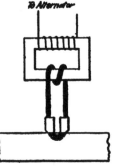

Fig. 7

by machine tools. The advantage of this method is that it allows the annealing to be done locally, only the part of the plate immediately surrounding the secondary terminals being softened.

ELECTROLYTIC FORGE

39. It was discovered by Lagrange and Hoho that, if a piece of iron or other metal a, Fig. 8, were connected to the negative pole and the circuit closed by dipping a into a conducting liquid b, the metal soon became heated to a white heat or even melted. In Fig. 8 the liquid b is held in a wooden tank c provided with a lead lining d. The metal to be heated, or the tongs in which it is held, is rested against the bar e, which forms the negative terminal. In America this process of heating metals to a welding heat has been developed principally by Burton. The solution generally used consists of 10 parts of sal soda and 1 part of borax dissolved in sufficient water to give a specific gravity of about 1.17. When the metal is plunged into the liquid, there is a sudden evolution of hydrogen at the submerged portion, and the envelope of gas introduces such a high resistance that intense heat is developed at the surface of the metal and soon brings it up to a temperature depending on the strength of the current and on the length of time that the metal is held in the liquid. In the Burton method of applying this process, the metal is brought in contact with the surface of the electrolyte so that the hydrogen that is evolved can ignite and aid in heating up the metal instead of passing off as it does when the metal is plunged beneath the liquid. At least 110 volts should be used for this method of heating, and, of course, direct current is required,

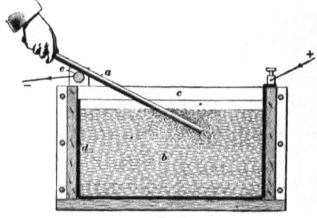

Fig. 8

since alternating current would not produce the necessary evolution of gas. The objects to be heated must be handled rapidly, as it is somewhat difficult to regulate the heat and there is danger of burning the metal. The amount of current required depends on the area of metal submerged. As the piece to be heated is dipped farther into the liquid, the current becomes greater because the resistance is reduced so that the forge, to a certain extent, regulates the amount of current automatically.

ELECTRIC FURNACES

40. When current is made to flow between carbon terminals that are separated a short distance, a so-called electric arc or flame is formed, and this arc constitutes the source of highest temperature known. The temperature of the arc is estimated to be in the neighborhood of 3,500° C., and in it the most refractory substances are easily melted. This high temperature has been made use of in a number of metallurgical operations, and electric furnaces of many different styles have been devised. The **electric furnace,** in addition to high temperature, has the advantage that the heat is not obtained by ordinary combustion of carbon, so that the furnace does not have to be supplied with air in order to effect the heating. Also, with the ordinary furnace,

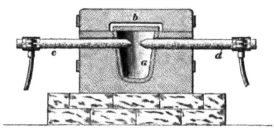

FIG. 9

a considerable amount of heat is carried off with the products of combustion, but in the electric furnace the loss of heat is small and a very high temperature can be obtained.

41. Fig. 9 shows a simple form of furnace suitable for experimental work. A crucible a of refractory material is surrounded by firebrick. The crucible is also covered by a slab of fireclay b, and entering from either side are the carbon electrodes c, d between the ends of which the arc is formed, thus heating up the mass of material in the crucible. The arc can be started by first turning on the current, then sliding one of the electrodes in until it touches the other, and immediately withdrawing it a short distance; or, a small rod of carbon, say about $\frac{1}{16}$ inch in diameter, may be placed between the points. When the current is turned on, this rod is soon burned out and the arc established. In some furnaces the space between the electrodes is filled with finely divided graphite of carbon, so that the heating is effected by incandescent particles of carbon rather than by a regular electric arc. A small furnace, such as that shown in Fig. 9, may be used for quite a wide range of experimental work.

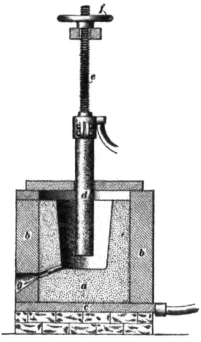

FIG. 10

42. Fig. 10 shows a furnace designed by Willson. In this case the crucible *a* is of carbon and forms one electrode. It is surrounded by masonry *b* and rests on a plate *c* to which one terminal is attached. The other electrode *d* is suspended from the screw thread *e*, so that it can be raised or lowered by turning the wheel *f*. A hole is provided at *g* through which the product of the furnace, if in liquid form, can be drawn off, the hole being plugged with clay when not in use. Calcium carbide, from which acetylene gas is obtained, can be made in either of the above furnaces by heating a mixture of lime and carbon. Electric furnaces have been designed for many different electrochemical and electrometallurgical processes, but further treatment cannot be given here, as it is beyond the scope of this Course and belongs properly to electrometallurgy or electric smelting.

ELECTROMAGNETS AND SOLENOIDS

43. **Electromagnets** and **solenoids** are used for a number of purposes in practical work, and the object here is to give a general outline of the principles involved in their design, and incidentally point out the limitations in the use of these devices. There seems to be a great deal of misunderstanding as to the amount of work that can be done by means of magnets and solenoids, but it will be seen from the following that where the object is to overcome a resistance over a considerable range, they are very inefficient devices. Where the magnet is used simply to lift objects, the armature is in direct contact with the magnet and a strong pull can be obtained with a comparatively small expenditure of current. When, however, the pull has to be exerted through a distance, there is an air gap in the magnetic circuit and the power consumption for a given pull is enormously increased.

LIFTING MAGNETS

44. The lifting power or adhesive force of a magnet is called its *tractive force*, or simply *traction*. **Lifting magnets** are employed to a considerable extent in rolling mills for handling steel plates and other articles that are not readily handled by ropes or slings. The lifting magnet allows the work to be handled more rapidly than if slings or ropes were used, and thus saves time. In most cases the object to be lifted constitutes the armature of the magnet. The magnet is lowered on to the object by means of a crane, the article is lifted and carried wherever desired, and the tractive effort that a given magnet will exert depends very largely on the nature of the contact between the poles and the armature or object to be lifted. It is evident that the contact surface may often be very rough and uneven, and this will materially affect the lifting power. If, on the other hand, the magnet makes contact with an armature having a true surface, the lifting power will approach more nearly to the calculated value, which is based on the assumption that there is perfect contact between the armature and the object to be lifted. If a magnet is to be used for general lifting work, where the nature of the contact is indeterminate, it is almost useless to attempt to calculate the ampere-turns necessary to give the required lifting effort. In such cases the only safe way is to design a magnet and determine the necessary ampere-turns experimentally. This can easily be done by winding on a temporary coil of a known number of turns, and then increasing the current through it until the required pull is obtained. The product of the current and number of turns will give the ampere-turns required, and a permanent coil can then be designed to give this number of ampere-turns. In case the magnet makes good contact with its armature, the ampere-turns required for a given tractive effort or the tractive effort corresponding to a given number of ampere-turns can be calculated with a fair degree of accuracy by using Maxwell's formula.

FORMULAS FOR LIFTING MAGNETS

45. Maxwell's Law of Traction.—The relation between the tractive force, the area of contact of the polar surfaces, and the magnetic density at the contact surfaces, was first demonstrated by Maxwell. The formula expressing this relation is as follows:

$$P_1 = \frac{\mathfrak{B}^2 A_1}{8\pi} \qquad (1)$$

where P_1 = total pull, in dynes, exerted between the contact surfaces;

$\mathfrak{B}$ = magnetic density at the contact surfaces, in lines per square centimeter;

A_1 = area of contact, in square centimeters;

$\pi = 3.1416$.

In this formula A_1 is the total polar contact. For example, in a horseshoe magnet, A_1 would be the combined area of the two poles and P the total pull exerted by the magnet. Maxwell's formula may be reduced to the form:

$$P = \frac{\mathbf{B}^2 A}{72{,}134{,}000} \qquad (2)$$

where P = pull, in pounds;
        $\mathbf{B}$ = magnetic density at pole face, in lines per square inch;
        A = polar area, in square inches.

Formula 2 should be carefully noted. It shows that the tractive force of a magnet increases directly as the total area of surface in contact with the armature, and as the square of the density of the lines of force across that surface. The formula assumes that the distribution of lines is uniform throughout the entire contact surface. In actual practice it is impossible to obtain this result on account of magnetic leakage and other causes.

In all electromagnets designed for traction there will be two contact surfaces, one at the north pole of the magnet and the other at the south pole; or, in other words, the total lines of force developed in the magnetic circuit are used twice in producing the traction of the magnet. If the two contact surfaces are symmetrical and equal in area, the total tractive force of the magnet will be twice the result obtained by considering one contact surface alone; but if the contact surfaces are unlike, the tractive force exerted by each surface must be calculated separately, and the two results thus obtained added together.

The most economical electromagnet designed for traction is one that will lift the greatest load in proportion to its own weight. To accomplish this result, the following must be considered: The magnetic circuit in the magnet and armature should be as short as possible; the sectional area of the magnetic circuit should be uniform and large in proportion to the over-all dimensions; the iron or steel used in the magnet should have a high permeability; the magnetic density of the contact surface should not be much over 100,000 lines of force per square inch, for, if the density is pushed higher than this, the reluctance of the magnetic circuit will be increased, and this increases the weight of the copper used in the magnetizing coils.

From formula **2**, it is evident that the pull p exerted per square inch of contact surface is

$$p = \frac{B^2}{72{,}134{,}000} \qquad (3)$$

because the pull per square inch is equal to the total pull divided by the total contact area A.

Rule.—*The tractive force of an electromagnet in pounds per square inch is equal to the square of the magnetic density at the contact surface divided by 72,134,000.*

46. For convenience in making calculations, Table XI, giving the pull per square inch of polar surface corresponding to different magnetic densities, is here given.

TABLE XI

Magnetic Density B. Lines per Square Inch	Pounds Pull per Square Inch	Magnetic Density B. Lines per Square Inch	Pounds Pull per Square Inch
10,000	1.39	45,000	28.07
11,000	1.68	50,000	34.66
12,000	2.00	55,000	41.94
13,000	2.34	60,000	49.91
14,000	2.72	65,000	58.57
15,000	3.12	70,000	67.93
16,000	3.55	75,000	77.98
17,000	4.01	80,000	88.72
18,000	4.49	85,000	100.16
19,000	5.00	90,000	112.29
20,000	5.55	95,000	125.11
25,000	8.66	100,000	138.63
30,000	12.48	105,000	152.84
35,000	16.98	110,000	167.74
40,000	22.18	115,000	183.34

47. Formula **2** may be changed so as to give the magnetic density at the contact surfaces necessary for a given tractive force, the area of contact being known. In this case the formula becomes

$$\mathbf{B} = 8{,}493\sqrt{\frac{P}{A}} \qquad (4)$$

Rule I.—*In an electromagnet the density of lines of force at the contact surface is equal to 8,493 times the square root of the tractive force in pounds divided by the area in square inches.*

The total area of contact surface required for a given pull with a given density is given by the formula

$$A = \frac{P \times 72{,}134{,}000}{\mathbf{B}^{\cdot}} \qquad (5)$$

Rule II.—*In an electromagnet, the total polar area required for a given pull with a given density at the contact surfaces is found by multiplying the pull, in pounds, by 72,134,000 and dividing the product by the square of the density.*

DESIGN OF LIFTING MAGNETS

48.   General Construction.—Lifting magnets are made in a variety of designs, and their shape depends, to a considerable extent, on the class of work for which they are intended.

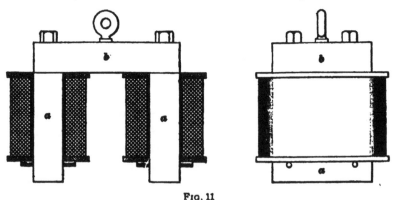

Fig. 11

Where magnets are used for lifting purposes, they must be constructed so that the windings will be protected from mechanical injury. The style of magnet shown in Fig. 11

is much used, though, of course, it would be provided with a metal shell or other protection for the coils. The cores *a, a* and yoke *b* are of wrought iron or mild steel, and ·are rectangular in shape. Fig. 12 shows a bipolar lifting magnet designed by Mr. E. B. Clark* and intended for lifting steel sheets. The construction is evident from the figure, (*a*) showing the details and (*b*) a perspective view. This magnet is capable of lifting 6 tons with 6¼ amperes

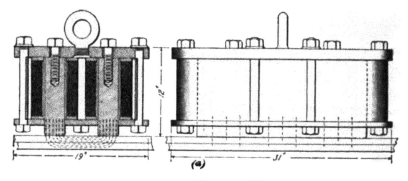

(*b*)

FIG. 12

at 250 volts, and its total weight is 1,200 pounds. Fig. 13 shows another style of magnet that has been used considerably, principally because it is iron clad; the magnetizing coil is well protected, and only one coil is needed to provide the excitation. The central pole piece *a* is circular in cross-section, while the other pole piece *b* is in the form of an annular ring. If the magnet proper *M* is made in one casting, the coil *c c* is wound on a form, after which it is thoroughly insulated by wrappings of cloth, mica, or tape, then slipped over the core of the magnet and held in position by a ring of brass or other non-magnetic metal *r*. The connections to the coil are made to terminal wires passing from the coil up through

*American Electrician, Vol. XIII, No. 2.

holes in the top of the magnet. By designing the magnet low and large in diameter, the magnetic circuit can be made exceedingly short in proportion to its sectional area, thus realizing one of the conditions of an economical design. The hooks h are provided so that slings may be attached in case the magnet becomes disabled. The magnet of the design illustrated was capable of lifting 12 tons and took a current of 13 amperes at 250 volts. The total weight of the magnet was 1,235 pounds. It has been found, however, that this circular form of magnet does not give good results for some kinds of work, and that, especially for plate lifting, the rectangular form of core is much better, because the circular outline of the pole pieces allows a plate, which always tends to bend or droop down, to tear away much more easily than from a rectangular pole piece.

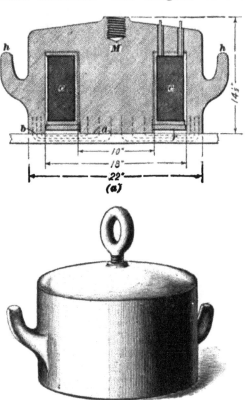

(a)

(b)

Fig. 13

For this reason the rectangular cores are very much used, even though they require a greater length of copper, for a given area of core section, than circular cores.

49. Fig. 14 (a) and (b) shows a type of magnet also designed by Mr. E. B. Clark for lifting plates. This magnet is made up of twelve small cores attached to a plate that forms the yoke. Each core is provided with a coil and is made from ,bar iron 1 inch thick by 5 inches wide. The cores are 3½ inches long. Each coil has 644 turns of No. 17 wire, and the magnet lifts 4 tons when taking a current of 4.4 amperes at 250 volts. The magnet weighs only 300 pounds, so that it is capable of lifting nearly 27 times its own weight. It is thus seen that this magnet can lift a greater load in proportion to its weight than either of the styles shown in Figs. 12 or 13. Fig. 15 shows a style of

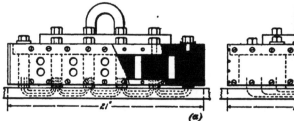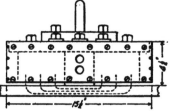

(a)

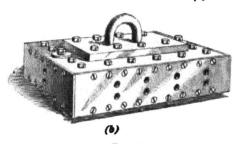

(b)

Fig. 14

protected terminal suitable for lifting magnets. It is designed so that there is little danger of the wire a, which connects to the winding, being broken. The wire b, which connects to the line, may become broken, because it is exposed, but this is easily remedied by removing the brass cap C and making a new connection to the stud d.

50. Calculations regarding lifting magnets are always more or less unsatisfactory, for a number of reasons. In the first place, the nature of the contact between magnet and armature is always variable and calculations based on the assumption of a perfect contact are likely to lead to unsatisfactory results. Again, the amount of magnetic leakage between the magnet cores is hard to estimate, and the quality of the iron used in the magnet may not be accurately known. Experimental data relating to the type of magnet to be designed are therefore necessary if the results are to be other than approximate. In what follows we will illustrate the general method of calculating an ordinary lifting magnet of the horseshoe type.

51. A horseshoe type lifting magnet similar to that shown in Fig. 11 is to be designed for a tractive force of 1 ton and is to operate on 220 volts. The cores and yoke are to be of rectangular cross-section, and we will assume that the magnetic qualities of the iron are represented by the curve for wrought iron shown in Fig. 16, which is here repeated from *Dynamos and Dynamo Design*, for convenient reference.

The first thing to be decided on is the density at the pole face. This should be taken fairly high in order to secure a large pull per square inch. It cannot be forced too high without causing a large amount of magnetic leakage and requiring an excessive number of ampere-turns. A fair value for the

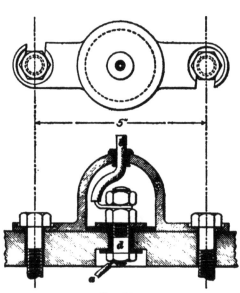

Fig. 16

density at the pole faces is 100,000 lines per square inch. The pull per square inch corresponding to this density is 138.63 pounds. The total pull is to be 2,000 pounds, or 1,000 pounds per pole, so that the area of each pole should be $\frac{1,000}{138.63} = 7.21$ square inches. In order to allow some margin, we will make the pole face area 8 square inches, with dimensions 2 inches by 4 inches. The total magnetic flux entering or leaving a pole face will be $100,000 \times 8 = 800,000$ lines.

47—37

The flux in the magnet cores will be greater than that at the pole faces by the amount of the magnetic leakage.   The

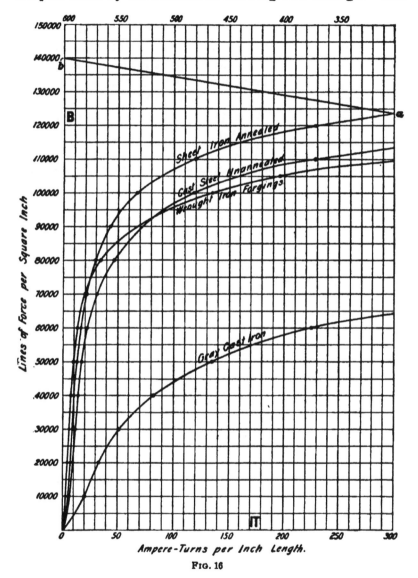

Fig. 16

coefficient of leakage depends considerably on the extent to which the armature or object to be lifted is saturated and may also change with the character of contact between

magnet **and** armature.   With a thoroughly good path
through the armature, there would not be so great a tendency for the magnetism to leak across between the poles.
In order to make allowance for leakage, we will take the
leakage coefficient as 1.3 and calculate the magnet for a
flux of 800,000 × 1.3 = 1,040,000 lines in the cores.

In order to keep down the density in the cores, we will
make them 2½ inches by 4½ inches.   This will give a cross-
sectional area of 2½ × 4½ = 11¼ square inches, and the density
in the cores will be $\dfrac{1,040,000}{11.25}$ = 92,400 lines per square inch.
If the cross-section of the cores were not made greater than

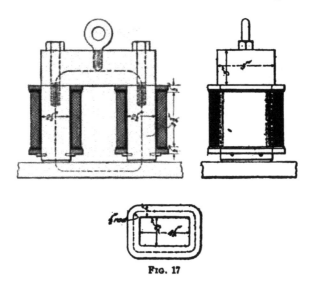

the area of the pole faces, the density in the cores would be
very high and a large number of ampere-turns would be
required to force the magnetism through them.   The cores
will be chamfered down, as shown in Fig. 17, thus reducing
the core cross-section to the pole-face area.

52. Calculation of Ampere-Turns.—In order to
estimate the ampere-turns, we must take some trial dimensions for the magnetic circuit and will therefore try those
shown in Fig. 17, making the length of the cores about

three times their thickness and the distance between the centers of cores $7\frac{1}{2}$ inches. If these dimensions do not work out well, they can easily be modified later. The dotted line indicates the average path of the magnetic flux.

We will make the yoke 3 inches by 5 inches in order to bring down the density in it and thereby lessen the required magnetizing power. The density in the yoke will be $\frac{1,040,000}{15} = 70,000$ lines per square inch, approximately.

The ampere-turns per inch length of the magnet cores in which the density is 92,400 will be about 80, as can be seen by referring to the magnetization curves. The length of the path in each core is $6\frac{1}{2}$ inches, so that the ampere-turns required for the cores are $2 \times 6\frac{1}{2} \times 80 = 1,040$.

The ampere-turns per inch length of the path in the yoke will, for a density of 70,000, be about 20. The length of the path in the yoke will be $5 + \frac{1}{2}(3.1416 \times 2\frac{1}{2}) = 8.93$ inches, or, say, 9 inches. The length of path through the armature will be about the same, so that the ampere-turns required for yoke and armature may be taken as $2 \times 9 \times 20 = 360$.

The air gap between armature and magnet remains to be accounted for. If there were a perfect fit between the two, this air gap might be almost neglected, but in actual operation there is always a certain amount of dirt or scale or else the surface is more or less uneven. In order to allow for these, we will calculate the magnet as if there were an air gap of $\frac{1}{16}$ inch between pole pieces and armature. The total length of air gap would then be $\frac{1}{8}$ inch and the density at the poles is to be 100,000. The ampere-turns for the air gap are then $\frac{1}{8} \times 100,000 \times .313 = 3,900$, approximately.

The total ampere-turns to be supplied by the two magnet coils are $1,040 + 360 + 3,900 = 5,300$.

53. Heating of Magnet Coils.—When operating commercial lifting magnets, the supply voltage is usually fixed and the most economical way is to design the winding so that the magnet can be connected directly across the circuit without any danger of overheating, and without the intervention of any resistance. In order that a coil may not reach a dangerously high temperature, it must be designed so that its outside surface can dissipate the heat. The temperature which a coil attains depends not only on the amount of energy dissipated in it, but also on the facility with which it can get rid of its heat to the surrounding air. Now, the amount of heat that a coil can dissipate depends on its exposed surface and the difference in temperature between the coil and the air. It is evident, then, that for a given rise in temperature, a coil cannot dissipate as much energy in a hot engine or boiler room as it could in a cooler location. The outside dimensions of the coil must, therefore, be fixed so that the energy can be dissipated without causing the coil to heat more than 50° to 75° F. above the temperature of the room. Since the coils of lifting magnets are usually wound rather deep, radiation is not easily effected, and, as a rule, from 2 to 5 square inches of outside coil surface must be allowed for each watt dissipated in the coil. If the magnet is used in a fairly cool location, from 2 to 3 square inches per watt will be sufficient, but if used in a hot engine room, rolling mill, or similar location, from 4 to 5 square inches should be allowed. For example, if a magnet required 200 watts for its operation, the outside surface of the coils should be from 800 to 1,000 square inches if the magnet is to be used in a very warm location, or from 400 to 600 square inches if used in a fairly cool place. The surface here meant is the outside surface of the coils not including the end surfaces. All the above allowances are made on the assumption that the magnet is to be capable of continuous operation. If it is used only intermittently, the radiating surface may be very much less for the same rise in temperature. When a magnet is used for intermittent work, it has time to get rid of the heat between the intervals when it is used. Under such circumstances an allowance of 1 square inch of radiating surface per watt should be sufficient to prevent overheating. If, however, there is any liability of the magnet being used continuously, it must be designed so as to give ample radiating surface; otherwise, the winding may be burned out or at least permanently injured.

54.   Calculation of Winding.—If the voltage on which the magnet is to operate is fixed, and the ampere-turns required are known, then the size of wire that must be used can at once be determined if the mean or average length of a turn is known.

Let $I T$ = total ampere-turns required;

$\quad I$ = amperes;

$\quad T$ = total number of turns;

$\quad R$ = total resistance of coils;

$\quad E$ = E. M. F. applied to coils, i. e., the line E. M. F., because no resistance is used in series with the magnet;

$\quad l$ = average length of 1 turn of wire expressed in feet;

circ. mils = cross-sectional area of the wire in circular mils.

Since the wire on the magnet will be hot under normal working conditions, we will take the resistance per mil foot as 12 ohms.

$$R = \frac{12\ Tl}{\text{circ. mils}} \qquad (6)$$

If l'' is the length of a turn in inches, then

$$R = \frac{Tl''}{\text{circ. mils}} \qquad (7)$$

. because Tl is the total length in feet of the wire on the magnet.

Also, from Ohm's law, we have

$$I = \frac{E}{R}$$

or $$I = \frac{E}{\dfrac{Tl''}{\text{circ. mils}}}$$

or $$I = \frac{E \times \text{circ. mils}}{Tl''}$$

and $$\text{circ. mils} = \frac{Tl''I}{E} \qquad (8)$$

We may change the fraction on the right to read $\dfrac{I\,T\,l''}{E}$.

Now, the product of I and T appearing in this fraction is known, because it is equal to the ampere-turns, although the values of I and T separately are not known; hence,

$$\text{circ. mils} = \frac{I\,T\,l''}{E}. \qquad (9)$$

Rule.—*The cross-section, in circular mils, of the wire required for a magnet designed to operate on a given voltage E is equal to the ampere-turns multiplied by the mean length of a turn, in inches, and divided by the voltage.*

If the coil were operated at a low temperature, the resistance of a mil foot of copper would be somewhat less than 12 ohms, but this is a good average value to use in making calculations.

Coming back to the example given above, we can use formula **9** to determine the size of the wire, but before doing so it will be necessary to assume a trial depth of the winding. We will try a depth of 1 inch, which is somewhat less than half the thickness of the core. The average length of a turn is as indicated in Fig. 17 and will be $2 \times 4\frac{1}{2} + 2 \times 2\frac{1}{4}$ $+ 3.1416 \times 1 = 17.1416$ inches, or, say, $17\frac{1}{4}$ inches, approximately. The two coils of the magnet are connected in series across the circuit; hence, $I\,T$ in formula **9** must be the ampere-turns supplied by both coils. We then have

$$\text{circ. mils} = \frac{5,300 \times 17\frac{1}{4}}{220} = 415$$

No. 24 B. & S. comes nearest to this, having 404 circular mils, so we will adopt this size. The diameter of No. 24 double cotton-covered magnet wire is about .028 inch over the insulation; hence, the number of wires per inch will be $\dfrac{1}{.028} = 35$, approximately, and the number that can be placed in each square inch of winding space $35^\bullet = 1,225$, or, say, 1,200, in order to allow a little for clearance.

55. So far we have taken no value for the current. We will assume that this magnet is to be used rather steadily and that an allowance of 2 square inches of coil surface per watt will be sufficient to get rid of the heat. The cores are $6\frac{1}{2}$ inches long, so that allowing $\frac{3}{4}$ inch for projection of the core beyond the spools and $\frac{1}{2}$ inch at each end for spool heads and insulation, the actual length of winding space would be $6\frac{1}{2} - (\frac{3}{4} + 2 \times \frac{1}{2}) = 4\frac{3}{4}$ inches. The outside perimeter of the coil will be $2 \times 4\frac{1}{2} + 2 \times 2\frac{1}{2} + 3.1416 \times 2 = 20.28$ inches. The outside surface of each coil is $20.28 \times 4\frac{3}{4} = 96.3$ square inches. The total surface of both coils is 192.6 square inches, so that the allowable number of watts dissipated will be 96.3 and the allowable current $\frac{96.3}{220} = .44$ ampere, nearly. With a current of .44 ampere, the number of turns required will be $\frac{5,300}{.44} = 12,000$, approximately, or 6,000 on each spool. The space occupied by 6,000 turns will be $\frac{6,000}{1,200} = 5$ square inches. The winding space allowed was $4\frac{3}{4}$ inches by 1 inch, thus giving 4.75 square inches of winding space, which is hardly sufficient to accommodate the calculated winding. It will therefore be necessary to make the winding space a trifle deeper or a little longer. The latter would be preferable because additional radiating surface would be obtained without adding to the depth of the winding, and the slight increase in length of the magnet cores would not increase the reluctance of the magnetic circuit appreciably.

56. It will be noticed that in making the foregoing calculations, two rather indefinite quantities had to be assumed, namely, the coefficient of magnetic leakage and the nature of the contact, the latter being taken as equivalent to a $\frac{1}{16}$-inch air gap. The calculations have been given not with the intention of assuming that the lifting power as shown by test will agree exactly with that calculated, but to indicate the general method of applying magnetic calculations to the design of an electromagnet and incidentally to point out why such magnets do not admit of exact calculation. In calculating such magnets, one must be prepared to expect results shown by the test to be considerably different from those calculated unless in the calculations allowances have been made that are based on experimental data obtained from magnets of similar type.

The only magnets that can be calculated with even a fair degree of accuracy are those in which there is a closed magnetic circuit or where the flux is confined to a well-defined path in which the density of the lines is known. The lifting power, for example, of a straight-bar electromagnet cannot be determined except from experimental data, because the reluctance and distribution of the lines in the return path through the air are not known. In fact, the reluctance and distribution are influenced quite largely by the shape of the pole pieces.

———

MAGNETS WITH MOVABLE ARMATURES

———

SHORT-RANGE MAGNETS

57. Electromagnets designed for attracting their armatures through a distance can be divided into two classes, namely, short- and long-range magnets. **Short-range magnets** are used in places where the armature is required to move rapidly through a short distance, exerting comparatively little force; as, for example, in telegraph apparatus, electric bells, arc lights, etc. Such magnets are usually of the horseshoe type, as shown in Fig. 18, which represents an electromagnet

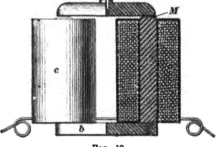

FIG. 18

for a telegraph relay. In this particular magnet the cores are made of two round bars of soft iron M, $\frac{3}{8}$ inch in diameter and 2 inches long. The cores are screwed into a yoke of soft iron b, $\frac{3}{8}$ inch wide by $\frac{1}{4}$ inch thick and 2 inches long. The magnetizing coils are wound over vulcanized rubber bobbins or spools, and contain, all told, about 8,500 convolutions, or turns, of insulated copper wire .009 inch in diameter. The total resistance of the wire in the two magnetizing coils is about 150 ohms. A vulcanized rubber shell or cover *c* is slipped over each coil when wound to protect it from dust and bruises.

58. Fig. 19 represents another form of magnet used for rapid vibrations of the armature. The cheapness of winding

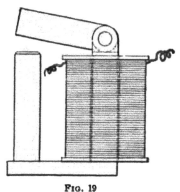

FIG. 19

only one coil instead of two and its simplicity of construction recommend it for a large variety of practical uses. The principal disadvantage is the large amount of magnetic leakage caused by an unbalanced magnetic field.

The magnitude of the force which short-range electromagnets are usually required to exert is comparatively small; in most cases the armature moves only a fraction of an inch against the tension of a light helical spring; consequently, it is unnecessary to calculate the magnetic circuit and the force of attraction. The size and amount of wire to be used for the magnetizing coils depend on the local conditions, and the most satisfactory results are obtained by experimental trials in each particular case.

SOLENOIDS

59. For some kinds of work a magnet is required to exert a pull through a considerable range of movement. Generally speaking, the magnet is not an efficient device for exerting a strong pull through a long range, and in many cases it would be found much more economical to use an electric motor operating through suitable gearing. A long range of movement implies a long air gap in the magnetic circuit, and this in turn necessitates a very large magnetizing force. However, magnets are simple in construction and not likely to get out of order, hence they are used for many different lines of work, such as operating electrically controlled switches, automatic motor starters, semaphores, etc.

The ordinary horseshoe type of magnet is not well adapted for long-range work. If the armature is moved away from the poles, a large number of the lines of force leak between the pole pieces without passing through the armature at all, and the pull is enormously reduced, both on this account and because the introduction of the air gap into the magnetic circuit cuts down the magnetic flux. More-

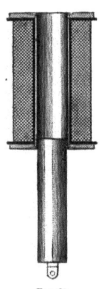

over, the ampere-turns required for a given pull through a given range cannot be calculated with any degree of exactness, because the leakage cannot easily be determined. This is also true in the case of the plain solenoid and plunger shown in Fig. 20. A solenoid of this kind will exert a pull through a considerable range, but the greater part of the magnetic circuit is through air, and the device is therefore very inefficient, giving but a very small pull considering the power required for its excitation. The maximum pull is obtained when the core is about half way in the coil, but the number of ampere-turns required for a given pull cannot be calculated with any degree of accuracy. The only safe way is

Fig. 20

to determine the ampere-turns experimentally by winding a temporary coil.

60. Fig. 21 shows a style of coil and plunger magnet or solenoid that is much superior to the simple solenoid when a fairly strong pull must be obtained through considerable range. This type is sometimes called a "stopped" magnet or solenoid. The coils cc surround the cores, as shown, and core a is free to move. The air gaps are within the coils where the lines of force are kept together.

Fig. 22 shows another type of long-range magnet in which part of the magnetic circuit is outside of the magnet through the air.   This introduces magnetic leakage, as indicated by the dotted lines, and makes this form undesirable.

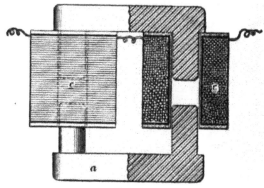

FIG. 21

The style of solenoid that has been used perhaps more than any other for long-range work is shown in Fig. 23. If desired, the yoke Y can be made cylindrical, so as to completely cover the coil; this construction is desirable where the magnet is exposed to moisture or mechanical injury, but

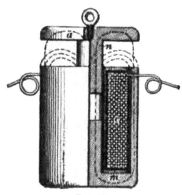

FIG. 22

otherwise it is better to leave the coil open, as shown, in order to promote ventilation. The fixed core A and the movable core B are of wrought iron. If the cross-section of the yoke be made about the same as that of the core, the density in the yoke will be about one-half that in the core, because the flux divides, half of it flowing through each side of the yoke.

A very small clearance is all that is necessary at b, and this can be provided by means of a thin brass bushing that serves also as a guide for the plunger. The advantages of this type of magnet are that it has only one air gap in the magnetic circuit; with the exception of the gap necessitated by the movement of the plunger, and the very small gap at *b*, the path of the magnetic flux is entirely through iron. Sometimes the end of the upper core is coned and the end of the lower core hollowed out to correspond; this shaping of the ends makes the pull somewhat more uniform throughout the stroke. If the solenoid is to draw its core in to the extreme limit, it is advisable to cover

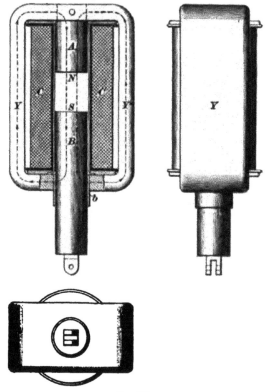

Fig. 23

one of the poles with a thin sheet of brass, in order to prevent sticking when the current is cut off.

61. A solenoid depends for its action on the principle that a piece of iron placed in a magnetic field tends to move so that the reluctance of the magnetic path will be made as small as possible; that is, the iron moves so as to increase the flux that passes through it. If the iron were so situated that its movement would not decrease the reluctance and thus increase the number of lines, there would be no force on the piece whatever. Consider the solenoid in Fig. 23; with the core in the position shown and with a given number of ampere-turns supplied by the magnetizing coil, there will be a certain magnetic flux between the ends of the core. Now, suppose the core moves in a short distance; the air gap in the magnetic circuit will be shortened and the flux increased. The average pull exerted on the core is proportional to this increase in magnetic flux, and the greater the increase in flux, the greater will be the pull. The pull cannot be calculated by the formulas previously given for lifting magnets, as they apply only when the two surfaces are in contact and do not give even approximate results when applied to a solenoid where there may be an air gap of several inches. The pull varies with the position of the core and increases quite rapidly as the air gap becomes smaller. The pull in any position depends on the ampere-turns and the *flux gradient* or the variation of the flux corresponding to unit movement of the plunger; for example, if a movement of $\frac{1}{4}$ inch caused an increase in the flux of 20,000 lines, the flux gradient would be 80,000 lines per inch, or 960,000 lines per foot. The pull exerted may be expressed as follows:

$$\text{pull (pounds)} = \frac{738}{10^{11}} \times \text{ampere-turns} \times \text{flux gradient} \qquad \textbf{(10)}$$

In this formula,* the flux gradient must be expressed in lines of force per linear foot.

This formula does not enable a solenoid to be designed for a given pull over a given distance, but it is very useful in checking the results that may be expected from a solenoid of certain dimensions and has been found to give results that agree with those obtained from solenoids actually constructed. The design of a solenoid is largely a cut-and-try operation; by assuming a number of dimensions and calculating the corresponding pull, an approximate design can

---

*Wm. A. Del Mar, American Electrician, Vol. XVI, No. 4.

soon be obtained.   Solenoids usually have a large air gap in the magnetic circuit at the beginning of the stroke, and they are, as a rule, required to give a certain pull at the beginning of the stroke.   Hence, in making rough, preliminary calculations in order to form some idea as to the probable dimensions of a solenoid for a given service, the reluctance of the iron parts of the magnetic circuit can be neglected in comparison with the reluctance of the large air gap.   If this is done, the magnetic gradient can be easily calculated, because with a fixed number of ampere-turns, a decrease in the length of the air gap is accompanied by a proportional increase in the flux.   The flux gradient in lines of force per linear foot can then be calculated by the formula

$$\text{flux gradient} = \frac{38.3 \, I T A}{D^2} \qquad (11)$$

where $I T$ = ampere-turns;

A = area of cross-section of gap (in square inches);

D = length of gap (in inches).

Substituting this value for the flux gradient in formula **10**, we have

$$\text{pull (pounds)} = \frac{28.3 \, (I T)^2 A}{10^2 \, D^2} \qquad (12)$$

It must be remembered that this formula neglects the reluctance of the iron part of the magnetic circuit, but it gives the approximate relation between the pull and the other quantities entering into the problem if the air gap reluctance is very large compared with that of the iron part of the magnetic circuit.   It is also assumed that all the flux passes through the space between the ends of the cores; this is practically the case with a solenoid like that shown in Fig. 23.

The application of formula **12** will be illustrated by the following example: Required a solenoid similar to that shown in Fig. 23 for a 3-inch stroke and a pull of 50 pounds at the beginning of the stroke; the solenoid to be wound for 110 volts.

It is first necessary to assume values for the area of the core, and the ampere-turns; the resulting pull is then obtained from formula **12**. The air gap in this solenoid is long and the pull is quite large; hence, it is plain that the core must have considerable area.   Let us assume a core $2\frac{1}{2}$ inches in diameter = 4.9 square inches, area, and a coil giving 15,000 ampere-turns as the first trial. D = length of air gap = 3 inches; hence,

$$\text{pull} = \frac{28.3 \times 15,000 \times 15,000 \times 4.9}{10^9 \times 3 \times 3} = 34.7 \text{ pounds, approximately}$$

This is not large enough, so we must increase either the diameter of the plunger, the number of ampere-turns, or both. Try a plunger $2\frac{3}{4}$ inches in diameter = 5.9 square inches, area, and use 16,000 ampere-turns on the coil. Then,

$$\text{pull} = \frac{28.3 \times 16,000 \times 16,000 \times 5.9}{10^9 \times 3 \times 3} = 47.5 \text{ pounds}$$

For a third trial we will leave the area of the core as in the last case and increase the ampere-turns to 17,000; then,

$$\text{pull} = \frac{28.3 \times 17,000 \times 17,000 \times 5.9}{10^9 \times 3 \times 3} = 53.6 \text{ pounds}$$

This is close enough to use as a further basis for the design.

With an air gap of 3 inches and 17,000 ampere-turns, the magnetic density in the gap will, since $IT = \mathbf{H} \, l \times .313$ and $\mathbf{B} = \mathbf{H}$ for air, be

$$\mathbf{B} = \frac{IT}{l \times .313} = \frac{17,000}{3 \times .313} = 18,100 \text{ lines per square inch, and the total}$$

flux equals $18,100 \times 5.9 = 106,800$ lines, approximately.

We will make the iron yoke of ample cross-section, say 6 square inches or 4 inches by $1\frac{1}{2}$ inches. Half of the flux passes through each side of the cast-iron yoke, as indicated in Fig. 23, so that the density will be $\dfrac{53,400}{6} = 8,900$ lines per square inch; about 18 ampere-turns per inch will be required to set up this density. Of course, the density will increase as the core moves in. About 9 ampere-turns per inch will be sufficient to set up the density of 18,100 lines per square inch in the wrought-iron cores.

If the central core is $2\frac{3}{4}$ inches in diameter and a trial winding space $1\frac{1}{2}$ inches deep be taken, the mean diameter of the coil will be about $4\frac{3}{4}$ inches, allowing for clearance and insulation, and the length of an average turn will be 14.9 inches. We have then,

$$\text{circ. mils} = \frac{I T \, l''}{E} = \frac{17,000 \times 14.9}{110} = 2,302$$

No. 18 B. W. G. has 2,401 circular mils area, so that this size will be used. In order to determine the number of turns, we must assume a value for the current. This is determined by the heating effect, and as the dimensions of the coil are not known, the radiating surface cannot be calculated. However, an allowance of 1,200 to 1,500 circular mils per ampere should be safe for a coil of this kind, assuming that it is exposed as shown in Fig. 23; if the coil were completely covered by the iron yoke, a large number of circular mils per ampere would be necessary. Allowing 1,200 circular mils per ampere fixes the current at 2 amperes, and the resistance must, therefore, be 55 ohms. The number of turns to give this resistance can be obtained from formula 7, transposed to read,

$$T = \frac{R \times \text{circ. mils}}{l''} = \frac{55 \times 2,401}{14.9} = 8,860 \text{ turns, approximately}$$

and ampere-turns = 8,860 × 2 = 17,720. The difference between this number of ampere-turns and the 17,000 for which the winding was originally calculated is due to the fact that a wire of exactly the calculated size could not be selected from the wire table. No. 18 B. W. G. single cotton-covered wire has a diameter over the insulation of .054 inch, and about 324 turns can be placed in each square inch of winding space; 8,860 turns would, therefore, require 27.4 square inches. Since the winding space was assumed 1½ inches deep, the required length of space would be 18.26 inches, say 18¼ inches. Allowing ⅜ inch at each end for the spool heads makes the distance between the inside of the iron yokes 19 inches. The length of the magnetic path through the iron yoke will be about 25 inches and through the wrought-iron cores about 16 inches, with the core at the beginning of the stroke. The ampere-turns for the yoke will, therefore, be 25 × 18 = 450, and for the cores 16 × 9 = 144, or 594 for the whole of the iron part; this is practically negligible when compared with the 17,000 ampere-turns allowed for the 3-inch air gap. There will be a small air gap when the plunger slides through the yoke. The yoke is 1½ inches thick, and the plunger 2¼ inches in diameter, or 8.6 inches in circumference, so that the area of the gap will be 8.6 × 1½ = 12.9 square inches. The density in the gap will be $\frac{106,800}{12.9}$ = 8,280 lines per square inch. The length will not be more than $\frac{1}{16}$ inch, so that the ampere-turns for this part of the circuit will be 8,280 × $\frac{1}{16}$ × .313 = 162. The total ampere-turns will, therefore, be 17,000 + 594 + 162 = 17,756. The coil as finally designed provides 17,720 ampere-turns, so that the solenoid should exert the required pull of 50 pounds.

62. In case a solenoid is to be designed for a short lift, formula **12** will not give close results, because the reluctance of the parts of the magnetic circuit, other than the air gap, will have a decided influence on the variation of the flux with the movement of the plunger. In such cases, formula **11** may be used to advantage, as follows: Adopt some trial dimensions for the magnetic circuit; also, assume a trial value for the flux, and from the known densities calculate the ampere-turns per inch for each part of the circuit with the plunger, say, in its lowest position. Assume that the coil provides this constant number of ampere-turns, and determine, by making a number of trial calculations, what the flux will be when the core is moved in a small distance, say ⅛ inch. The difference between the two fluxes is the increase in flux corresponding to a movement of ⅛ inch; 96 times this amount will give the flux gradient per foot to be used in the formula, and the resulting pull can be calculated. By making two or three sets of calculations, using different dimensions, the main features of the design can be decided on.

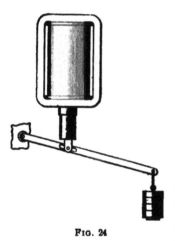

Fig. 24

63. From formula **12,** it is seen that with a given number of ampere-turns and a given size of core the pull is inversely proportional to the square of the length of the air gap. In other words, if the magnetizing force and size of core are left the same, halving the air gap will make the pull four times as great. This, of course, is not strictly true unless the reluctance of the iron part of the circuit is negligible. However, it shows that when a pull is required over a wide range it is advantageous to employ a lever, as indicated in Fig. 24, to reduce the stroke of the solenoid and work with a strong pull over a short range.

64. Sparking Caused by Magnets.—Very often it is found that when the current supplied to a magnet is turned off there is severe sparking at the switch, and in some cases the insulation of the magnet is punctured. When the current flowing through a magnet coil is suddenly interrupted, a high E. M. F. is induced by reason of the large self-induction. As the magnetic field threading through the coil collapses, an E. M. F. is set up in the turns, and since this

E. M. F. is very much higher than that of the line, there is
danger of its puncturing the coil insulation, or, in any event,
it makes a long arc at the switch, thus burning the contacts.   Either of two methods may be used for
obviating these effects.   The first is to wind a
coil of wire of a comparatively small number
of turns around the magnet and fasten the ends
of the coil together, thus making it form a
closed circuit.   When the current in the main
coil is broken, a heavy induced current is set
up in this auxiliary secondary coil and prevents
the production of the high primary E. M. F.
Another method, and one that is more widely
used, is to connect incandescent lamps, as
shown in Fig. 25.   When the switch is opened,
the induced E. M. F. sets up a current around
through the lamp, thus dissipating the energy
stored in the magnetic field without arcing at
the switch and without danger to the insulation.
If it is found that one lamp is not sufficient to stand the
induced voltage, two or more may be connected in series.

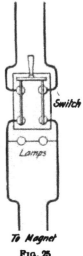

Line

Switch

Lamps

To Magnet

FIG. 25

A SERIES OF QUESTIONS
AND EXAMPLES

RELATING TO THE SUBJECTS
TREATED OF IN THIS VOLUME

It will be noticed that the Examination Questions that follow have been divided into sections, which have been given the same numbers as the Instruction Papers to which they refer. No attempt should be made to answer any of the questions or to solve any of the examples until that portion of the text having the same section number as the section in which the questions or examples occur has been carefully studied.

ELECTRIC POWER STATIONS
(PART 1)

EXAMINATION QUESTIONS

(1) Give a brief description of the Cochrane feedwater heater.

(2) (*a*) What is one of the chief advantages of the use of condensed steam for boiler-feed purposes? (*b*) What is the principal disadvantage?

(3) (*a*) What is the Porter-Clark process of treating feedwater? (*b*) Give a brief description of the Kennicott water softener.

(4) How can the load line representing the daily output of a proposed station be predetermined with a fair degree of accuracy?

(5) What effect has a small amount of kerosene oil introduced into the feedwater supplied to a boiler?

(6) What area of tube surface should be provided in a feedwater heater for a 250-horsepower engine if the steam is supplied at a temperature of 215° and the water heated from 65° to 195°? Brass tubes are to be used in the heater.

Ans. 60.76 sq. ft.

(7) Name some of the substances that commonly exist as impurities in feedwater.

(8) (*a*) What is meant by the efficiency of a boiler? (*b*) What efficiency is usually attained in properly proportioned boilers with clean heating surfaces?

§ 28

(9) (*a*) What is a so-called fuel economizer? (*b*) Give a brief description of the Green economizer.

(10) Point out the difference between open and closed feedwater heaters.

(11) How should doors used in power stations be constructed and hung?

(12) State the conditions that should be met as far as possible by the location selected for an electric power station.

(13) How should the foundations for the generating units in a station be constructed with regard to the walls of the building and to each other?

(14) (*a*) Why should advantage be taken of all sources of waste heat to increase the temperature of the water supplied to boilers? (*b*) What are the principal sources of waste heat in connection with a steam-power plant?

(15) When a foundation is to rest on soil of a yielding nature and where piling is unnecessary or impracticable, how should the foundation be arranged in order to secure uniform settlement?

(16) State the various boiler-feeding appliances in the order of their economy.

(17) (*a*) How may feedwater be purified by means of live steam? (*b*) Describe, briefly, the Hoppes purifier.

(18) What load will a pile driven in firm soil support if it is driven by a 1,500-pound hammer dropping 25 feet and the driving continued until the last blow moves the pile .5 inch? **Ans. 25 tons**

ELECTRIC POWER STATIONS

(PART 2)

EXAMINATION QUESTIONS

(1) (*a*) What two classes of steam boilers are suitable for electric power stations? (*b*) Name at least three kinds belonging to each class.

(2) If a boiler has a surface ratio of 50 and if 12 pounds of coal is burned per square foot of grate per hour, about how many pounds of feedwater, supplied at a temperature of 150° F., can the boiler convert into steam at 100 pounds pressure, per hour, per square foot of grate surface?

Ans. 118 lb.

(3) (*a*) Name two or three methods of producing forced draft. (*b*) What are some of the advantages and disadvantages of forced draft?

(4) A chimney 140 feet high is to carry off the gases of combustion from a battery of boilers aggregating 400 horsepower; what should the inside diameter of the chimney be?

Ans. 3 ft. 11 in.

(5) (*a*) Into what two general classes may mechanical stokers be divided? (*b*) Point out the difference between the two classes.

(6) What is meant by the surface ratio of a boiler?

(7) What is the difference between a forced draft and an induced draft?

(8) What are the least heights of chimney suitable for burning with natural draft: (*a*) bituminous slack? (*b*) anthracite pea? (*c*) anthracite buckwheat?

§ 29

(9)   About how many pounds of air are required, theoretically, for the combustion of 1 pound of: (*a*) anthracite coal? (*b*) bituminous coal? (*c*) dry wood? (*d*) petroleum?

(10)   A plant is to be equipped with engines of the Corliss compound condensing type: (*a*) about how many square feet of boiler heating surface should be allowed per horsepower? (*b*) How many square feet per horsepower should be allowed if automatic cut-off medium-speed engines are used?

(11)   Why are round chimneys better than square or octagonal chimneys?

(12)   (*a*) How is draft pressure usually measured? (*b*) In what two ways is the draft pressure expended?

(13)   What are some of the advantages claimed for mechanical stokers?

(14)   What points should be considered in the selection of boilers for a power plant?

(15)   How many pounds of air will be carried, per hour, by a chimney 180 feet high and having an effective flue area of 16 square feet?   The temperature of the gases is 300° F. above the temperature of the outside air at 60° F.

Ans. 209,196 lb.

(16)   (*a*) About how many pounds of anthracite can be burned per square foot of grate surface? (*b*) How many pounds of bituminous coal?

(17)   (*a*) Name two methods of producing induced draft. (*b*) What are some of the advantages and disadvantages of induced draft?

(18)   If a stack is capable of discharging 15,000 pounds of air per hour when the difference between the temperature of the gases supplied to the stack and the outside atmosphere at 60° F. is 200°, what amount can it discharge for a difference in temperature of 400°?   Ans. 16,666 lb.

ELECTRIC POWER STATIONS
(PART 3)

EXAMINATION QUESTIONS

(1) Why does the use of compound engines reduce the loss from cylinder condensation?

(2) What are the advantages and disadvantages of the tandem compound engine as compared with the cross-compound?

(3) What are the main advantages gained by the use of compound engines, operated with condensers, over simple engines operated without condensers?

(4) What are the limits of piston speed for long-stroke and short-stroke engines?

(5) How should the efficiency of the engine be related to that of the generator, in order to secure the best economy?

(6) What steam pressures are commonly used with the various types of engines?

(7) Name four considerations that govern the type of engine to be selected for any given station.

(8) What is the allowable limit in the angular variation, during a single revolution, of an engine direct-connected to an alternator?

(9) (a) For what purpose is a condenser used? (b) Why is it impossible to maintain a perfect vacuum in a condenser?

(10) Why are steam turbines of the Parsons or Curtis types more easily adapted to the driving of alternators than direct-current dynamos?

(11) For what purpose is an air pump used in connection with a condenser?

(12) What approximate saving, in pounds of steam per horsepower-hour, can be effected by using compound engines with condensers over simple, high-speed, non-condensing engines?

(13) Point out the difference between a surface condenser and a jet condenser.

(14) About what relation should exist between the tube area of a surface condenser and the amount of exhaust steam to be condensed?

(15) (a) Explain the action of the barometric column or siphon condenser. (b) What are some of the advantages of this type of condenser?

(16) Why is the induction type of condenser not well adapted for stations where the load is of a variable character?

(17) If the temperature of the injection water supplied to a condenser is 50° F. and the temperature of the discharge 100° F., how many units weight of injection water will be required per unit weight of steam condensed?   Ans. 21.8.

(18) (a) For what purpose are cooling towers used? (b) On what factors does the efficiency of a cooling tower depend?

(19) Name three methods for obtaining the necessary air circulation through cooling towers.

(20) State some of the points to be observed in connection with the installation and operation of surface condensers.

ELECTRIC POWER STATIONS
(PART 4)

EXAMINATION QUESTIONS

(1) Give an example of a good method of attaching pipe flanges to pipes for pressures up to 165 pounds per square inch.

(2) (*a*) State some of the covering materials suitable for the live-steam piping of power stations. (*b*) State some of the materials that are not suitable.

(3) Calculate the number of pounds of steam per minute that can be delivered through 300 feet of 4-inch pipe with an initial pressure of 120 pounds, the loss of pressure to be limited to 2 pounds, $D = .304$.   Ans. 90.91 lb. per min.

(4) (*a*) What are the advantages of using receivers in connection with live-steam piping? (*b*) How is the capacity of the receiver determined?

(5) How many horsepower can be obtained, theoretically, from a stream giving a flow of 15,000 cubic feet per minute under a head of 40 feet?   Ans. 1,136 H. P.

(6) State the main points that should be observed when designing electric power station buildings.

(7) What will be the loss in head, expressed in feet, in a 40-inch pipe, 1,000 feet long, through which water is flowing at the rate of 8 feet per second? _   Ans. 5.92 ft.

(8) Why are duplicate systems of steam piping not usually installed in modern power plants?

(9)   What two kinds of waterwheel are generally used for electric power plants?

(10)   What is the most satisfactory kind of gasket for use with live-steam piping?

(11)   (*a*) What is meant by the term hunting or racing as applied to a waterwheel?   (*b*) How is the Lombard governor arranged so as to prevent hunting?

(12)   Why is it advisable to use brass pipe for feedwater connections where the expense is warranted?

(13)   (*a*) Explain the operation of a turbine waterwheel. (*b*) Into what classes are turbines usually divided, and point out the distinction between them?

(14)   What essential. features must be provided for in the building of dams for water-power plants?

(15)   Why is it advisable to use long pipe bends, rather than elbows, in steam-piping systems?

(16)   Under what conditions are impulse wheels generally used in preference to turbines?

(17)   State the important points that must be taken into consideration when determining the advisability of a water-power installation.

TELEGRAPH SYSTEMS
(PART 1)

EXAMINATION QUESTIONS

(1) (a) What is electric telegraphy? (b) What two kinds of signals may be used?

(2) Why is a relay used in a long line circuit instead of a sounder?

(3) (a) How many intermediate offices may be put on one line? (b) What is the objection to very many offices on one line?

(4) What precaution must be taken in regard to the switches C and C' in Fig. 1?

(5) (a) What telegraph codes are in common use? (b) Where is each code used?

(6) Draw a sketch showing how to connect sounders, relays, and batteries on a line having four offices and twelve cells, the cells being equally divided among the four offices.

(7) (a) What is the Morse closed-circuit system? (b) Why is it called a closed-circuit system?

(8) (a) In the closed-circuit system, can batteries be used at intermediate as well as at terminal stations? (b) If batteries are inserted in a line circuit at several offices, must the same number of cells be used at each office? (c) If batteries are used at terminal and intermediate offices, how must they be connected in the line circuit so as to assist one another?

§ 46

(9)   What is the Morse open-circuit system and why so named?

(10)   Why should all telegraph instruments connected in series in the same line be similar in construction and equal in resistance?

(11)   Why will lightning jump across a thin, or narrow, space filled with air, mica, or paraffined paper, in order to reach the ground, in preference to going along the regular path through the wire coils on the instruments?

(12)   Why are telegraph magnets designated by their resistance, as, for instance, a 20-ohm sounder, when the important feature is really the number of turns of wire?

(13)   (*a*) What is a telegraph repeater?   (*b*) Why are repeaters needed?

(14)   (*a*) What is an automatic repeater?   (*b*) Why is an operator needed to attend these repeaters?

(15)   (*a*) What is an artificial line?   (*b*) Why is it used in differential and polar duplex systems?

(16)   (*a*) What is duplex telegraphy?   (*b*) What is diplex telegraphy?

(17)   State four ways in which the two line circuits in the Wood repeater may be used and also the corresponding positions of the switches *M*, *g*.

(18)   (*a*) What is a polarized relay?   (*b*) What is the object to be kept in view in winding differential neutral, or polar, relays?

(19)   Why should the sending be heavy, or firm, on circuits containing repeaters?

(20)   (*a*) Are all circuits normally closed or open in a Milliken repeater?   (*b*) When only the key at the western station is opened, why do the eastern relay and the transmitter on the same side remain inactive, that is, closed?

(21) (*a*) On what principle does the bridge duplex system of telegraphy depend? (*b*) What parts in the system are arranged the same as for the differential duplex? (*c*) What instrument in the ordinary Wheatstone bridge has the same position as the relay of this system?

(22) How is the polar duplex balanced?

(23) What is a continuity-preserving pole changer?

(24) Explain, briefly, the fundamental principles in virtue of which two messages may be sent in the same direction over the same wire at the same time.

(25) Explain the action of repeating from east to west through a Wood button repeater.

TELEGRAPH SYSTEMS

(PART 2)

EXAMINATION QUESTIONS

(1) (*a*) On what changes in the line currents does the quadruplex system of telegraphy depend for its action? (*b*) How, in this system, does each key govern its own sounder without affecting the others?

(2) In the district telegraph service, what is meant by a return-call box?

(3) What is meant by the short-end and long-end batteries in quadruplex telegraphy?

(4) In the Jones quadruplex system, what is the use of the induction coil, and how is the production of false signals by the neutral relay avoided?

(5) In the battery quadruplex system, how is the current reversed in direction and how is it changed in strength?

(6) What positions of the four keys will cause no currents to flow in the line in the Jones quadruplex?

(7) (*a*) How are the sounders arranged in connection with the neutral relays in the battery quadruplex system? (*b*) Why are the sounders arranged in the way you describe?

(8) What kinds of wire are used in telephone and telegraph line work?

(9) (*a*) Describe two methods of joining telegraph or telephone wires. (*b*) Which is the better for hard-drawn copper wire? (*c*) Why?

§ 47

(10) (*a*) What are the terms by which the various grades of iron wire are distinguished? (*b*) Which is the best grade for ordinary line wires and why?

(11) (*a*) Into what two classes may paper cables be divided? (*b*) What advantages has each?

(12) What advantages has copper over iron for line wires?

(13) (*a*) What methods are available for reducing the electrostatic capacity of the conductors in cables? (*b*) Which of these methods is most effective and why?

(14) What is an artificial cable?

(15) (*a*) What is a coherer, and how does the ordinary coherer show the presence of Hertzian waves? (*b*) What is a serious objection or defect of most wireless telegraph systems?

(16) (*a*) In the Muirhead cable duplex system, what are used in the arms of the bridge in place of or in addition to resistances? (*b*) What is connected in series with the receiving instrument?

(17) What is meant when it is stated that the receiving station of a wireless-telegraph system is in tune, resonance, or syntony with a particular transmitting station?

(18) What is the use of the petticoat of an insulator?

(19) (*a*) In submarine telegraph cable systems, how is the transmitting key arranged? (*b*) Explain the principle on which the receiving instrument works. (*c*) What constitutes the dots and dashes with this instrument?

(20) How are the call boxes associated with any one circuit usually connected in district-messenger telegraph systems?

TELEPHONE SYSTEMS

(PART 1)

EXAMINATION QUESTIONS

(1) (*a*) What is meant by the packing of a telephone transmitter? (*b*) How may it be remedied?

(2) (*a*) What is a grounded circuit? (*b*) What is a complete metallic circuit? (*c*) What are the advantages of the latter over the former?

(3) (*a*) What is cross-talk? (*b*) How is it caused in line wires?

(4) What are the three most common causes of trouble in telephone instruments and their line circuits?

(5) Explain the action of a battery transmitter.

(6) What is a party line?

(7) Name and describe, briefly, the two kinds of party lines in common use.

(8) (*a*) What is an automatic shunt for a series magneto-generator? (*b*) Why is an automatic shunt used around a series magneto-generator?

(9) How does the action of a battery transmitter differ from that of a magneto-transmitter?

(10) (*a*) How is sound produced? (*b*) Explain how it is transmitted through the air.

(11) (*a*) Why does the diaphragm of a receiver vibrate when an alternating current traverses the receiver coil? (*b*) If a diaphragm of a receiver is made to vibrate by an

external force, such as vibrations in the air, why are currents produced in the receiver coil if the circuit is closed but no battery used?

(12) What problems must be met in telephone-line construction that do not occur in ordinary line work?

(13) (*a*) Name the characteristics of sound. (*b*) Define each. (*c*) Give examples of various differences in familiar sounds.

(14) If a microphone transmitter and battery are connected in a local circuit with the primary winding of an induction coil and a receiver in series with the secondary winding, what kind of currents are produced: (*a*) in the primary? (*b*) in the secondary circuits when the transmitter is operated? (*c*) why?

(15) (*a*) What are the sounds made by the vocal organs in producing speech termed? (*b*) Are these sounds produced by simple or complex waves?

(16) If a non-inductive, but high-resistance, coil is connected in parallel with an inductively wound but relatively low-resistance coil: (*a*) which coil will offer the least opposition to a direct steady current? (*b*) which coil to a high-frequency alternating current?

(17) (*a*) What source of induction is the most troublesome in telephone lines, electromagnetic or electrostatic? (*b*) How may disturbances due to induction on line circuits be eliminated?

(18) What are some of the mechanical defects to be guarded against in telephone receivers?

(19) To what, besides induction, may the cross-talk produced in a telephone cable be due?

(20) What is a selective signaling party-line system?

TELEPHONE SYSTEMS
(PART 2)

EXAMINATION QUESTIONS

(1) Explain the action of a telephone-switchboard drop.

(2) What is a clearing-out drop?

(3) Show, by a diagram, how a line drop may be associated with a spring jack.

(4) What is a listening key and for what purpose is it used?

(5) (*a*) What is a supervisory signal? (*b*) What kind of signals are used for this purpose?

(6) By what means may calls be received in telephone exchanges?

(7) Show, by a diagram, how a lamp may be used as a line signal.

(8) Under what conditions does a line test busy in the central-energy multiple switchboard described in this Section?

(9) What is a line cut-off relay?

(10) (*a*) What is a house system? (*b*) By what other names is it frequently called?

(11) (*a*) What is a telephone exchange? (*b*) What two kinds of switchboards may be used for large exchange systems?

(12) What are the general requirements of switchboard drops?

(13)   (*a*) For what purpose is the night alarm used in telephone exchanges?   (*b*) Describe its operation.

(14)   Mention three remedies that may be applied to prevent cross-talk between switchboard drops.

(15)   (*a*) What advantage does the multiple switchboard have over all forms of switchboard systems?   (*b*) What disadvantage has it?

(16)   What is the one feature that distinguishes multiple switchboards from all other forms?

(17)   Why is some form of busy test necessary in multiple switchboards?

(18)   What is meant by the statement that signaling from a subscriber to the central office is done automatically?

(19)   What is a line-pilot lamp and for what is it used?

(20)   (*a*) What is a central-energy telephone system? (*b*) By what other name is it known?

APPLIED ELECTRICITY

EXAMINATION QUESTIONS

(1) What type of dynamo is generally used for electroplating?

(2) A rheostat is to be made up of tinned-iron wire wound into spirals and mounted on an iron frame. The box is to have a resistance of 120 ohms and be capable of carrying a maximum current of 1.3 amperes. Give the size of wire, amount, and arrangement that you would consider suitable for a rheostat of this capacity.

(3) Explain the Thomson electric welding process.

(4) How may incandescent lamps be connected so as to take up the inductive discharge when a magnet is switched off and thus prevent arcing at the switch and strains on the magnet insulation?

(5) Why is alternating current employed for electric welding in preference to direct current?

(6) Give the composition of a solution suitable for silver plating.

(7) (a) Why is it difficult to design efficient electromagnets or solenoids to give a strong pull through a wide range of movement? (b) If a solenoid has to overcome a resistance over a long range, why is it advisable to design the solenoid for a strong pull through a short range and then increase the movement, with corresponding reduction in the force exerted, by means of levers?

§ 50

(8) How many watts will it require to raise the temperature of a room 8 ft. × 10 ft. × 9 ft. from 50° F. to 70° F. in 20 minutes, assuming that none of the heat escapes?

Ans. 216 watts

(9) A lifting magnet requires 7,500 ampere-turns for its excitation. The mean length of a turn is 26 inches. What size of wire should be used for the winding, assuming that the magnet is to be connected across 250 volts?

Ans. No. 21 B. & S.

(10) Give the composition of a cyanide copper-plating solution.

(11) How many amperes at 250 volts must be supplied in order to heat 3 gallons of water from 40° F. to 200° F. in $\frac{1}{2}$ hour, assuming that all the heat supplied passes into the water? Ans. 9.4, nearly

(12) If a two-pole electromagnet has an area of 25 square inches at each pole, and if the density at the pole face is 95,000 lines per square inch, how many pounds tractive force will the magnet exert? Ans. 6,255.5 lb.

(13) In electroplating, to which pole of the battery or dynamo are the objects to be plated always connected?

(14) A two-pole electromagnet is to be designed for a tractive force of 1,500 pounds. If the pole-face density is 105,000, what should be the area of each pole face?

Ans. 4.91 sq. in.

(15) A wire is strung in still air and is maintained at a temperature of 50° C. above that of the surrounding air. How many watts will it take to supply the loss by convection from this wire if the wire is $\frac{1}{4}$ mile in length?

Ans. 3,498 watts

(16) On what quantities does the pull exerted on the core of a solenoid, at any point in its stroke, depend?

A KEY

The Keys that follow have been divided into sections corresponding to the Examination Questions to which they refer, and have been given corresponding section numbers. The answers and solutions have been numbered to correspond with the questions. When the answer to a question involves a repetition of statements given in the Instruction Paper, the reader has been referred to a numbered article, the reading of which will enable him to answer the question himself.

To be of the greatest benefit, the Keys should be used sparingly. They should be used much in the same manner as a pupil would go to a teacher for instruction with regard to answering some example he was unable to solve. If used in this manner, the Keys will be of great help and assistance to the student, and will be a source of encouragement to him in studying the various papers composing the course.

ELECTRIC POWER STATIONS
(PART 1)

---

(1)   Give a brief description by referring to Fig. 15.

(2)   (a) The water resulting from the condensed steam contains no vegetable or mineral impurities.

(b)   It contains cylinder oil carried over from the engine with the exhaust steam.   This oil must be removed before the water is suitable for boiler-feed purposes.   See Art. **78.**

(3)   (a) Give a short description of the process obtained from Art. **60.**

(b)   See Arts. **59** and **60.**

(4)   Give a short description of the method of obtaining the probable load line as given in Arts. **13** and **14.**

(5)   It loosens the boiler scale and prevents it from adhering firmly to the tubes.   See Art. **63.**

(6)   From formula **2**, Art. **71,** $S = .248 \log \dfrac{T_s - T_i}{T_s - T_a}$; in this case $T_s = 215°$, $T_i = 65°$, $T_a = 195°$; hence, $S = .248 \log \dfrac{215 - 65}{215 - 195} = .217$ sq. ft. per H. P. if copper tubes are used.   In this case the tubes are to be of brass; hence, the area of the tube surface per horsepower will be $.217 \times 1.12$, and for 250 H. P. $.217 \times 1.12 \times 250 = 60.76$ sq. ft.   Ans.

(7)   Carbonate of lime, sulphate of lime, chloride of calcium, calcium nitrate, magnesium carbonate, magnesium chloride, magnesium sulphate, sodium sulphate, sodium carbonate, iron, silica, carbonic acid, and suspended matter.   See Arts. **42** to **55.**

(8)   (a) The ratio of the heat utilized in evaporating the water to the heat supplied by the combustion of the fuel.

(b)   From 75 to 80 per cent.   See Art. **7.**

(9)   (a) Fuel economizers are devices designed to abstract the heat from the gases of combustion on their way from the boiler to the stack

and impart this heat to the boiler feedwater before it enters the boiler. See Art. **73**.

(*b*)   Give a brief description by referring to Fig. 19.

(10)   In open heaters, the feedwater is in contact with the atmosphere; in closed heaters, the feedwater is not in contact with the atmosphere. See Art. **66**.

(11)   Give a brief description of the Underwriters' style of door, as described in Art. **34**.

(12)   Give a short statement of the conditions as laid down in Art. **3**.

(13)   The foundations should be absolutely independent of the walls of the building; and where foundations are to be built within a short distance of each other for several units, these foundations should be bonded together so as to form one solid structure. See Art. **24**.

(14)   (*a*) Because all heat supplied to the water from waste sources represents so much coal saved. The waste heat costs as much in proportion as the useful heat and the most effective means of utilizing the waste heat is to impart it, as far as possible, to the boiler feedwater. See Art. **64**.

(*b*)   Exhaust steam from the engines, exhaust steam from pumps or other steam-driven auxiliary appliances, and heat carried up the stack with the gases of combustion. See Art. **65**.

(15)   The foundation should be supported on a grillage made of iron beams supported on concrete, which serves to distribute the weight over a large area. See Art. **31**.

(16)   Give a list of the methods, as contained in Art. **79**.

(17)   (*a*) By heating the feedwater to a high temperature by means of live steam, the scale-forming impurities are precipitated from the water before it is introduced into the boiler. The scale-forming substances are precipitated because they are unable to remain in solution when the water is raised above a certain temperature. See Art. **62**.

(*b*)   See Art. **63**.

(18)   From formula 1, $L = \dfrac{2\,WH}{p+1}$. In this case $H = 25$, $W = \dfrac{1,500}{2,000} = .75$, $p = .5$; hence,

$$L = \frac{2 \times .75 \times 25}{.5 + 1} = 25 \text{ tons.}$$

ELECTRIC POWER STATIONS
(PART 2)

(1) (*a*) Fire-tube and water tube.

(*b*) The return tubular boiler, the vertical tubular boiler, and the Sederholm boiler are examples of the fire-tube class. The Babcock and Wilcox, the Heine, the Stirling, and the Climax are examples of the water-tube type. See Art. **30.**

(2) This question can be answered by referring to Table V. In the column at the left, giving coal burned per square foot of grate per hour, find the number 12 and opposite this find the number 118 in the body of the table, in the column headed 50, which is the surface ratio in this case. The boiler can therefore convert 118 lb. of water into steam per hour for each square foot of grate surface on which 12 lb. of fuel is burned per hour, assuming that the water is supplied at a temperature of 150° F. and converted into steam at 100 lb. pressure.

(3) (*a*) By blowing air under the grate by means of fans; by using steam blowers, such as the McClave Argand blower; and by use of steam jets under the grates. See Arts. **20, 21,** and **22.**

(*b*) Forced draft may be used with boilers already installed in order to increase their evaporative capacity and also make it possible to burn a cheaper grade of fuel than was originally intended. The chief disadvantages are that ashes accumulate rapidly on the heating surfaces; there is a tendency for the fires to burn unevenly; soot and ashes are liable to blow out whenever the fire-doors are opened, unless the draft is first shut off; and ashes cannot be conveniently removed without first shutting off the draft. There is also the power required to maintain the draft to be considered. See Art **24.**

(4) This problem is solved by means of the rule given in Art. **12,** from which the effective area $= \dfrac{400}{3.33 \times \sqrt{140}} = \dfrac{400}{3.33 \times 11.83} = 10.15$ sq. ft. The relation between the diameter and area is $.7854 \, d^2 =$ area; hence, $.7854 \, d^2 = 10.15$, or $d = \sqrt{\dfrac{10.15}{.7854}} = 3.6$ ft. nearly, or, say,

3 ft. 7 in. To this must be added 4 in. to allow for friction, thus making the actual diameter 3 ft. 11 in. Ans.

(5) (*a*) Overfeed stokers and underfeed stokers.

(*b*) The difference between the two classes is indicated by their names. In the overfeed stokers the coal is fed on top of the coal already on the grate, while in the underfeed stokers it is fed into the mass of burning coal from the underside. Compare descriptions of Roney stoker, Art. **52**, and American stoker, Art. **56**.

(6) The surface ratio of a boiler is the ratio of the heating area to grate area, or it is the number of square feet of heating surface per square foot of grate surface. See Art. **42**.

(7) In the forced draft, air is introduced under the grate bars and forced through the fire by means of blowers or other appliances. In the induced draft, fans or other devices are introduced in the flues beyond the boiler so as to create a suction and draw air through the coal. See Arts. **19, 20, 21,** and **25**.

(8)   (*a*) 100 ft.
     (*b*) 130 ft.
     (*c*) 150 ft.   See Art. **10**.

(9)   (*a*) 12.13.
     (*b*) 12.06.
     (*c*) 7.68.
     (*d*) 15.65.   See Table I.

(10)   (*a*) From $7\frac{1}{2}$ to 8 sq. ft.
      (*b*) From $11\frac{1}{2}$ to 12 sq. ft.   See Art. **40**.

(11) Because the round chimney offers least resistance to wind pressure. Also a round flue offers less resistance to the passage of the gases than a square flue of equal area. See Art. **14**.

(12) (*a*) By means of a draft gauge, which consists of a U-shaped tube containing water. One side of the tube is connected with the flue and the other is open to the atmosphere. The draft pressure is expressed in inches of water as indicated by the difference in the levels of the columns of water in the two legs of the tube.

(*b*) It is expended in imparting the velocity to the air and in overcoming the frictional resistances offered by the grate, fuel, flue passages, and chimney. See Art. **6**.

(13) Give a statement of the advantages as enumerated in Art. **58**.

(14) Give a statement of the considerations as enumerated in Art. **38**.

(15) This problem is solved by means of the rule given in the latter part of Art. **13.** Referring to curve A, Fig. 2, the ordinate corresponding to a temperature of 300° F. is .975. The effective flue area is 16 sq. ft. and the height of the chimney 180 ft., hence applying the rule, we have

lb. of air carried per hr. = .975 × 1,000 × 16 × $\sqrt{180}$ = 209,196. Ans.

(16) (*a*) From 5 to 28 lb.

      (*b*) 8 to 15 lb. See Art. **45.**

(17) (*a*) Induced draft may be produced by using steam blowers in the stack or by introducing fans in the flue passages between the boilers and the stack. See Arts. **25, 26,** and **27.**

(*b*) The chief advantages are that the amount of air required for perfect combustion can be closely regulated; an intense draft can be secured at less first cost than it would require to build a good chimney; and the draft can be controlled independently of atmospheric conditions so as to burn low-grade fuel and increase the evaporation capacity of the boilers. The chief disadvantages are the daily cost of operation and maintenance. See Art. **28.**

(18) Curve A, Fig. 2, shows that for a temperature 200° above the outside air, the ordinate of curve A is .9 and for a temperature 400° above the air, the ordinate is 1. These ordinates represent the relative quantities of air discharged at the two temperatures. If, therefore, 15,000 lb. is discharged per hour at 200° the weight discharged per hour at 400° will be 15,000 × $\frac{1}{.9}$ = 16,666 lb. Ans. See Art. **13.**

ELECTRIC POWER STATIONS
(PART 3)

---

(1) Because the range of temperature in any one cylinder is reduced, consequently the entering steam does not come in contact with surfaces that are as much below it in temperature as would be the case if the steam were expanded in a single cylinder. See Art. **7**.

(2) The tandem compound engine is cheaper than the cross-compound, for an equal number of horsepower, but the low-pressure cylinder is hard to get at for examination or repairs. The cross-compound engine admits of a somewhat higher rated speed and exerts a more uniform turning effort on the engine shaft. See Art. **9**.

(3) The use of a compound engine in connection with a condenser allows a high ratio of expansion and effects a considerable saving in steam. The higher steam economy thus obtained permits the installation of less boiler capacity and thus reduces the outlay for boilers. With simple engines used with condensers the possible benefits of a high ratio of expansion cannot be realized because of the large loss due to cylinder condensation. See Art. **13**.

(4) About 700 to 800 ft. per min. for long-stroke engines and 600 ft. per min. for short-stroke engines. See Art. **16**.

(5) There is always a certain load for which the engine gives its maximum steam economy and there is a certain output for which the generator gives its highest efficiency. The engine and generator should, therefore, be so selected that they reach their maximum efficiencies simultaneously. See Art. **19**.

(6) 100 to 120 lb. for single-expansion engines, 140 to 160 lb. for compound condensing engines, 180 to 190 lb. for triple-expansion condensing engines. See Art. **20**.

(7) An abstract of Art. **21** is required.

(8) One-sixtieth of the pitch angle between the centers of adjacent poles. Since the distance between centers of adjacent poles corresponds to $180°$ of phase difference, the allowable variation is $\frac{180°}{60} = 3°$ of phase difference either way, or $6°$ altogether. See Art. **28**.

(9) (*a*) To remove the back pressure that would otherwise act on the piston of the engine if the steam was exhausted into the atmosphere. In other words, the object of the condenser is to create a partial vacuum behind the piston and allow the steam to be expanded down to a lower pressure and temperature than would be possible if it were exhausted into the atmosphere.

(*b*) Because there is always some leakage of air, and a certain amount of air is mixed with the steam. The condensed steam also emits a certain amount of water vapor, which reduces the vacuum. See Arts. 37 and 38.

(10) Because these turbines run at high speed and it is somewhat difficult to build direct-current dynamos, for very high speeds, that will run without sparking at the commutator. See Art. 31.

(11) The air pump is used to remove the mixture of condensing water, air, vapor, and condensed steam from the condenser. Sometimes a pump is arranged so that it removes air and vapor only, in which case it is often referred to as a dry-air pump. See Art. 38.

(12) From 12½ to 20 lb. of steam per indicated horsepower-hour. See Art. 40.

(13) In the surface condenser, the condensing water and condensed steam are discharged separately. The cooling water circulates through a nest of small tubes, around which the exhaust steam passes, and is condensed by coming into contact with the cold walls of the tubes. In the jet condenser, the exhaust steam passes into a chamber where it is condensed by coming into direct contact with a jet of cold water. The water of condensation and the cooling water are thus mixed in the jet condenser. See Arts. 43 and 48.

(14) There should be not less than 1 sq. ft. of tube surface for each 10 lb. of exhaust steam condensed per hour. See Art. 44.

(15) (*a*) Give an abstract of Arts. 53 and 54.

(*b*) The cost of this condenser is low, and in cases where the water is supplied from an elevation or under pressure, the condenser can be operated without any pumping. See Art. 56.

(16) Because, when the load is variable, the amount of exhaust steam varies, and there might be times when there would not be sufficient exhaust steam to move the condensing water unless the condensers were continually adjusted to suit the changing conditions. See Art. 61.

(17) From the rule given in Art. 68, we have units weight of injection water required per unit weight of steam $= \dfrac{1{,}190 - 100}{100 - 50} = 21.8$. That is, the weight of the injection water required will be 21.8 times the weight of steam condensed. Ans.

(18) (*a*) They are used to cool the water discharged from a condenser so that it can be used over and over for condensing purposes, thus enabling a plant to use condensing engines even though the water supply be limited. See Art. **74.**

(*b*) Give an abstract of Art. **76.**

(19) See Art. **79.** Short descriptions of the three methods mentioned in this article are required.

(20) Give a brief outline of the points enumerated in Art. **47.**

ELECTRIC POWER STATIONS
(PART 4)

---

(1) The pipe is provided with threaded flanges with the threads carefully cut. The end of the pipe should be peened into the flange as shown in Fig. 6. See Art. **14.**

(2) (*a*) and (*b*). See Art. **31.**

(3) From formula **1**, we have

$$W = 87 \sqrt{\frac{D(p_1 - p_2)d^5}{L\left(1 + \dfrac{3.6}{d}\right)}}$$

In this case, the initial pressure p_1 is 120 lb. and the density D of the steam corresponding to this pressure is .304 lb. per cu. ft. The loss of pressure in the pipe is $p_1 - p_2 = 2$ lb. The diameter d is 4 in. and $L = 300$ ft.; hence,

$$W = 87 \sqrt{\frac{.304 \times 2 \times 4^5}{300\left(1 + \dfrac{3.6}{4}\right)}} = 90.91 \text{ lb. per min. } \textbf{Ans.}$$

(4) (*a*) The receiver allows smaller steam piping than would otherwise be required; it tends to maintain a steady flow of steam through the piping; and provides a cushion that takes up the reaction caused by the sudden stopping of the flow of steam when the admission valves close and thus prevents vibrations from being transmitted through the piping system. The use of the receiver also tends to maintain a constant steam pressure at the engines.

(*b*) The volume of the receiver should be from three to four times the volume of the high-pressure engine cylinder to which the steam is delivered. See Art. **19.**

(5) The theoretical horsepower can be calculated by means of formula **3.** In this case, the volume V is 15,000 cu. ft. per min. and the head H is 40 ft.; hence,

$$\text{theoretical H. P.} = \frac{62.5 \times 15,000 \times 40}{33,000} = 1,136. \text{ Ans.}$$

§ 31

(6) See Art. **85.** A statement of the points given in this article is required.

(7) The loss in head can be calculated by means of formula **4.** In this case, the length L = 1,000 ft., v = 8 ft. per sec., d = 40 in., k = .0037 since the pipe is 40 in. in diameter.

$$h = \frac{k\,L\,v^2}{d} = \frac{.0037 \times 1,000 \times 8^2}{40} = 5.92 \text{ ft.} \quad \text{Ans.}$$

(8) Because the duplication makes the piping complicated and it is not necessary if proper care is taken in the selection of the piping and fittings. See Art. **7.**

(9) Turbines and impulse wheels. See Art. **66.**

(10) The corrugated copper gasket. See Art. **22.**

(11) (a) It is a periodic surging of the speed above and below the normal whenever there is a change in load. It is caused by the governor not responding at once to the changes in speed and on that account opening the gates too wide or closing them too much for the load that the wheel is carrying. See Art. **81.**

(b) The movements of the waterwheel governor instead of being controlled simply by a centrifugal governor, are modified in such a manner by the movement of the gate itself that the tendency of the flyball governor to overshoot the mark is counteracted. See Art. **84.**

(12) Because hot water corrodes iron pipe rapidly so that it is soon rusted out. See Art. **26.**

(13) (a) An abstract of Art. **67** is required.

(b) Point out the distinction between the different classes of turbines as given in Art. **68.**

(14) State the features as given in Art. **60.**

(15) Long bends permit expansion and contraction of the piping system and offer less obstruction to the flow of the steam. See Arts. 2 and **9.**

(16) Impulse wheels are particularly suited to high heads and are very generally used where the head is from 100 feet upwards. See Art. **78.**

(17) The student is required to state, briefly, the considerations given in Art. **48.**

TELEGRAPH SYSTEMS
(PART 1)

---

(1) (*a*) See Art. **1.**

(*b*) Visible and audible signals may be used.

(2) Because it is not practicable to obtain in a long line circuit a current of sufficient strength to operate a sounder, but a current strong enough to operate a relay may be readily obtained; and the relay is able to open and close a local circuit containing a sounder and a battery that can furnish sufficient current to work the sounder. See Art. **3.**

(3) (*a*) See Art. **8.**

(*b*) See Art. **8.**

(4) These switches must both be kept closed at all times, except at the one key where an operator is sending.

(5) The Continental, or Universal, code is used all over Europe and practically all over the world on submarine cables. The Morse alphabet and numerals and the Phillips punctuation code is used all over the United States and Canada, except on submarine cables, on which the Continental code is used.

(6) The arrangement would be the same as shown in Fig. 3, except that there will be one more intermediate office. Note that there should be three cells at each office, the main-line batteries at all the offices being connected in series.

(7) (*a*) The Morse closed-circuit system is one in which all line batteries are in series in the line circuit, which is normally closed. Thus, current is flowing through the whole circuit at all times except when the key is opened at some one station in order to send a space. The circuit is closed, even when no messages are being sent.

(*b*) Because normally, even when no messages are being transmitted, the circuit is closed.

(8) (*a*) Yes.

(*b*) Not necessarily.

(*c*) All batteries in the line circuit must be in series with one another.

§ 46

(9) The Morse open-circuit system is one in which the batteries are so arranged as to be cut out of the circuit except when signals are being sent. A battery is required at each office. It is so named because all batteries are normally on open circuit, current flowing over the line only when dots and dashes are being transmitted.

(10) See Art. **18.**

(11) See Art. **24.**

(12) Because the resistance can be determined so much easier than the number of turns. See Art. **17.**

(13) (*a*) A telegraph repeater may be defined as an arrangement of apparatus for repeating signals from one main line into another main line. It is controlled by the sending operator at the end of one main line, and, in turn, controls the second main line, and, hence, the relay at the far end of it. The repeater itself is located at about the middle point of the distance covered.

(*b*) The following three causes combine to limit the length of line over which it is practical to signal without using repeaters: First, as a line increases in length, the effective current decreases on account of leakage until it becomes so small that satisfactory signals cannot be transmitted, no matter how much the battery power is increased. Second, as a line increases in length, the resistance increases, and, consequently, the E. M. F. must be correspondingly increased, assuming that the insulation remains perfect; if it does not, the E. M. F. must increase faster than the resistance. But it is impractical to use over about 400 volts, and 300 is usually considered very high for single working. Third, as a line increases in length, the electrostatic capacity increases until the latter seriously diminishes the speed of signaling. See Art. **37.**

(14) (*a*) An automatic repeater is one that will automatically repeat in either direction; that is, it does not require an operator at the repeater to turn a switch when the direction of sending is to be reversed.

(*b*) An operator is needed to adjust the instruments and care for the batteries. Besides doing other work, one operator may look after a number of repeater sets.

(15) (*a*) An artificial line is a branch circuit to the ground having the same resistance and capacity as the line. The resistance and the capacity must be properly arranged so that the artificial line will not only have the same resistance and capacity, but will also charge and discharge at the same rate as the line.

(*b*) It is used in order to make the current from the home battery divide equally between the line and artificial line circuits, so that it will not energize the differentially wound relay or relays at the home station.

(16) (a) See Art. 44.

(b) See Art. 44.

(17) With the switch g closed and k connecting a with d, the west line may repeat into the east line; with g closed and k connecting b with c, the east line may repeat into the west line; with g closed and k connecting c and d, the east line and the west line may be used independently; with g open and k connecting c and d, the west line and the east line are connected straight across without using the apparatus as a repeater.

(18) (a) A polarized relay is one that requires the direction of the current flowing through it to be reversed in order to move the armature from one stop to another. See Arts. 52 and 53.

(b) See Art. 51.

(19) The sending should be heavy or firm, that is, the signals should be somewhat prolonged, because the current requires time to rise from zero to its maximum on account of the electrostatic capacity of the line and the inductance of the relays, and because time is also required for the various armatures to move across the gaps between the two stops.

(20) (a) Closed.

(b) When the key at the western station is opened, the western relay R_1 of the repeater, Fig. 18, opens the local circuit through the magnet S_1 of the transmitter T_1, because M_1 does not release its armature. The transmitter T_1 breaks two contacts, one slightly before the other. The contact x_1, that is broken first, opens a local circuit through the extra magnet M on the opposite, or eastern, side, causing this extra magnet to release its armature and allow the spring s to hold the contact at y closed. Thus, the transmitter T on the same, or eastern, side and the western line is held closed. The second contact a_1 that is broken at the western transmitter T_1 opens the eastern main line, but the armature g of the eastern relay R is not released, although the eastern relay R is demagnetized, because the spring s is stronger than s_1 and no current is flowing in M. Consequently, the local circuit controlled by the armature g is not opened at the contact, or front stop, y. Hence, the armature of the eastern relay R, as long as only the western key is being operated, remains against its front stop y and, therefore, keeps the transmitter T on the same, or eastern, side closed, and, consequently, the western line is closed at a while the western key is being operated. Moreover, the circuit through the extra magnet M_1 is kept closed at x by the eastern transmitter, thus allowing the western relay R_1 to have full control of its armature g_1.

(21) (a) On the principle of the Wheatstone bridge.

(b) The battery, key, and artificial line.

(c) The galvanometer.

(22) This may be briefly answered as follows: The polar, Stearns, and bridge duplex and quadruplex systems are balanced by properly adjusting the various instruments and the resistance and capacity of the artificial lines until the latter are equal to and discharge at the same rate as the main line; the systems being satisfactorily balanced when sending on the home key, will not interfere in any way with the signals that are being received at the home office, but made at the distant office. See also Art. 63.

(23) A continuity-preserving pole changer is a device for reversing the direction of a current in one part of a circuit without opening the circuit.

(24) One message is sent by operating a pole changer that controls the direction of the current, and is received by a polar relay at the distant station. Another message is sent by operating a transmitter that controls the strength of the current, and is received at the distant station by a neutral relay. The transmitter is located at the same station as the pole changer.

(25) To repeat from the east line to the west line (see Fig. 17) the ground switch g is closed and the lever k is placed from b to c. When the eastern key is closed a current flows from $+B_1$ through $b-o_2-c-k-l-G-$ the ground to the eastern station–the eastern line and R_1 to $-B_1$. This current causes R_1 to hold its local circuit closed, thereby causing the sounder S_1 to close the circuit of the battery B through the relay R–western line–western station–ground–$G-l-k-b-o_1$–contacts of S_1-2-B. Thus, the signal is repeated over the western line, and the circuit controlling the sounder S is closed. When the eastern key is opened, the eastern relay R_1 and then the sounder S_1 release their armatures, thereby causing the western line to be opened between the contact stop and armature of S_1. Since there is now no current in R, the sounder S will release its armature. No circuit is controlled, however, by the armature of the sounder S, as there is a break between a and k. The sounder S, therefore, acts merely as a reading sounder. See Arts. 39 and 40.

TELEGRAPH SYSTEMS

(PART 2)

---

(1) (*a*) On changes, both in the direction and in the strength of the current.

(*b*) One key at each end of the line governs the direction of the current, and the other key its strength. At each end there are two differentially wound relays, one of which (the neutral relay) is operated only by the changes in the strength of the current, while the other (the polar relay) is operated only by the changes in the direction of the current. The key at one end governing the strength of the current, therefore, produces no effect on the polar relay at the other end, because the latter responds only to changes in the direction of the current, but this key does operate the neutral relay, because the latter responds to changes in the strength of the current. The key governing the direction of the current operates only the polar relay at the other end, because the latter is affected by changes in the direction, but not by changes in the strength of the current. The home relays are not operated by the home keys, because both relays are differentially wound and not affected by the changes in strength or direction of the currents produced at the home station.

(2) See Art. **28.**

(3) See Art. **2.**

(4) See Arts. **10** and **11.**

(5) In the battery quadruplex system, the current is reversed in direction by means of an instrument called a pole changer. The pole changer in one position connects the positive pole of the battery toward the line circuit, and the negative pole of the battery to the ground; when the key controlling the pole changer is operated the connections are changed at the pole changer so as to connect the negative pole of the battery toward the line and the positive pole to the ground, thus reversing the direction in which the current flows toward the line circuit. The pole changer used in the battery quadruplex reverses the direction of the current without opening the circuit, and,

§ 47

although it momentarily short-circuits the battery, this does no particular harm where gravity cells are used, because the gravity cells have sufficient internal resistance to prevent the flow of an excessively large current. The current is changed in strength by means of a so-called continuity-preserving transmitter. In one position of this transmitter a certain number of cells is connected in the circuit, and in the other position of the transmitter a different number of cells is connected in the circuit; thus, the strength of the current in the circuit is varied by varying the number of cells connected in the circuit. The transmitter is called a continuity-preserving transmitter because it increases or decreases the number of cells connected in the circuit without opening the circuit at any time, and although it momentarily short-circuits some of the cells, this does no particular harm for the same reason as given above.

(6)   This condition exists for four combinations of the four keys, which are as follows: All four keys open, all four keys closed, the pole-changer keys Pk and Pk_1 closed and the transmitter keys Tk and Tk_1 open at both terminal stations, and finally the pole-changer keys Pk and Pk_1 open and the transmitter keys Tk and Tk_1 closed at both stations. See Fig. 2.

(7)   (a) Two sounders are used, the first sounder being closed on the back stop of the neutral relay and the second sounder being closed on the back stop of the first sounder. See Art. 8.

(b)   The sounders are arranged in this manner so as to eliminate the production of false signals. For when the magnetization of the neutral relay passes through zero as the current through the relay is reversed in direction, there is a tendency for the relay to release its armature, but this cannot produce a false signal unless the time of no magnetism in the relay is sufficient to allow the armature to touch the back stop of the relay, because the first sounder, which is called a repeating sounder, cannot close the circuit of the second, or reading, sounder unless the armature of the relay touches its back stop. Thus, a slight fluttering of the relay armature can produce no false signal unless it is released for a sufficient length of time to cross the gap between the front and back stops of the relay, and, also, to allow the armature of the repeating sounder to at least partially cross, in turn, the gap between its front and back stops.

(8)   Galvanized-iron, steel, and hard-drawn copper wire.   See Art. 49.

(9)   (a) See Art. 54.

(b)   The McIntire sleeve joint.

(c)   It does not require the use of solder to maintain good contact, the joint is made more easily, and the heat required to solder a joint generally weakens hard-drawn copper wire.

(10) (*a*) "Extra Best Best," "Best Best," "Best," and "Steel."

(*b*) "Extra Best Best" is considered the best for ordinary line wires, because it has the least resistance and is the most uniform in quality, being both tough and pliable.

(11) (*a*) Saturated- and dry-core cables.

(*b*) Dry-core cables have a low electrostatic capacity per mile and a very high insulation resistance. They are, however, very susceptible to injury by moisture and should only be used where a very low electrostatic capacity is an absolute necessity, as in telephone cables. The saturated-core cables have a higher electrostatic capacity per mile than the dry-core cables, but have the advantage of not being so susceptible to moisture, and are, therefore, more desirable for telegraph purposes.

(12) Copper wire possesses at least six times as great a conductivity as iron, and is non-corrosive. Copper wire is, therefore, far more durable than galvanized-iron or steel wire.

(13) (*a*) First, the wires may be placed farther apart; second, the wires may be made smaller in diameter; third, an insulating medium having a lower inductivity may be used.

(*b*) The third. The first method makes the cable too bulky; the second may be used to some extent, but if carried to an extreme, the resistance of the wires is made too high and the tensile strength too small. The method involving the use of an insulating material having a low inductivity affects neither the strength nor the size of the cable appreciably, and is, therefore, the most desirable.

(14) An artificial cable is a branch circuit to the ground having the same electrical properties as the line; that is, it should have the same resistance, capacity, and leakage as the real cable. These properties must be properly arranged so that the artificial cable will not only have the same resistance and capacity, but will also charge and discharge at the same rate as the line. In order to accomplish this, the capacity of the artificial cable must be distributed more or less throughout its length, so that its capacity shall be distributed in a manner similar to that in the cable.

(15) (*a*) A coherer is a device for detecting the presence of electromagnetic, or Hertzian, waves. It is sensitive to electromagnetic waves, because the latter causes the resistance of the ordinary type of coherer to decrease enormously.

(*b*) A serious objection made against most wireless-telegraph systems is the fact that with the methods generally used at present, two independent communications cannot be received at the same station readily, if at all; and every receiver placed within the radius of action of a transmitter is liable to be acted on by the waves sent out by the one transmitter. Hence, if two transmitters are simultaneously operated, complete interference or a confusion of the two sets of signals is very liable to be the result. A second serious objection, or defect, is the fact that the distance between the stations cannot be indefinitely increased. Both of these objections may, of course, be overcome in the future.

(16)　(*a*) Condensers.

　　(*b*) A condenser.

(17)　The receiving station of a wireless-telegraph system is said to be in tune, resonance, or syntony with a particular transmitting station when it will respond only to electromagnetic waves of the particular frequency emitted from this transmitting station. If the frequency of the waves is greater or less than those for which the receiver is tuned, the latter will not respond to them.

(18)　It makes the surface resistance between the wire and the pin large. See Art. 60 and Fig. 15.

(19)　(*a*) The transmitting key is so arranged that by pressing one lever, the plus pole of the battery is connected to line while the minus pole is connected to earth; and by pressing the other lever, the minus pole is connected to line while the plus pole is connected to earth.

(*b*) The receiving instrument is usually a siphon recorder. It consists of a coil of wire swinging between the poles of a permanent magnet. When the current flows through this coil in a certain direction it swings in one direction, and in the opposite direction when the current is reversed in direction. A glass siphon has one end dipping into a vessel of ink and the other end, whose movement is controlled by the coil, spurts ink on a strip of paper that is drawn past the end of the siphon by a clockwork or electric motor. This spurting of the ink is caused by the vibration of the glass siphon, which is due to an electromagnet resembling a vibrating bell in its action. The ink causes a record of the movement of the coil to be made on the paper tape.

(*c*) A movement of the siphon in one direction indicates a dot, and in the other direction a dash.

(20)　The call boxes associated with any one circuit in district-messenger telegraph systems are usually connected in series in a complete metallic-line circuit. A ground is only used as a return circuit in a case of a fault on the line, or for producing the return call in return-call boxes.

TELEPHONE SYSTEMS

(PART 1)

(1) (*a*) and (*b*) See Art. **24.**

(2) (*a*), (*b*), and (*c*) See Art. **61.**

(3) (*a*) and (*b*) See Art. **63.**

(4) See Art. **50.**

(5) The battery transmitter has two or more points in loose contact. As the diaphragm vibrates, the pressure between the points in contact varies; as the pressure increases, they make better contact (that is, more points are caused to touch one another), and hence the resistance across the contact decreases; as the pressure decreases, the contact becomes poorer (that is, fewer points touch one another), and hence the resistance across the contact becomes greater. This variation in resistance causes a variation in the current that flows through the transmitter circuit from the battery. Since the diaphragm vibrates in unison with the air waves produced by talking, the pressure produced by the diaphragm varies in unison with the vibration of the diaphragm, the resistance in unison with the pressure, and the current in unison with the resistance, it therefore follows that the undulations produced in the strength of the current are in unison with the sound waves. This assumes that there is no loss in quality in the transformation, which is not quite true. See Arts. **11** and **12.**

(6) See Art. **41.**

(7) Series and bridged. For descriptions of series and bridged party lines see Arts. **43** and **44.**

(8) (*a*) A device for normally maintaining a short circuit around the winding of the armature of a series-generator, but which opens this short circuit automatically when the generator handle is turned.

(*b*) It removes the resistance of the generator armature from the line circuit, so that it is not in the path of incoming currents from the other stations.

(9) The magneto-transmitter serves as a generator of electricity, because the movement of the iron diaphragm in front of the pole pieces varies the reluctance of the path of the lines of force, and hence increases and decreases the number, or distribution, of the lines of force through the receiver coil, thereby inducing an alternating E. M. F. in the receiver coil, and hence an alternating current will flow if the circuit is closed. The battery transmitter serves simply as a valve for controlling the flow of current from a battery. This it does by varying the resistance of the circuit, and hence a variable, or undulating, direct current (that is, one that changes in strength but not in direction) flows in the transmitter circuit.

(10) (*a*) and (*b*) See Art. **2.**

(11) (*a*) See latter part of Art. **6.**
      (*b*) See Art. **7.**

(12) The problems of eliminating cross-talk and other troubles due to induction.

(13) (*a*) Loudness, pitch, and timbre.
      (*b*) and (*c*) See Art. **3.**

(14) (*a*) The current in the primary circuit is an undulating direct current.
      (*b*) The current in the secondary circuit is an alternating current.
      (*c*) The current in the primary circuit is an undulating direct current, because the transmitter merely varies the resistance of the circuit in which a direct current from a battery flows; hence, only the strength of the current is varied; its direction always remains the same. The lines of force set up in the iron core of the induction core increase and decrease in number as the strength of the current in the primary increases and decreases, but the direction of the lines of force is not reversed. However, increasing the number of lines of force will induce a current in the secondary in one direction, while decreasing the number will induce a current in the opposite direction. Therefore, the current in the secondary circuit flows in one direction, while the primary current is increasing, and in the other direction while the primary current is decreasing. It is, therefore, an alternating current.

(15) (*a*) Articulate speech.
      (*b*) By exceedingly complex waves.

(16) The inductive but low-resistance coil will offer the least opposition to a direct steady current, and the non-inductive but high-resistance coil the least opposition to the high-frequency alternating current. See Arts. **15** and **16.**

(17) (*a*) Electrostatic induction.
      (*b*) By using two line wires from each circuit, and so arranging

the two wires that the average distance of each from any other wire is the same. This is accomplished with overhead bare wires by transposing the two wires forming a circuit properly and sufficiently often, whereas in cables the two wires forming a circuit are twisted together, thus effecting a complete transposition of the wires every few inches.

(18) See Art. **17.**

(19) To leakage (poor insulation) between the various conductors. It is generally caused by dampness.

(20) See Art. **45.**

TELEPHONE SYSTEMS

(PART 2)

(1) See Art. **2.**

(2) A clearing-out drop is generally some form of electromagnetically operated signal used to inform an operator when one or both subscribers that are connected together, desire to be disconnected. See Art. **13.**

(3) See the jack *J'* and drop *D'*, Fig. 4.

(4) A listening key is a switching device that is used to connect an operator's telephone set across a cord circuit. See the key *K'*, Fig. 4.

(5) (*a*) Supervisory signals are lamps or electromagnetic signals associated with the cord circuit and adapted to keep the operator informed as to the condition of a subscriber's circuit as long as a plug remains in the jack associated with that subscriber's line. The signal always shows, when the plug is in the jack, whether the subscriber's receiver is on or off the hook.

(*b*) Sometimes electromagnetic signals resembling annunciators, or line drops, are used, but more frequently miniature incandescent lamps ($\frac{1}{8}$ candlepower) are used.

(6) By means of electromechanical annunciators or drops; these being provided with a shutter that is moved by an electromagnet connected in the line circuit; or by means of a miniature incandescent lamp connected in a local circuit that is controlled by a relay whose coil is connected in the line circuit.

(7) See the line-relay and line-lamp circuits associated with one of the lines in Fig. 11.

(8) The line will test busy at any section of the multiple switchboard as long as a plug remains inserted in some jack associated with that line. See Art. **34.**

(9) A relay adapted to open the circuit of a subscriber's line at such a point that the spring jacks will be left connected with a subscriber's line, but the other apparatus in the switchboard circuit, such

as the line relay, will be cut off from the circuit. The line cut-off relay usually has its coil included in a local circuit that is completed when a plug is inserted in a jack belonging to that line. See Fig. 11.

(10) (*a*) A system adapted for use in factories and business and private houses, and involving the use of at least one more line wire than there are stations. All these line wires are connected at each station to a switch that enables the party at any station to connect his telephone with the line belonging to any other station and to call up that station without the assistance of any operator at any central exchange.

(*b*) The intercommunicating and speaking-tube system.

(11) (*a*) A telephone exchange is a combination of a number of telephones with their line circuits, signals, and switching devices, by means of which any telephone may be connected with any of the other telephones.

(*b*) Transfer and multiple switchboards.

(12) Drops should be so constructed as not to be liable to get out of adjustment. Both poles of the electromagnet should be presented to the armature. The armature should come near to the pole pieces without touching them, and all iron used in the construction of drops should be of the softest possible grade. It is usually desirable that the drops shall occupy but little space, and that they may be easily removed from and replaced in the switchboard when repairs are necessary.

(13) (*a*) To attract the attention of the operator at night or such other times when her presence is not required continuously at the board.

(*b*) The drop, in falling, closes an auxiliary local circuit, containing a battery and a vibrating bell. The closing of this circuit causes the bell to ring until the operator restores the shutter after answering the call. In one central-energy system, the night-bell circuit is controlled by a relay in series with all the line-pilot lamps; whenever any subscriber calls, one line-pilot lamp lights, the night-bell relay is energized, and the night bell rings; on inserting a plug in the jack of the calling subscriber, the line-pilot-lamp circuit is opened, the night-bell relay releases its armature, and the night bell stops ringing.

(14) First, the drop may be cut out of the circuit during conversation; second, the drops, if left in circuit, may be placed far enough apart not to affect one another; and third, a magnetic shield, or shell, may be placed around each drop.

(15) (*a*) By means of the multiple switchboard any operator can connect any two subscribers on the entire board without the assistance of any other operator, while in all the transfer systems, at least two operators must take part in most of the connections.

(*b*) It is more complicated and expensive.

(16) There is a spring jack on each section of the board connected to each line circuit; thus, if a board has 30 sections, each line will have 31 spring jacks, one on each section, except at the answering position for this line, where there are usually 2 jacks.

(17) Since each line is provided with a spring jack on each section of the board, it follows that any line may have a connection made with it at any one of the sections. In order to prevent a connection with the same line at more than one section at the same time, means are provided whereby the first connection that is made with the line will establish certain conditions that will indicate to operators at other sections the fact that the line is already in use. The provision made to indicate whether a line is in use is called the "busy test." The fact that any one of the 90 operators at a 30-section board can see only one of the 30 jacks belonging to each line makes a busy test absolutely necessary.

(18) It means that the removal of the receiver from the hook will give a signal which the operator will recognize as a desire on the part of the subscriber for a connection or other service; that is, the subscriber does not have to press any button, or operate any device; all that is necessary for him to do is to take down the receiver.

(19) A line-pilot lamp is a miniature incandescent lamp (usually, however, a little larger and more powerful than a line lamp) whose circuit is usually controlled by a relay (called the line-pilot relay), so connected that current flowing through any line lamp at any one operator's position on a central-energy multiple switchboard, will also flow through the relay coil, close the relay, and cause the line-pilot lamp at that operator's position to light. A line-pilot lamp enables the supervising operator to observe how well an operator is answering the calls received at her position. Furthermore, with the arrangement of the central-energy multiple switchboard shown and described in this Section, the line-pilot lamp will light when a subscriber takes down his receiver, even if the line lamp associated with that subscriber's line is burnt out. Hence, the subscriber is not indefinitely cut off from the exchange or his line hung up, as it is termed, because the lighting of the line-pilot lamp without the lighting of the line lamp indicates that some line lamp is burned out and it is a simple matter to locate and replace the faulty line lamp.

(20) (a) A central-energy system is one in which all current for operating both the transmitters and the line-signaling devices at the exchange is supplied from the exchange, thus doing away with the primary battery and magneto-generator required at each subscriber's station in most of the older magneto-switchboard systems.

(b) It is also known as the "common-battery" system, because one battery located at the central office is used in common to supply current for all line circuits terminating in that exchange.

APPLIED ELECTRICITY

(1)   A compound-wound machine designed to give a large current output at low voltage. See Art. 5.

(2)   The rheostat has to be capable of carrying 1.3 amperes; hence, according to Table IX, No. 23 wire would be used, as this size is capable of handling 1.49 amperes when mounted on an iron frame. The total resistance is to be 120 ohms, and if the wire is wound on a .4-in. mandrel, the resistance per inch of spiral will be .735 ohm, and $\frac{120}{.735}$ = 163 in. of spiral will be needed. If each spiral is made about 8 in. long, the rheostat could be made up of, say, 21 such spirals connected in series.

(3)   In the Thomson welding process, the metals to be welded are butted together and sufficient current passed through the junction to bring the metal to a welding heat. See Art. 33.

(4)   See Art. 64 and Fig. 25.

(5)   Because for electric welding, very large currents at low pressure are required, and such currents are easily obtained by generating alternating current at comparatively high pressure and then stepping-down to large current at low pressure by means of a special transformer. See Art. 33.

(6)   The following solution may be used for silver plating: Silver chloride, 3 oz.; cyanide of potassium, 10 to 12 oz.; water, 1 gal. See Art. 14.

(7)   (*a*). Because a long range of pull is equivalent to a long air gap in the magnetic circuit, and if a strong pull is to be exerted with a magnet of reasonable dimensions, a fairly high density must be used. A large number of ampere-turns would be needed to set up the magnetic flux, and the winding would be expensive.

(*b*)   See Art. 63.

(8)   The number of cubic feet to be heated is $8 \times 10 \times 9 = 720$. The temperature is to be raised $70 - 50 = 20°$ F. For each cubic foot an expenditure of 18 joules will raise the temperature 1°. Hence, to

raise the temperature of 720 cu. ft. 20° would require an expenditure of 720 × 20 × 18 = 259,200 joules. The air is to be heated in 20 min., or 60 × 20 = 1,200 sec.; hence, work must be done at the rate of $\frac{259,200}{1,200}$ = 216 joules per sec., or 216 watts must be supplied, since 1 watt is equal to 1 joule per sec. See Art. **20.**

(9) From formula **9,** we have

$$\text{circ. mils} = \frac{I\,T\,l''}{E}$$

In this case, $I\,T$ = ampere-turns = 7,500, l'' = 26, E = 250; hence,

$$\text{circ. mils} = \frac{7,500 \times 26}{250} = 780$$

No. 21 B. & S., 810 circular mils, is the nearest size.

(10) A cyanide copper bath may be made as follows: Carbonate of copper, 1 lb.; carbonate of potash, 6½ oz.; cyanide of potassium, 2 lb.; water, 3 gal. See Art. **8.**

(11) The number of joules required to raise the temperature of 1 gal. of water 1° F. is 8,803. In this case, 3 gal. are to be raised 200 − 40 = 160° F.; hence, the number of joules required will be 3 × 160 × 8,803 = 4,225,440. The heating is to be done in ½ hr., or 1,800 sec.; hence, the joules per second or watts will be $\frac{4,225,440}{1,800}$ = 2,347. At 250 volts, 2,347 watts are equivalent to $\frac{2,347}{250}$ = 9.4 amperes, nearly. Ans. See Art. **21.**

(12) With a pole face density of 95,000 lines per sq. in. (see Table XI), 125.11 lb. pull per sq. in. will be exerted. The pull at each pole will, therefore, be 125.11 × 25 = 3,127.75 lb., and the total pull will be twice this, or 6,255.5 lb. Ans. See Arts. **45** and **46.**

(13) To the negative pole. See Art. **2.**

(14) For a pole face density of 105,000 lines, the pull per sq. in. will be 152.84 lb. The total tractive force is to be 1,500 lb., or 750 lb. per pole; hence, the area of each pole face must be $\frac{750}{152.84}$ = 4.91 sq. in. Ans.

(15) The loss by convection is about .053 joule per second (or watts) per foot per degree centigrade difference in temperature. In this case, the length is $\frac{5,280}{4}$ = 1,320 ft., and the difference in temperature is 50° C , so that the loss will be .053 × 1,320 × 50 = 3,498 watts. Ans. See Art. **26.**

(16) See formula **10,** Art. **61.** The student is required to give an explanation of this formula.

INDEX

NOTE.—All items in this index refer first to the section and then to the page of the section. Thus, "Combustible 29 1" means that combustible will be found on page 1 of section 29.

INDEX

Lightning Source UK Ltd.
Milton Keynes UK
UKHW012306140219
337323UK00011B/400/P